Communications
in Computer and Information Science 753

Commenced Publication in 2007
Founding and Former Series Editors:
Alfredo Cuzzocrea, Xiaoyong Du, Orhun Kara, Ting Liu, Dominik Ślęzak,
and Xiaokang Yang

Editorial Board

More information about this series at http://www.springer.com/series/7899

Leonid Sokolinsky · Mikhail Zymbler (Eds.)

Parallel Computational Technologies

11th International Conference, PCT 2017
Kazan, Russia, April 3–7, 2017
Revised Selected Papers

 Springer

Editors
Leonid Sokolinsky
South Ural State University
Chelyabinsk
Russia

Mikhail Zymbler
South Ural State University
Chelyabinsk
Russia

ISSN 1865-0929   ISSN 1865-0937 (electronic)
Communications in Computer and Information Science
ISBN 978-3-319-67034-8   ISBN 978-3-319-67035-5   (eBook)
DOI 10.1007/978-3-319-67035-5

Library of Congress Control Number: 2017953411

Printed on acid-free paper

This Springer imprint is published by Springer Nature
The registered company is Springer International Publishing AG
The registered company address is: Gewerbestrasse 11, 6330 Cham, Switzerland

Preface

This volume contains a selection of the papers presented at the 11th International Scientific Conference on Parallel Computational Technologies, PCT 2017, held during April 3–7, 2017, in Kazan, Russia.

The PCT series of conferences aims at providing an opportunity to discuss the future of parallel computing, as well as to report the results achieved by leading research groups in solving both scientific and practical issues using supercomputer technologies. The scope of the PCT series of conferences includes all aspects of high performance computing in science and technology such as applications, hardware and software, specialized languages, and packages.

The PCT series is organized by the Supercomputing Consortium of Russian Universities and the Federal Agency for Scientific Organizations. Originated in 2007 at the South Ural State University (Chelyabinsk, Russia), the PCT series of conferences has now become one of the most prestigious Russian scientific meetings on parallel programming and high-performance computing. PCT 2017 in Kazan continued the series after Chelyabinsk (2007), St. Petersburg (2008), Nizhny Novgorod (2009), Ufa (2010), Moscow (2011), Novosibirsk (2012), Chelyabinsk (2013), Rostov-on-Don (2014), Ekaterinburg (2015), and Arkhangelsk (2016).

All papers submitted to the conference were scrupulously evaluated by three reviewers on the relevance to the conference topics, scientific and practical contribution, experimental evaluation of the results, and presentation quality. PCT's Program Committee selected the 24 best papers to be included in this CCIS proceedings volume.

We would like to thank the Russian Foundation for Basic Research for their continued financial support of the PCT series of conferences, as well as respected PCT 2017 sponsors, namely platinum sponsors, Intel and RSC Group, gold sponsor, NVIDIA, silver sponsor, Hewlett Packard Enterprise, and track sponsor AMD.

We would like to express our gratitude to every individual who contributed to the success of PCT 2017. Special thanks to the Program Committee members and the external reviewers for evaluating papers submitted to the conference. Thanks also to Organizing Committee members and all the colleagues involved in the conference organization from Kazan Federal University, the South Ural State University, and Moscow State University. We thank the participants of PCT 2017 for sharing their research and presenting their achievements as well.

Finally, we thank Springer for publishing the proceedings of PCT 2017 in the Communications in Computer and Information Science series.

May 2017

Leonid Sokolinsky
Mikhail Zymbler

Organization

The 11th International Scientific Conference on Parallel Computational Technologies, PCT 2017, was organized by the Supercomputing Consortium of Russian Universities and the Federal Agency for Scientific Organizations, Russia.

Steering Committee

Berdyshev, V.I.	Krasovskii Institute of Mathematics and Mechanics, Yekaterinburg, Russia
Ershov, Yu.L.	United Scientific Council on Mathematics and Informatics, Novosibirsk, Russia
Minkin, V.I.	South Federal University, Rostov-on-Don, Russia
Moiseev, E.I.	Moscow State University, Russia
Savin, G.I.	Joint Supercomputer Center, Russian Academy of Sciences, Moscow, Russia
Sadovnichiy, V.A.	Moscow State University, Russia
Chetverushkin, B.N.	Keldysh Institute of Applied Mathematics, Russian Academy of Sciences, Moscow, Russia
Shokin, Yu.I.	Institute of Computational Technologies, Russian Academy of Sciences, Novosibirsk, Russia

Program Committee

Sadovnichiy, V.A. (Chair)	Moscow State University, Russia
Dongarra, J. (Co-chair)	University of Tennessee, USA
Sokolinsky, L.B. (Co-chair)	South Ural State University, Russia
Voevodin, Vl.V. (Co-chair)	Moscow State University, Russia
Zymbler, M.L. (Academic Secretary)	South Ural State University, Russia
Ablameyko, S.V.	Belarusian State University, Republic of Belarus
Afanasiev, A.P.	Institute for Systems Analysis RAS, Russia
Akimova, E.N.	Krasovskii Institute of Mathematics and Mechanics UrB RAS, Russia
Andrzejak, A.	Heidelberg University, Germany
Balaji, P.	Argonne National Laboratory, USA
Boldyrev, Y.Ya.	Saint-Petersburg Polytechnic University, Russia
Carretero, J.	Carlos III University of Madrid, Spain
Gazizov, R.K.	Ufa State Aviation Technical University, Russia

Gergel, V.P.	Lobachevsky State University of Nizhny Novgorod, Russia
Glinsky, B.M.	Institute of Computational Mathematics and Mathematical Geophysics SB RAS, Russia
Goryachev, V.D.	Tver State Technical University, Russia
Il'in, V.P.	Institute of Computational Mathematics and Mathematical Geophysics SB RAS, Russia
Kobayashi, H.	Tohoku University, Japan
Kunkel, J.	University of Hamburg, Germany
Labarta, J.	Barcelona Supercomputing Center, Spain
Lastovetsky, A.	University College Dublin, Ireland
Ludwig, T.	German Climate Computing Center, Germany
Lykosov, V.N.	Institute of Numerical Mathematics RAS, Russia
Mallmann, D.	Julich Supercomputing Centre, Germany
Michalewicz, M.	A*STAR Computational Resource Centre, Singapore
Malyshkin, V.E.	Institute of Computational Mathematics and Mathematical Geophysics SB RAS, Russia
Modorsky, V.Ya.	Perm Polytechnic University, Russia
Shamakina, A.V.	High Performance Computing Center in Stuttgart, Germany
Shumyatsky, P.	University of Brasilia, Brazil
Sithole, H.	Centre for High Performance Computing, Republic of South Africa
Starchenko, A.V.	Tomsk State University, Russia
Sterling, T.	Indiana University, USA
Taufer, M.	University of Delaware, USA
Turlapov, V.E.	Lobachevsky State University of Nizhny Novgorod, Russia
Wyrzykowski, R.	Czestochowa University of Technology, Poland
Yakobovskiy, M.V.	Keldysh Institute of Applied Mathematics RAS, Russia
Yamazaki, Y.	Federal University of Pelotas, Brazil

Organizing Committee

Nurgaliev, D.K. (Chair)	Kazan Federal University, Russia
Khramchenkov, M.G. (Co-chair)	Kazan Federal University, Russia
Mosin, S.G. (Co-chair)	Kazan Federal University, Russia
Vasiliev, A.V. (Secretary)	Kazan Federal University, Russia
Antonov, A.S.	Moscow State University, Russia
Antonova, A.P.	Moscow State University, Russia
Banderov, V.V.	Kazan Federal University, Russia
Elizarov, A.M.	Kazan Federal University, Russia
Frolov, N.A.	Kazan Federal University, Russia
Nikitenko, D.A.	Moscow State University, Russia
Ostrovskikh, N.O.	South Ural State University, Russia

Contents

Supercomputer Simulation

High Performance Architectures, Tools and Technologies

An AlgoView Web-visualization System for the AlgoWiki Project

Alexander S. Antonov$^{(\boxtimes)}$ and Nikita I. Volkov

Lomonosov Moscow State University, Moscow, Russia
asa@parallel.ru, volkovnikita94@gmail.com

Abstract. There are countless ways to define an algorithm structure, which are mostly organized by flow of data, by executed tasks or by data decomposition. The so-called information graph provides a combination of these patterns. A possibility to investigate visually the information graph of a particular algorithm is, therefore, an adequate tool that helps to understand the algorithm itself, determining its resource of parallelism and figuring out how to code it better for parallel computing systems. In this paper, we present our approach to the information graphs visualization system, where online availability and low computational cost are the primary goals.

Keywords: Information graph · Parallel form · AlgoWiki · AlgoView · Web · Visualization · Algorithm

1 Introduction

1.1 The AlgoWiki Project

The AlgoWiki project has been on the scene since 2014. Its main purpose is to describe algorithms with special attention to their logical structure, resource of parallelism [1,2], possible parallel implementations, their scalability and efficiency [13]. Attention is also paid to less-known characteristics, such as data locality [3]. Powered by the MediaWiki engine and available online by reference [4], the whole project has already been developed steadily for three years, with more than 30 articles written specifically by its authors, as well as hundreds of articles created by its visitors. It is supported by the Russian Science Foundation and officially led by Dr. J. Dongarra, from the University of Tennessee.

1.2 Why is the Information Graph Important

A very important part of an algorithm description is the so-called information graph, since it allows to show the algorithm's parallel structure both in detail and at the macro level only. An algorithm structure can be defined in countless

The results were obtained at the Lomonosov Moscow State University with the financial support of the Russian Science Foundation (agreement N 14-11-00190).

L. Sokolinsky and M. Zymbler (Eds.): PCT 2017, CCIS 753, pp. 3–13, 2017.
DOI: 10.1007/978-3-319-67035-5_1

ways that follow a certain organization pattern. The *information graph* is defined in [1] as an acyclic graph where vertices stand for operations in the algorithm and edges stand for data flows. Its structure is defined by the algorithm itself and the amount of input data. Therefore, the number of vertices in an information graph is not limited in any way, while the number of edges can reach that of a fully connected graph. Such a graph provides by definition a combination of organization by flow of data, by executed tasks and by data decomposition patterns. Graph edges give a complete overview of data flows within the algorithm, while vertices and the common graph structure explain data decomposition and executed tasks in detail. A very similar term, namely DAG, is used to describe the behavior of an algorithm realization in numerous articles (see, for instance, [5]).

1.3   Related Work

Software capable of graph visualization exists for sure. Good examples of such software are [6,7]. But where would common users normally look for an algorithm description nowadays, in case they need one? Definitely, on the Internet. And it is not common to install something specific on one's laptop when willing to read a Wiki article. A notable example in this field is a whole family of molecular visualization systems, such as [8]. Furthermore, common graph visualization software is not designed specifically for displaying information graphs, in which keeping the regular structure of the original algorithm intact is a matter of primary importance. In face of this fact, it is appropriate to present an important definition from [9]: a *parallel form* of an information graph, in which all graph vertices are divided into a number of levels. If two vertices belong to the same level, then the operations they represent may be executed simultaneously. The parallel form is actually a basic restriction of graph representation aimed towards handling its structure carefully.

1.4   Setting up the Goal

Thus, building a desired system involves addressing a set of lesser tasks. The first one is to develop a method of presenting information graphs in 3-dimensional space with minimal obscurement of their regular structure, so that one can comprehend it as easily as possible. This is followed by the creation of a tool capable of performing a transformation sequence from some well-known and developed representation of an algorithm to the actual 3D-model, so that source data for the tool to work would already exist, and it would not be necessary to build them from scratch. Finally, the online-availability goal defines that a displayer should not require large computational resources from the device it is run on. However, we do not bother much about amounts of Internet connection bandwidth required for the tool to work smoothly, since such approach follows a common tendency: increasing bandwidth of networks and increasing popularity of cloud computing.

2   Proposed Solution

In this section, we successively address the tasks described in the Introduction section by giving a detailed description of the methods developed. The whole system is divided into two independent tools. The exporter tool generates a set of 3D-models by means of parsing an information graph description obtained from the reference realization of an algorithm. Doing so beforehand, we relieve users of the task of performing this costly operation. The displayer tool combines the pre-generated set into an interactive 3D-representation available online through any browser with WebGL support. A simple and well-known mechanism is used to integrate html pages containing such 3D-representations directly into Wiki articles.

The builder and displayer software tools mentioned in this section will be depicted in detail in the Implementation part.

2.1   A Method of Organizing a 3-Dimensional Graph Representation

To start with, a method that makes an automatically built 3-dimensional representation of an information graph as easy to comprehend as possible should be developed. In our previous research within the AlgoWiki project, a set of recommendations was given in order to make 2-dimensional manually-built information graph visualizations more uniform and representative. Those recommendations included such things as explicit indication of data flow directions, using different colors for vertices containing different operations, visualizing vertices standing for macro operations (subgraphs of the original information graph, each represented as a single vertex in order to simplify the entire visualization) with circles of different sizes, and many others. Basic recommendations that make common sense are used in the AlgoView system.

But the most important thing is a method we used to create 2-dimensional projections of 3-dimensional representations manually. A 3-dimensional Cartesian space and a grid in it are introduced. With the use of macro operations, many information graphs can be represented with a combination of loop nests not deeper than three levels each. So, each graph vertex is placed in a grid node with coordinates based on iteration parameters of loops in the nest for the given node, while graph edges just connect grid nodes. This approach was further developed in our builder tool to work with loop nests of four and more levels. We also address a problem of graph edges possibly interfering with each other by using square Bezier curves to calculate exact line-positions. Since such a curve is defined by three points, and we have information only about the two grid nodes being connected, the exact position of the third point also depends on the iteration parameters of the loops. The axes of the Cartesian space are added directly into the information graph 3-dimensional representation to make orientation easier.

2.2   Building 3-Dimensional Graph Models

An exact definition of "well-known and developed algorithm representation" should be given in order to address the second task mentioned in the Introduction section. In this paper, we deem the source code of a reference algorithm realization to be such a representation. We propose building a system that performs a series of transformations: source code → inner representation → XML representation → a set of 3D-models → 3D-representation. Assuming that transformations here are indexed starting with zero, the displayer tool is responsible for the third transformation, the builder is responsible for the second, an Algo-Graph tool created by our colleagues responds for the first, and performing the zeroth step is among our future plans and projects. Thus, creating a set of 3D-models from an XML file of known structure is the main transformation described in this work.

```xml
<?xml version="1.0" encoding="UTF-8"?>
<algograph version="0.0.1">
 <algorithm num_ext_params="1" num_groups="2">
  <ext_param name="n"/>
  <loop>
   <group id="0" number="3" depth="1" num_or_parts="1" num_oper="1">
    <or_part id="0" num_and_parts="1">
     <and_part id="0"><![CDATA[]]></and_part>
     <statement id="0" num_inputs="1">
      <input id="0" num_solutions="0"/>
     </statement>
    </or_part>
   </group>
  </loop>
 </algorithm>
</algograph>
```

Fig. 1. XML file describing a simple code fragment

The main idea of the builder tool is to handle loop nests defined in an XML file by the previously described method. The XML file has the following structure (see Fig. 1). It describes one algorithm: a sequence of loops or groups; a loop consists of a sequence of loops or groups; a group stands for a set of vertices (solutions) and edges (dependencies); exact coordinates for vertices depend on exact iteration parameters of the loops a group belongs to according to the previously described method with 3-dimensional Cartesian space; an edge coordinates are based on the coordinates of the vertices it connects. Thus, the builder tool generates a list of coordinates for vertices and edges after parsing the given XML file. To create a set of 3D-models from that list, we use the following method. A set of functions is created to build geometrical 3-dimensional primitives, like spheres, cones, etc. This primitives can be transformed by passing parameters to appropriate functions, and these parameters are taken from the generated list for each edge and each vertex.

2.3   Achieving Online Availability

Creating a set of 3-dimensional models from such an XML file is a very costly operation, that is why it is done in advance without any participation of the client's device. The generated set of models is uploaded to the same web server that stores the web pages used for the final 3-dimensional representation. WebGL technology is used to display the uploaded models on the appropriate web page, so the 3-dimensional representation is accessible via the client's browser if it supports WebGL (and modern browsers do [10]). The displayer tool is basically a set of Java scripts attached to the web page that are responsible for loading pre-generated models into it.

3   AlgoView Toolset

3.1   The Builder Tool Description

The whole builder tool is written in C++ and is platform-independent. It, however, uses external libraries for certain auxiliary jobs and, thus, cannot be called a complete creation of ours. The RapidXML library [11] is used to make parsing the XML file into a DOM-tree easier. The Exprtk library [12] is used to evaluate expression strings contained in XML files. Our own source code is organized in several files as well.

The XML file name and output file names are used as input parameters. The builder tool works in several steps.

Initializing. First of all, objects of Para_parser and Builder classes are created. Input parameters are passed to these objects.

Parsing the DOM-tree. The Build_params method from Para_parser class does the heavy lifting here. The input XML file is, first of all, parsed into a DOM-tree by means of RapidXML. Then all the tree nodes are operated recursively. By the end of this stage, our Para_parser class object has two lists containing, respectively, the coordinates of the graph vertices and pairs of graph vertices (the first vertex is the beginning of an edge and the second is its end). We also plan to add information about loop variables for each vertex created, which can be later used for calculating exact edge-positions in 3-dimensional space, as well as for adding new interactive features for 3-dimensional representation that we do not have yet (see Sect. 4 for more details). The exact form of this information is still a matter of some concern.

Building 3D models in JSON format. The Build_JSON method from the Builder class takes pointers to both lists as parameters. All the list nodes are parsed in sequence, with 3D-model generation being called for each node. The Builder class has private methods used to create geometric primitives both for vertices and edges; these methods take coordinates stored in the list nodes as parameters.

3.2   The Displayer Tool Description

The output of the builder tool is, as we already know, a set of 3D-models, which are handled by the displayer tool to create a final representation. Each web page is used to display a single information graph and is supported by the mentioned set of Java scripts and source files. These include JS libraries, textures, auxiliary 3D-models (for example, axis models), Java scripts used to interact with the representation, etc. The whole set of files contains library files, auxiliary source files and actual elements of the displayer tool.

Displaying an information graph representation in a browser window is an easy task compared to building a set of 3D-models from an XML file. It is done in several steps.

Initializing. Whenever some client decides to open one of our web pages in his browser, the InitGUI() function is called, creating a simple GUI to interact with the information graph representation. The webGLStart() function creates the canvas used to display the representation; a GL type object is initialized, as well as shaders and textures; an appropriate set of 3D-models is loaded into memory; reaction to user actions is set up. Then the tick() function is called, so the system will work in real time.

Animating. The tick() function is responsible for drawing the current frame and calculating parameters for the next frame (both implemented in separate procedures: DrawScene() and Animate()).

3.3   Integrating the System into the AlgoWiki Project

This is a simple but important part. The problem is that the MediaWiki engine used by the AlgoWiki project does not support WebGL technology. So we had to figure out a method for integrating our web pages into AlgoWiki. Fortunately, the IFrame MediaWiki widget was designed for performing such actions. All we had to do was to install IFrame on AlgoWiki, and add the appropriate pages by its means.

4   Experiments and Results

Here we describe our achievements in creating 3-dimensional representations of information graphs. We have built some simple representations of hypothetic graphs for testing purposes, and moved on to creating actual representations of algorithms described in the AlgoWiki project. For now, automatic building does not support creating macro operations at all, since XML files do not contain that kind of information. Also, the aesthetic side is affected in certain cases. To fix that, the builder tool can also be run in a manual mode in which it creates vertices and edges according to commands given by the user. A basic representation displays the coordinate axes, followed by all the graph vertices and edges without paying attention to the exact nature of specific operations

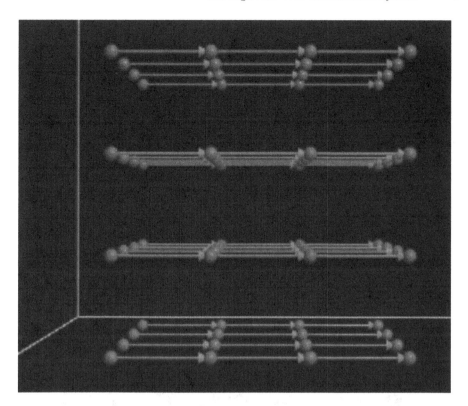

Fig. 2. Parallel matrix multiplication information graph, n = 4

and data flows within the information graph. In simple cases, this is sometimes not an issue, as shown in Fig. 2. Here a basic matrix multiplication is shown, all operations being ternary: $a + b * c$, where a is known from the previous operation in the same sequence, and b and c are elements of the input matrices. To address that, we plan to use different colors for different operations, and add auxiliary information about what each color means, like we did in the 2-dimensional representation. Another important idea one can deduce from the set of recommendations given for the information graph 2-dimensional representation was to display input data as vertices of different shapes. For the moment, we do not show them in the 3-dimensional representation.

An important part of our 3-dimensional representation concept is interactivity. That means several actions are available to users. Many "multidimensional" information graphs follow a set of structure patterns, different for each dimension. Thus, the most obvious feature is the adjustable point of view shown in Fig. 3 going along with some preset camera settings that can be used for displaying graph projections onto the coordinate planes, as one can see in Figs. 4 and 5.

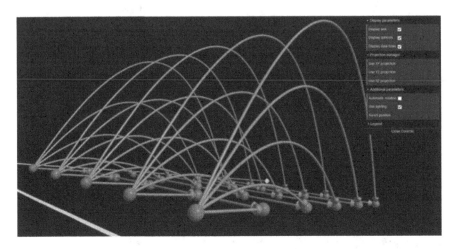

Fig. 3. Parallel matrix-vector multiplication information graph, n = 5; note the control panel in the upper right corner

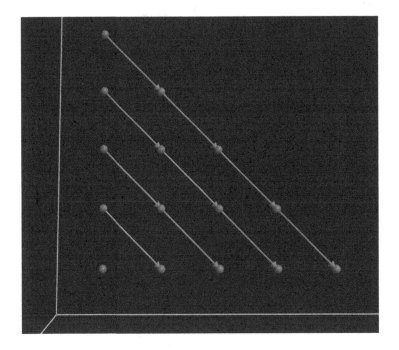

Fig. 4. Horner's scheme visualization, n = 6, XY projection

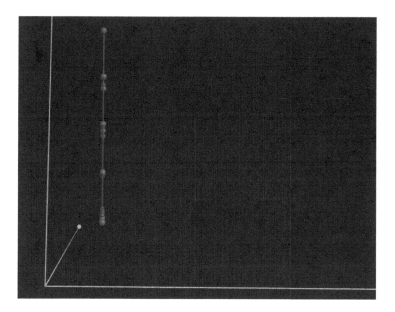

Fig. 5. Horner's scheme visualization, n = 6, YZ projection

Another option that may come in handy is the possibility to choose which parts of a 3-dimensional representation are currently displayed, as it can be noticed in Fig. 2 as well. This approach can ease the understanding of complicated structures common within information graphs, when it comes to something more intricate than matrix-vector operations. Addressing the parallel form of the information graph is also important, and the plan is to allow users to highlight a chosen level inside the graph, as well as to "walk" through levels to see how the algorithm execution goes. Like all the other features that are not present in the current version of 3-dimensional representation, it needs additional information stored in the vertex list described in Sect. 3.

The biggest issue we came across in building up 3-dimensional graph representations is the possible interference of graph edges. Assume we have a very simple information graph of an algorithm that represents a simple loop with n iterations where every iteration, except the first one, uses data from both the previous iteration and the first. Such a loop would result in a line of vertices with edges going from the first vertex to each one of the others, and neighbor vertices connected as well. This basically means that the edges would cover each other in a 3-dimensional representation. To avoid such an issue, the Bezier curve technique previously mentioned in Sect. 2 was introduced. But this technique itself does not completely resolve possible issues, as one can see in Fig. 6.

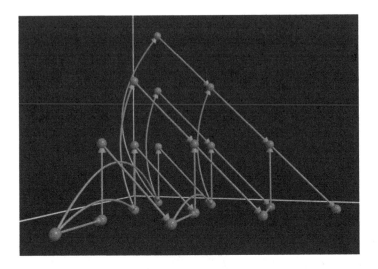

Fig. 6. Cholesky factorization information graph, n = 4

5   Conclusion

In this paper, we have reviewed our previously developed approach to the theoretical basis of creating 3-dimensional information graph representations. We have introduced the idea of a representation transformation sequence and have described a toolset implemented to automatically perform the two steps of the sequence that had not been worked out yet. Our toolset has been integrated into the AlgoWiki project, and some algorithm descriptions have been supplemented with our 3-dimensional representations. However, there is still work to do, since, as we have shown, our system sometimes does not work well in the aesthetic aspect, allowing data flows in the representation to interfere each other. Along with improvements to the tool responsible for transforming the source code into the inner representation, minimizing the number of such "difficult" algorithms stands as our primary goal. We have tested the system by creating 3D-representations of several sample algorithms described in AlgoWiki, and they have proved to be successful.

References

1. Voevodin, V.V., Voevodin, V.V.: Parallel Computing. BHV-Petersburg, St. Petersburg (2002). (in Russian)
2. Voevodin, V., Antonov, A., Dongarra, J.: Why is it hard to describe properties of algorithms? Proc. Comput. Sci. **101C**, 3–6 (2016). doi:10.1016/j.procs.2016.11.002
3. Alexander Antonov, A., Voevodin, V., Voevodin, V., Teplov, A.: A study of the dynamic characteristics of software implementation as an essential part for a universal description of algorithm properties. In: 24th Euromicro International Conference on Parallel, Distributed, and Network-Based Processing Proceedings, pp. 359–363 (2016). doi:10.1109/PDP.2016.24

4. Open Encyclopedia of Parallel Algorithmic Features AlgoWiki. https://algowiki-project.org/
5. Sharp, J.A. (ed.): Data Flow Computing: Theory and Practice. Ablex Publishing Corp, Norwood (1992)
6. The Open Graph Viz Platform. https://gephi.org/
7. Graph Visualization Software. http://www.graphviz.org/
8. An Open-Source web-application. http://molview.org/
9. Voevodin, V., Antonov, A., Dongarra, J.: AlgoWiki: an open encyclopedia of parallel algorithmic features. Supercomput. Front. Innov. **2**(1), 4–18 (2015). doi:10.14529/jsfi150101
10. Chong Lip Phang: Web Coding Bible (HTML, CSS, Javascript, PHP, SQL, XML, SVG, Canvas, WebGL et al.)
11. A RapidXML manual. http://rapidxml.sourceforge.net/manual.html
12. Exprtk manual. http://www.partow.net/programming/exprtk/
13. Antonov, A., Frolov, A., Kobayashi, H., Konshin, I., Teplov, A., Voevodin, V., Voevodin, V.: Parallel processing model for Cholesky decomposition algorithm in AlgoWiki project. Supercomput. Front. Innov. **3**(3), 61–70 (2016). doi:10.14529/jsfi160307

The Top50 List Vivification in the Evolution of HPC Rankings

Dmitry Nikitenko[✉] and Artem Zheltkov

Lomonosov Moscow State Uniersity, 1 Leninskie Gory, 119234 Moscow, Russia
dan@parallel.ru, artzlt@mail.ru

Abstract. The most productive and powerful supercomputers always arouse great interest being flagships of the HPC machines fleet. These giants always require new forms of presentation because of innovations in design. This also leads to evaluation and development of new ways of comparison and supercomputer ranking. There is a variety of known regional and world-level rankings based on different methods. Every ranking is rich in peculiarities in its development. In this paper we try to observe the evaluation of supercomputing rankings and discuss our vision of vivification of the oldest regional ranking - The Top50 list of most productive supercomputers of Russia and CIS and its features.

Keywords: Supercomputer · Model of supercomputer · HPC ranking · Performance and productiveness · Comparison of HPC systems · Methods of description for supercomputer design

1 Introduction

Researchers in the area of computational mathematics and adjacent areas have always been attracted to the choice of implementation methods which would better suit for a certain algorithm of calculations.

Computers have been evaluating fast and the diversity of architectures and technologies has been growing as well, multiplying the available variety of platforms and particular implementation hints.

It is quite natural that there appeared a need for measuring various platform capabilities even at relatively early stages of computing bringing various benchmarks to life. Later together with growing concurrence of platforms on the market it actually gave birth to appearance of HPC rankings.

Good news is that all these rankings altogether give us colorful multidimensional picture of HPC roadmap and trends, taken from a variety of viewing points at a time.

Bad news is that despite rich performance evaluation history, each ranking has its own weak sides. It can be seen taking form of insufficient architecture

The results were obtained in the Lomonosov Moscow State Uniersity with the financial support of the Russian Science Foundation (agreement No. 14-11-00190) and the Russian Foundation for Basic Research (project No. 16-07-01199).

L. Sokolinsky and M. Zymbler (Eds.): PCT 2017, CCIS 753, pp. 14–26, 2017.
DOI: 10.1007/978-3-319-67035-5_2

description, poor applicability to real-life applications seldom update period, low participants number, etc.

Of course, every list has its own goals, its own core audience and locality, but it is also interesting that all ranking engines have to be updated from time to time, for example, to start supporting heterogeneous systems with a number of computing unit and core types in the system.

It seems to be a good idea to look through the history of HPC rankings and peculiarities to develop a ranking design that would at least meet the HPC list requirements for close future and would be as useful for most users as it can be.

The paper has the following structure.

Section 2 covers a background of the project, reviewing existing HPC rankings and summarizing the state of the art as a basis for the project evaluation.

Section 3 is devoted to the implementation details of the proposed design of HPC system description for HPC rankings.

Section 4 introduces important features that become available with the proposed approach.

Section 5 addresses the forthcoming research work packages and our roadmap vision for the close future.

Broadly speaking in this paper we try to summarize our experience and vision of HPC rankings present, and to discuss approbation of new ideas on Top50 list of the most productive supercomputers of CIS and Russia.

2   Background

2.1   HPC Rankings

As of today, the most well-known lists are Top500, Green500 and Graph500, which are receiving applications for participation all over the world. There are also regional ratings, such as the Top50 of Russia and CIS, Top Supercomputers-India, The Irish Supercomputer List.

All these ratings include descriptions of supercomputer systems that may differ in structure and available characteristics; each list has its own method of describing and representing its participating systems.

All the rankings can be divided into two groups: global rankings (Table 1) that accept participating systems from all over the world and regional rankings (Table 2) that represent the situation in a certain region or a country.

2.2   HPC Rankings

Speaking about high performance computing, the most famous is a Top500 rating that is known to be as long living, as about fifty half-year editions old, with the first list announcement dated back to 1993 [1].

The Top500 rating is the largest ranking of supercomputer systems in the world. It is based on computer performance value, which is obtained via the HPL (High Performance Linpack [2]) benchmark. In this benchmark, a dense

Table 1. Global HPC rankings (Perf. – Performance, Eff. – Efficiency).

Global rankings	Top 500	Green 500	Graph 500	Green graph 500	HPCG	HPGMG
Gradation criteria	Perf.	Eff.	Perf.	Eff.	Perf.	Perf.
Benchmark	HPL	HPL	BFS	BFS	HPCG	HPGMG-FV
Units	FLOPS	FLOPS/W	TEPS	TEPS/W	FLOPS	DOF/S
Year 1st announced	1993	2007	2010	2013	2014	2014
Lists/Year	2	2	2	2	2	2
Editions issued on:	Jun Nov	Jun Nov	Jun Nov	Jun Nov	Jun Nov	Jun Nov
Number of systems as of 11.2016	500	500	216	30	101	11
Quality of system description	High	High	Fair	Poor	Poor	Poor
Heterogeneous architecture	No	No	No	No	No	No

Table 2. Global Top500 list and regional rankings.

Ranking title	Top 500	Top50 Russia	Top super-computers India	The Irish supercomputer list
Ranking region	World-wide	Russia and CIS	India	The island of Ireland
Gradation criteria	Perf.	Perf.	Eff.	Perf.
Benchmark	HPL	HPL	HPL	HPL
Year 1st announced	1993	2004	2008	2013
Lists/Year	2	2	2	2
Editions issued on:	June November	March/April September	June December	June November
Number of systems as of 11.2016	500	50	33	27
Quality of system description	High	High	Fair	Fair
Heterogeneous architecture	No	Yes	No	Partial

system of linear equations $Ax = f$ is being solved with an algorithm conforming to LU factorization with partial pivoting. The result of the benchmark is the number of floating point operations per time, referred to as FLOPS (Floating-point OPerations per Second).

Green500 list [3] is based on the same submitted to the Top500 list data, where the main criterion of comparison is the energy efficiency of systems. Participating systems are compared by performance obtained via HPL divided by the power consumption of system including its infrastructure during calculations. Thus, the metric for a given rating is the FLOPS/W ratio. The information about supercomputer systems for this list is the same as for the Top500. The Green500 ranking was first announced in 2007.

HPCG [4] is a computing systems rating, which is using the HPCG (High Performance Conjugate Gradients) benchmark to get performance of participating systems. Like HPL, this benchmark implements solution of systems of linear equation, but, unlike HPL, the conjugate gradients method is used for it. The resulting performance obtained by HPCG is also measured in FLOPS and turns out to be 1–2 orders of magnitude lower compared with the results of HPL. This rating was proposed by the creators of the Top500 ranking as an alternative to this list, because over time, the performance of computer systems on the HPL benchmark and real-world applications became significantly different. HPCG benchmark result can be considered as performance in the worst case, as "pessimistic" estimation of computing power of a system. The HPCG list appeared for the first time relatively recently in 2014.

The HPGMG rating is positioned as one of the Top500 list alternatives and it uses the HPGMG (High Performance Geometric MultiGrid) benchmark [5] for estimating systems performance. In this benchmark, solutions of systems of linear equations are also being found, but the multigrid methods are used. The result of this benchmark is measured in the number of solved differential equations in a unit of time or quantity of degrees of freedom per time DOF/s (Degrees Of Freedom per second). While HPL benchmark shows the best really achievable performance and HPCG benchmark - performance of a system in the worst case, the HPGMG benchmark was designed as a cross between these two, in order to estimate the applicability of computer systems for real applications. The benchmark is pretty young starting from 2014 just like HPCG, and the current list edition consists of just as much as 11 entries at present.

Graph 500 [6] is an another alternative supercomputer rating to Top500 and it is based on a performance value obtained via the DFS (Depth First Search) benchmark, which is oriented to process large data sets. The main task in this benchmark is to find edges in a very large graph, which is reflected in the title of the rating. The result of benchmark is the number of vertices being traversed divided by time measured in TEPS (Traversed Edges Per Second). Started in 2010, this ranking is the first large-scale successful attempt to move from performance to real productive capability measurement.

The Green Graph 500 rating [7] in relation to the rating Graph 500 is analogous to the list Green500 in relation to the Top500 rating, where performance divided by consumed energy is calculated, according to result of a supercomputer system in the Graph500 rating. Thus, the metric for rating Green Graph 500 is the TEPS/W ratio.

2.3   Regional HPC Rankings

Top50 supercomputer list of Russia and CIS [8] is a ranking of the 50 most powerful computing systems located on the territory of Russia and CIS countries. Performance is estimated using the HPL benchmark in accordance with the rules established in the Top500 rating.

List of Top Supercomputers-India [9] is a ranking of the most powerful high-performance systems in India. Performance value in this list is also made in accordance with the requirements of the Top500 list and measured via the HPL benchmark.

The Irish Supercomputer List [10] is a list of the supercomputers in Ireland with the highest performance. Performance of participating systems is measured using the HPL benchmark.

2.4   Regional HPC Rankings

As hardware capabilities of supercomputers and task scales grow rapidly, question of resource utilization is a hot topic. There are tons of conferences, discussions and publications on performance measurements and benchmarking. Let us highlight just several ideas that deal with HPC rankings.

- FLOPS of HPL at present are very much like Horse Power for the automobiles: nice for the marketing, but hardly represent real productivity. Though, the metric is and will be supported in close future because of rich historical HPC roadmap already covered.
- New rankings appear based on alternative benchmarks that more and more face productivity instead of performance as Top500 built on HPL no longer represents real-life computing for most applications.
- Global HPC rankings are very inert, but still have to renew engine one per 7–8 years with evolution of hardware design principles and new hot topics rising.
- Most lists try to avoid abundant data in supercomputer system description, especially global rankings.
- Regional, local rankings have rather low refreshment rate, usually no more than 10–20%. So systems in such rankings have a much longer lifetime, what is often close to the systems usual lifetime (7–10 years).

This information is valuable for us as it illustrates quite a difference in global and regional rankings [11–13]:

- Global rankings are up to the whole world scale trends exposure, but they do illustrate the top of the iceberg only.
- Local rankings are much closer to the real situation in HPC of a local area, as a significantly bigger part of all in-service systems is in the list.

Consequently, updating the engine of a local rating to support variety of metrics would potentially be close to an encyclopedia of real HPC capabilities of the specified geographical region, application area or computing pattern.

3   Implementation Basis

3.1   Disadvantages of Current Models for Describing Computer Systems

Speaking about presented information about supercomputer systems, the Top500 rating is one of the most detailed ratings, while, but even this rating does not reflect all the features of participating systems [14]. For instance, it does not provide a description of heterogeneous systems which incorporate groups of nodes of various types. Besides, coprocessors and accelerators are not separated and its description is not detailed enough (the total quantity and model of GPU or coprocessor unit are only given). Finally, there is no information about a topology of a system, its data storage system and its other additional characteristics. Descriptions of systems, represented in the Green500 and the Graph500 ratings, contain the same or less information, so the same can be said about them. The Top Supercomputers-India rating has a good detailed description of computation components of systems, however, this information is stored in a text format, and therefore cannot be structured well, so it becomes practically unsuitable for computing some statistic data.

The Top50 supercomputers of Russia and CIS rating also gives us a well-detailed description, and moreover it has more clearly defined structure, when compared to another regional rating the Indian supercomputer ranking, but some implementation features like separate models for certain HPC system types (clusters, SMPs, etc.) together with fixed hardware option descriptors doesn't allow using the ranking engine without plenty of temporary bug-fixing add-ons.

Based on analysis of information on participating systems, it can be assumed that the Top500 rating uses the only one entity to describe all the attributes of represented systems, or there is a main entity, which covers the majority of characteristics of given systems, and a number of small dictionary entities (countries, regions, processor series, etc.) which are connected to the main entity via the foreign key. The same can be said about the rest of ratings presented in this review.

Anyway, in these models, a set of properties of the described systems is strictly held, so it leads to incomplete description in case of appearance of systems with new types of components or makes it necessary to review and adjust model regularly for making it able to store new kind of information.

To sum up one can see a variety of supercomputer ratings and the diversity of information being represented in them. Systems are described with different levels of detail and in different formats. At the moment, there is no universal model, which can contain all the features being used in these ratings and have a possibility to extend the set of stored characteristics introducing new components and new peculiarities of supercomputer systems. This lack of such a model problem arises as often, as new features in HPC system design appear. The proposed model of describing computing systems is aimed at solving such a problem.

3.2   Proposed Method for Describing Supercomputer Systems

When building of the supercomputer model a computational node was separated as the main component and the basic element of description. This choice was made due to the fact that a compute node can be logically isolated in any computing device, as the component which is responsible for computation (in the trivial case, the device is represented as a single compute node, coinciding with this device itself).

A computer consists of a number of node groups with different types; nodes within the same group are considered to be the same in architecture. Each node can contain a number of components of various types (processors, co-processors, accelerators, etc.) and have several attributes (including, for example, RAM or HDD storage). Each component, in turn, has its own set of attributes and, generally speaking, may contain other components.

The computer itself can also have a list of attributes, which are applicable for a whole computing machine like operating system, etc.

This model is abstract enough to store all sorts of information and requires the following key entities:

- Object. An Object represents a component. Objects can be linked with other components via links of the type "contains". Computing node and a computer itself are objects.
- Attribute. An Attribute represents a characteristic of each computer and its components. Values of attributes are specified for objects they belong to.

Via these entities we can obtain unification of representation of systems components and its properties and achieve the requirement of universality and flexibility of the model. For further convenience, the model can be supplemented with other entities, which are specific for all types of the described systems.

Entities and relationships used in the model are presented in Fig. 1.

The general idea lies in the fact that not every type of element of a system needs to be described by a separate entity (or table) an opposite approach meaning detailed description of each element type would lead to the creation of enormous number of entities and that would cause difficulties when working with a model. Furthermore, such an approach limits a possible description model by initially created entities, and for describing new, not pre-specified components of the system, we will need to re-design the model each time they appear.

Lets illustrate an example of how it would be possible to take advantage of our approach. For example, in the Top500 rating there is no information about the system storage of a computer system.

In our proposed model, this information will be easy to add: the storage system will be presented as a separate device with its own characteristics, i.e., it will be described as an instance of an existing entity Object and its properties as instances of Attribute. Likewise, you can describe the memory hierarchy of supercomputer systems, describing their levels one-by-one.

The developed abstraction of the model will not limit its scope of application only to the ratings of supercomputers and will provide an opportunity to expand it to other areas. Via this model, for example, we can describe the characteristics of mobile devices or embedded systems, using a similar approach to the description of devices and its components.

It stands to mention the importance of structuring the data, which is adhered in the model. A structured approach to the description of supercomputer systems will allow performing deeper analysis of data and identifying trends in

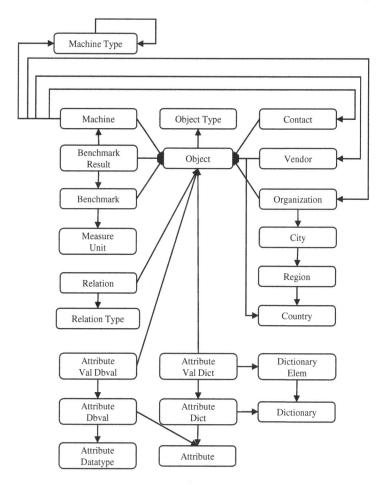

Fig. 1. Principle design of description for HPC systems [14].

evolution of the supercomputer area, as well as identifying the key factors of growth of the performance of systems. It is one of the main objectives of the emergence and existence of the supercomputer rankings. In addition, such structured descriptions can be used to automate the processing of new applications for participation in the ranking. Fully automation of this process is unlikely to be reached, but existence of a clear structure can be suitable for the most basic checks such as verifying the theoretical performance (Rpeak) of a system, the correctness of components characteristics, etc.

4 Approach Features

The developed core of the ranking engine is designed with the following topics in mind as important features of the project results.

4.1 Vivification of Ranking Data Analysis

In-depth analysis. Support for in-depth analysis that would allow studying the statistics and history of references. This would allow following short links to the specific sublists and stats.

In the regular ranking list, almost every part of system description is a link to the corresponding sublist and statistics (Fig. 2), for example:

- entries of all the vendor systems in the current list,
- systems using specified interconnect,
- systems having specified type of accelerators or processor model,
- systems hosted by the same site,
- and so on.

The link from the displayed HPC system name of its position drives to the detailed system information page, where more detailed information is provided, including history of appearing in HPC rankings. By the links from this history user can proceed to the description of the HPC system that corresponds to the exact ranking edition. Clicking on system components in detailed HPC system info mode shows detailed information on the selected component. For example, for a CPU it is vendor, family, model, number of cores, and frequency. The actual layout for the pages is under development now and it is planned to be complete in the close future.

The key idea is to give an answer to the most popular statistical questions in no more than 3 clicks from the main page.

Stats constructor. A needless part for efficient tool is a clear interface with a high usability, flexible sublist and stats generator functionality. Besides the quickly accessed sublists and stats directly from the ranking list, it is quite important to provide means for an all-round analysis. Not only statistics by a specified hardware option may be interesting, but their combinations as well. For example, the average available memory per core is quite an interesting parameter, and so on.

Speaking of building various statistics and their representation, Top500 list is ultimately leading as for now. Nevertheless, we plan to built even more flexible toolkit for the stats, especially keeping in mind possible alternative ranking engine implementation areas.

Visualization techniques. Speaking of visualization methods it is quite important to develop and support optional representation formats in addition to regular graphs and diagrams. For example, introducing methods of music charts where changing popularity of popular compositions is marked, can be useful and illustrative for the list itself.

Figure 2 illustrates the draft of the default main list layout. In the left column one can see current system position and changes indicators: is it a new or upgraded system (upgrade sign links to the description of previous system state), indicator of movement up or down comparing to the previous ranking edition (if the system position changed, user can see a pop-up message with the details like

№	Название Место установки	Узлов Проц. Ускор.	Архитектура: кол-во узлов [конфигурация узла] сеть: вычислительная / сервисная / транспортная	Rmax Rpeak (Tflop/s)	Разработчик
1 ▲ upgrade	«Ломоносов-2» Московский государственный университет имени М.В.Ломоносова	1280 1280 1280	1280 [1x Intel Xeon E5-2697v3, 64 GB RAM; Acc: 1x NVIDIA Tesla K40M] сеть: FDR Infiniband / Gigabit Ethernet / FDR Infiniband	1849.0 2575.87	Т-Платформы
2 ▽	«Ломоносов» Изменение: -1. Место в предыдущей редакции списка: #1 Московский государственный университет имени М.В.Ломоносова	6199 2130	4160 [2x Intel Xeon X5570, 12 GB RAM] 260 [2x Intel Xeon X5570, 24 GB RAM] 640 [2x Intel Xeon X5670, 24 GB RAM] 40 [2x Intel Xeon X5670, 48 GB RAM] 30 [2x IBM PowerXCell 8i, 16 GB RAM] 777 [2x Intel Xeon E5630, 12 GB RAM; Acc: 2x NVIDIA Tesla X2070] 288 [2x Intel Xeon E5630, 24 GB RAM; Acc: 2x NVIDIA Tesla X2070] 4 [4x Intel Xeon E7650, 512 GB RAM] сеть: QDR Infiniband / Gigabit Ethernet / Gigabit Ethernet	901.9 1700.21	Т-Платформы
3 new	«Политехник РСК Торнадо» Суперкомпьютерный центр, Санкт-Петербургский политехнический университет	712 1424 120	612 [2x Intel Xeon E5-2697v3, 64 GB RAM] 56 [2x Intel Xeon E5-2697v3, 64 GB RAM; Acc: 2x NVIDIA Tesla K40] 28 [2x Intel Xeon E5-2697v3, 128 GB RAM] 8 [2x Intel Xeon E5-2697v3, 128 GB RAM; Acc: 1x NVIDIA K1] 8 [2x Intel Xeon E5-2697v3, 128 GB RAM] сеть: FDR Infiniband / Gigabit Ethernet / Gigabit Ethernet	658.11 829.34	Группа компаний РСК

Fig. 2. Draft Top50 list layout with clickable components and extended position info.

previous position and number of positions gained or lost). Many other methods yet to be studied and implemented.

4.2   AlgoWiki

Extremely perspective and ambitious integration point managing data from the bases of real-life applications and algorithm implementations.

The idea is quite simple at a first glance. People run a popular algorithm implementations or an application packages with the specified parameters on various platforms and the results are stored into a database. This forms the basis for portability, efficiency and scalability analysis and study of suitability of this or that HPC system (or its part) for the specific application execution.

The question of all-round formal description of the studied algorithms should be taken care of at first. From this point of view AlgoWiki [15–17] is a perfect project to work in cooperation with.

4.3   OctoTron

Every system that participates in the ranking should have a page with full information on it available. It should contain full-detailed configuration, history of positions in popular ranking, benchmarking results, even photos and interesting facts that can be provided by the system holder or a vendor. Such a description would be much wider if we could show not only the configuration itself, but real topologies of supercomputer components.

In this, we see one of the most noticeable results of the OctoTron project a formal model of supercomputers that can be built experimentally according to the real installation peculiarities [18]. Asserted to the data on a specific machine configuration it would form a complete description of the HPC system design regarding hardware.

Such aggregated information on system configuration would also be very useful being available through the systems for the support of any computing center [19–21], especially for the largest HPC centers [22].

4.4 Mobile Linpack

The proposed implementation principles allow to apply the engine in similar adjacent areas. For example, it is planned to be tested as an engine for the Mobile Linpack project [23]. This list implements HPL testing of mobile devices very much alike Top500 and Top50 does.

5   Conclusion and Future Work

In the conclusion we would like to share our plans for the future. The project roadmap for now includes following work packages.

- Implementation of the existing engine full functionality by April, 2017 to meet the upcoming revision of the Top50 list with all existing functionality of present engine to ensure the operation capabilities of the new engine are fine.
- Presentation the draft of renewed Top50 rating at the Parallel Computing Technologies 2017 international conference.
- Establishing a tradition of having a specialized poster or exhibition section participation with major up to the date list highlights at the key supercomputing-oriented events in Russia Russian Supercomputing Days and Parallel Computing Technologies conference series.
- Gathering feedback on obtained announced results and analyzing it with the following implementation of the analysis results.
- Making full-functioning statistics subsystem operating by summer to have a full-service engine as a point of discussion by the key European HPC event ISC2017 International Supercomputing Conference.
- Research on integration with AlgoWiki project.
- Test-driving the developed engine on The Mobile Linpack ranking.
- Russian Supercomputing Days 2017 official introduction of the totally renewed vivificated Top50 ranking of the most powerful and productive HPC systems of CIS and Russia.

We hope that our integral approach and the revived Top50 rating engine will bring fresh breath to the image of HPC ratings and the whole scientific community will benefit from that both in Russia and abroad.

References

1. Top500 Supercomputing Sites Timeline. https://www.top500.org/timeline
2. Dongarra, J.J., Luszczek, P., Petitet, A.: The LINPACK benchmark: past, present and future. Concurr. Comput. Pract. Exp. **15**, 803–820 (2003). doi:10.1002/cpe.728
3. The Green500 List. http://www.green500.org
4. HPCG Benchmark. http://www.hpcg-benchmark.org
5. HPGMG High-performance Geometric Multigrid. https://hpgmg.org
6. The Graph 500 List. http://www.graph500.org
7. Green Graph 500. http://green.graph500.org
8. Top50 Supercomputers of Russia and CIS. http://top50.supercomputers.ru
9. Top Supercomputers India. http://topsupercomputers-india.iisc.ernet.in
10. The Irish Supercomputer List. http://www.irishsupercomputerlist.org
11. Nikitenko, D.A.: Top50 rating as an indicator of progress in HPC area. In: High-Performance Parallel Computing on Cluster Environment (HPC-2010): Proceedings of the X International Conference (Perm, Russia, 2010), vol. 2, pp. 258–260. Publishing of the Perm National Research Polytechnic University, Perm (2010). http://hpc-education.unn.ru/files/conference_hpc/2010/files/2010_144(2).pdf
12. Nikitenko, D.A.: Top50 rating: where are we sneaking out? In: Supercomputers. no. 4 (16), pp. 60–61 (2013). http://supercomputers.ru/images/stories/Supercomputers_16-2013.pdf
13. Antonov, A.S., Nikitenko, D.A., Sobolev, S.I.: The 18th edition of the Top50 list of most powerful computers of Russia: expectations and prospects. In: Parallel Computational Technologies (PCT 2013): Proceedings of the International Scientific Conference (Ufa, Russia, 1–2 April 2013), pp. 258–260. Publishing of the South Ural State University, Chelyabinsk. http://omega.sp.susu.ru/books/conference/PaVT2013/short/161.pdf
14. Zheltkov, A.A.: Methods for describing high-performance computing systems and formation of supercomputer rankings. In: Russian Supercomputing Days: Proceedings of the International Conference (Moscow, Russia, 26–27 September 2016), Moscow State University, pp. 897–911 (2016)
15. Voevodin, V.V., Antonov, A.S., Dongarra, J.J.: Why is it hard to describe properties of algorithms? Procedia Comput. Sci. **101C**, 3–6 (2016). doi:10.1016/j.procs.2016.11.002
16. Antonov, A.S., Voevodin, V.V., Voevodin, V.V., Teplov, A.M.: A study of the dynamic characteristics of software implementation as an essential part for a universal description of algorithm properties. In: 24th Euromicro International Conference on Parallel, Distributed, and Network-Based Processing Proceedings, 17th–19th February 2016, pp. 359–363 (2016). doi:10.1109/PDP.2016.24
17. Antonov, A.S., Frolov, A.V., Kobayashi, H., Konshin, I.N., Teplov, A.M., Voevodin, V.V., Voevodin, V.V.: Parallel processing model for cholesky decomposition algorithm in AlgoWiki Project. Supercomput. Front. Innov. **3**(3), 61–70 (2016). doi:10.14529/jsfi160307
18. Antonov, A.S., Nikitenko, D.A., Shvets, P.A., Sobolev, S.I., Stefanov, K.S., Voevodin, V.V., Voevodin, V.V., Zhumatiy, S.: An approach for ensuring reliable functioning of a supercomputer based on a formal model. In: Wyrzykowski, R., Deelman, E., Dongarra, J., Karczewski, K., Kitowski, J., Wiatr, K. (eds.) PPAM 2015. LNCS, vol. 9573, pp. 12–22. Springer, Cham (2016). doi:10.1007/978-3-319-32149-3_2

19. Nikitenko, D.A., Voevodin, V.V., Zhumatiy, S.A.: Resolving frontier problems of mastering large-scale supercomputer complexes. In: Proceedings of the ACM International Conference on Computing Frontiers (CF 2016), pp. 349–352. ACM, New York (2016). doi:10.1145/2903150.2903481

20. Nikitenko, D.A., Voevodin, V.V., Zhumatiy, S.A.: Octoshell: large supercomputer complex administration system. In: Russian Supercomputing Days International Conference, Moscow, Russian Federation, 28–29 September 2015, Proceedings, CEUR Workshop Proceedings, vol. 1482, pp. 69–83 (2015). http://ceur-ws.org/Vol-1482/069.pdf

21. Nikitenko, D.A., Zhumatiy, S.A., Shvets, P.A.: Making large-scale systems observable - another inescapable step towards exascale. J. Supercomput. Front. Innov. **3**(2), 72–79 (2016). doi:10.14529/jsfi160205

22. Voevodin, V.V., Zhumatiy, S.A., Sobolev, S.I., Antonov, A.S., Bryzgalov, P.A., Nikitenko, D.A., Stefanov, K.S., Voevodin, V.V.: Supercomputer Lomonosov practices. In: Open Systems, no. 7, pp. 36–39 (2012). https://www.osp.ru/os/2012/07/13017641/

23. Mobile Linpack. http://linpack.hpc.msu.ru

Reconfigurable Computer Based on Virtex UltraScale+ FPGAs with Immersion Cooling System

I.I. Levin[1], A.I. Dordopulo[1(✉)], A.M. Fedorov[1], and A.A. Gulenok[2]

[1] Scientific Research Centre of Supercomputers and Neurocomputers Co. Ltd.,
Taganrog, Russia
levin@superevm.ru, scorpio@mvs.tsure.ru, ss24@mail.ru
[2] Scientific Research Institute of Multiprocessor Computer Systems
at Southern Federal University, Rostov-on-don, Russia
andrei_gulenok@mail.ru

Abstract. The paper covers the reconfigurable computer systems based on high integration field programmable gate arrays (FPGAs) of Xilinx Virtex UltraScale+ series, where immersion liquid cooling system is used for cooling of all electronic components. The original engineering and technological solutions for thermal interface, cooling liquid, radiators and the entire reconfigurable computer system are considered. Owing to these features it is possible to achieve unprecedented layout density, up to 128 FPGAs for a 3U computational module placed into a standard 19″ computer rack. The paper presents results of prototyping and experimental verification of energy-efficient computational module with immersion liquid cooling system. On the base of the new computational module it is possible to achieve the performance of 1 Pflops in a standard 47U rack with total power consumption not more than 150 kW. The designed immersion liquid cooling system has power reserve for already produced and for next-generation high integration FPGA series, resistance to leaks and their consequences, and compatibility with traditional water cooling systems based on industrial chillers.

Keywords: Reconfigurable computer systems · Immersion liquid cooling system · FPGAs · Liquid cooling · Computational module · High-performance computer systems · Real and specific performance · Energy efficiency

1 Introduction

One of perspective approaches to achieve high real performance of a computer system is adaptation of its architecture to a structure of a solving task for creation of a special-purpose computer device which hardwarily implements all

The project has been funded in part by the scholarship of the President of the Russian Federation for young scientists and graduate students (SP-173.2016.5).

L. Sokolinsky and M. Zymbler (Eds.): PCT 2017, CCIS 753, pp. 27–41, 2017.
DOI: 10.1007/978-3-319-67035-5_3

computational operations of the information graph of the task with the minimum delays. Here, versatility of solving tasks, i.e. possibility of modification of the solving task or of its algorithm, is the necessary requirement to the computer system, the same as its capability of special-purpose application. It is possible to eliminate these contradictions, combining creation of a special-purpose computer device with a wide range of solving tasks, within a concept of reconfigurable computer systems (RCS) based on FPGAs that are used as a principal computational resource [1].

RCS, which contain large FPGA computational fields, are used for implementation of computationally laborious tasks from various domains of science and technique, because they have a number of considerable advantages in comparison with clusterlike multiprocessor computer systems: high real and specific performance, high energetic efficiency, etc. So, special-purpose reconfigurable computer systems such as Janus [2,3] and Janus2, which are used for spin glass calculations and provide more than 100-time speedup in comparison with commercial cluster systems. The supercomputer Anton [4], designed on basis of ASIC, speeds up execution of molecular dynamic tasks in more than 1000 times.

The scientific team of Scientific Research Centre of Supercomputers and Neurocomputers (SRC SC & NC, Taganrog, Russia) designs and produces supercomputerlike RCS. The principal computational resource of these RCS is not microprocessors, but a set of FPGA chips, united into computational fields by high speed data transfer channels. Spectrum of designed and produced products is rather wide: from completely stand-alone small-size reconfigurable accelerators (computational blocks), desktop or rack computational modules (Rigel, based on Xilinx Virtex-6, Taygeta, based on Xilinx Virtex-7 FPGAs) to computer systems which consist of several computer racks, placed in a specially equipped computer room (RCS-7).

The main distinctive feature of the RCS, produced in SRC SC & NC, is high board density and high (not less than 90%) filling of FPGA chips, that, as a result, provide high specific energetic efficiency of such systems [5].

Practical experience of maintenance of large computer complexes based on RCS proves that air cooling systems have reached their heat limit. Continuous, at least double increasing of circuit complexity and 1.5-time increasing of the clock rate of each new family of Xilinx FPGAs lead to considerable growth of power consumption and also lead to growth of the maximal temperature on chip. So, for the XC6VLX240T-1FFG1759C FPGAs of a computational module (CM) Rigel-2 the maximum overheat of FPGAs relative to the indoor temperature of $25\,^\circ$C in an operating mode and with power of 1255 W, consumed by the CM, is $33.1\,^\circ$C, i.e. the maximum temperature of the FPGA chip of the CM Rigel-2 is $58.1\,^\circ$C. For the XC7VX485T-1FFG1761C FPGAs of the CM Taygeta the maximum overheat of FPGAs relative to the indoor temperature of $25\,^\circ$C in an operating mode and with power of 1661 W, consumed by the CM, is $47.9\,^\circ$C, i.e. the maximum temperature of the FPGA of the CM Taygeta is $72.9\,^\circ$C. If we take into account that the permissible temperature of FPGA functioning is $65...70\,^\circ$C, then it is evident, that for normal and safe maintenance of the CM Taygeta it is necessary to have specially cooled air of $15\,^\circ$C in a conditioned rack.

According to the obtained experimental data, conversion from the FPGA family Virtex-6 to the next family Virtex-7 leads to growth of the FPGA maximum temperature on 11...15 °C. Therefore further development of FPGA production technologies and conversion to the next FPGA family Virtex Ultra Scale (about 100 million equivalent gates, power consumption not less than 100 W for each chip) will lead to growth of FPGA overheat on additional 10...15 °C. This will shift the range of their operating temperature to 80...85 °C, which means that their operating temperature exceeds the permissible range of the FPGA operating temperature (65...70 °C), and hence, this will have negative influence on their reliability when chips are filled up to 85–95% of available hardware resource. This circumstance requires a quite different cooling method which provides keeping of growth rates of the RCS performance for the designed advanced Xilinx FPGA families: Virtex UltraScale, Virtex UltraScale+, Virtex UltraScale 2, etc.

2 Liquid Cooling for Reconfigurable Computer Systems

Development of computer technologies leads to design of computer technique which provides higher performance, and hence, more heat. Dissipation of released heat is provided by a system of electronic element cooling, that transfers heat from the more heated object (the cooled object) to the less heated one (the cooling system). If the cooled object is constantly heated, then the temperature of the cooling system grows and in some period of time will be equal to the temperature of the cooled object. So, heat transfer stops and the cooled object will be overheated. The cooling system is protected from overheat with the help of cooling medium (a heat-transfer agent). Cooling efficiency of the heat-transfer agent is characterized by heat capacity and heat dissipation. As a rule, heat transfer is based on principles of heat conduction, that require a physical contact of the heat-transfer agent with the cooled object, or on principles of convective heat exchange with the heat-transfer agent, that consists in physical transfer of the freely circulating heat-transfer agent.

To organize heat transfer to the heat-transfer agent, it is necessary to provide heat contact between the cooling system and the heat-transfer agent. Various *radiators* – facilities for heat dissipation in the heat-transfer agent are used for this purpose. Radiators are set on the most heated components of computer systems. To increase efficiency of heat transfer from an electronic component to a radiator, a *heat interface* is set between them. The heat interface is a layer of heat-conducting medium (usually multicomponent) between the cooled surface and the heat dissipating facility, used for reduction of heat resistance between two contacting surfaces. Modern processors and FPGAs need cooling facilities with as low as possible heat resistance, because at present even the most advanced radiators and heat interfaces cannot provide necessary cooling if an air cooling system is used.

Till 2013 air cooling systems were used quite successfully for cooling supercomputers. But due to growth of performance and circuit complexity of microprocessors and FGAs, used as components of supercomputer systems, air cooling systems have practically reached their limits for designed advanced supercomputers, including hybrid computer systems. Therefore the majority of vendors of computer technique consider liquid cooling systems as an alternative decision of the cooling problem. Today liquid cooling systems are the most promising design area for cooling modern high-loaded electronic components of computer systems.

A considerable advantage of all liquid cooling systems is heat capacity of liquids which is better than air capacity (from 1500 to 4000 times), and higher heat-transfer coefficient (to 100 times growth). To cool one modern FPGA chip, 1 m^3 of air or 0.00025 m^3 (250 ml) of water per minute is required. Transfer of 250 ml of water requires much less of electric energy, than transfer of 1 m^3 of air. Heat flow, trans-ferred by similar surfaces with traditional velocity of the heat-transfer agent, is in 70 times more intensive in the case of liquid cooling than in the case of air cooling. Additional advantage is use of traditional, rather reliable and cheap components such as pumps, heat exchangers, valves, control devices, etc. In fact, for corporations and companies, which deal with equipment with high packing density of components operating at high temperatures, liquid cooling is the only possible solution of the problem of cooling of modern computer systems. Additional possibilities to increase liquid cooling efficiency are improvement of the initial parameters of the heat transfer agent: increasing of velocity, decreasing of temperature, providing of turbulent flow, increasing of heat capacity, reducing of viscosity.

Heat transfer agent of liquid cooling systems of computer technique is liquid such as water or any dielectric liquid [6–8]. Heated electronic components transfer heat to the permanently circulating heat transfer agent liquid, which, after its cooling in the external heat exchanger, is used again for cooling of heated electronic components. There are several types of liquid cooling systems. Closed loop liquid cooling systems have no direct contact between liquid and electronic components of printed circuit boards. In open loop cooling systems (liquid immersion cooling systems) electronic components are immersed directly into the cooling liquid [7]. Each type of liquid cooling systems has its own advantages and disadvantages.

In closed loop liquid cooling systems all heat-generating elements of the printed circuit board are covered by one or several flat plates with a channel for liquid pumping. So, for example, cooling of a supercomputer SKIF-Avrora is based on a principle "one cooling plate for one printed circuit board". The plate, of course, had a complex surface relief to provide tight heat contact with each chip. Cooling of a supercomputer IBM Aquasar is based on a principle "one cooling plate for one (heated) chip". In each case the channels of the plates are united by collectors into a single loop connected to a common radiator (or another heat exchanger), usually placed outside the computer case and/or rack or even the computer room. With the help of the pump the heat transfer agent is pumped through the plates and dissipates heat, generated by the computational elements, by means of the heat exchanger. In such system it is necessary to provide access of the heat transfer agent to each heat-generating element of the calculator, what means a rather complex "piping system" and a large number of pressure-tight connections. Besides, if it is necessary to provide maintenance of the printed circuit boards without any serious demounting, then the cooling system must be equipped with special liquid connectors which provide pressure-tight connections and simple mounting/demounting of the system.

In closed loop liquid cooling systems it is possible to use water or glycol solutions as the heat transfer agent. However, leak of the heat transfer agent can lead to possible ingress of electrically conducting liquid to unprotected contacts of printed circuit boards of the cooled computer, and this, in its turn, can be fatal for both separate electronic components and the whole computer system. To eliminate failures the whole complex must be stopped, and the power supply system must be tested and dried up. Control and monitoring systems of such computers always contain multiple internal humidity and leak sensors. To solve the leak problem a method, based on negative pressure of liquid in the cooling system, is frequently used. According to this method, water is not pumped in under pressure, it is pumped out, and this practically excludes leak of liquid. If air-tightness of the cooling systems is damaged, then air ingresses the system but no leak of liquid happens. Special sensors are used for detection of leaks, and modular design allows maintenance without stopping of the whole system. However, all these capabilities considerably complicate design of hydraulic system.

Another problem of closed loop liquid cooling systems is a dew point problem. In the section of data processing the air is in contact with the cooling plates. It means that if any sections of these plates are too cold and the air in the section of data processing is warmer and not very dry, then moisture can condense out of the air on the plates. Consequences of this process are similar to leaks. This problem can be solved ether by hot water cooling, which is not effective, or by control and keeping on the necessary level the temperature and humidity parameters of the air in the section of data processing, which is complicated and expensive.

The design becomes even more complex, when it is necessary to cool several components with a water flow proportionally to their heat generation. Besides branched pipes, it is necessary to use complex control devices (simple T-fittings and four-way fittings are not enough). An alternative approach is use of an industrial device with flow control, but in this case the user cannot considerably change configuration of cooled computational modules.

Advantages of closed loop liquid cooling systems are:

– use water or water solutions as the heat transfer agent which are available, have perfect thermotechnical properties (heat transfer capacity, heat capacity, viscosity), simple and comparatively safe maintenance;
– the large number of standard mechanisms, nodes and details for water supply systems, which can be used;
– great experience of maintenance of water cooling systems in industry.

However, closed loop liquid cooling systems have a number of significant disadvantages, which restrict their widespread use:

- difficulties with detection of the point of water leakage;
- catastrophic consequences that are the result of leakages not detected in time;
- technological problems of leakage elimination (a required power-off of the whole computer rack, that is not always possible and suitable);
- required support of microclimate in the computer room (a dew point problem);
- a problem of cooling of all the rest components of the printed circuit board of the RCS computational module. Even slight modification of the RCS configuration requires a new heat exchanger;
- a problem of galvanic corrosion of aluminum heat exchangers or a problem of mass and dimensions restrictions for more resistant copper heat exchangers (aluminum is three times as lighter than copper);
- air removal from the cooling system that is required before starting-up and adjustment, and during maintenance;
- complex placement of the computational modules in the rack with a large number of fittings required for plug-in of every computational module;
- necessity of use of a specialized computer rack with significant mass and dimension characteristics.

In open loop liquid cooling systems the heat transfer agent is the principal component, a dielectric liquid based, as a rule, on a white mineral oil that provides much higher heat storage capacity of the heat transfer agent, than the one of the air in the same volume. According to their design, such system is a bath filled with the heat transfer liquid (also placed into a computer rack) and which contains printed circuit boards and servers of computer equipment. The heat, generated by electronic components, is dissipated by the heat transfer agent that circulates within the whole bath. Advantages of immersion liquid cooling systems are simple design and capability of adaptation to changing geometry of printed circuit boards, simplicity of collectors and liquid connectors, no problems with control of liquid flows, no dew point problem, high reliability and low cost of the product.

The main problem of open loop liquid cooling systems is chemical composition of the used heat transfer liquid which must fulfil strict requirements of heat transfer capacity, electrical conduction, viscosity, toxicity, fire safety, stability of the main parameters and reasonable cost of the liquid.

Open loop liquid cooling systems have the following advantages:

- insensibility to leakages and their consequences, capability of operating even with local leakages of the heat transfer agent;
- insensibility to climate characteristics of the computer room;
- solution of the problem of cooling of all RCS components, because the print-ed circuit board of the computational module is immersed into the heat transfer agent;
- capability of modification of the configuration of the printed circuit board of the computational module without modification of the cooling system;
- simplicity of hydraulic adjustment of the system owing to lack of complex system of collectors;

- possibility of use of standard mechanisms, nodes and details, produced for hydraulic systems of machine industry, and know-how of maintenance of electrical equipment that uses dielectric oils;
- increasing of the total reliability of the liquid cooling system.

Disadvantages of open loop liquid cooling systems are the following:

- necessity of an additional pump and heat exchange equipment for improvement of thermotechnical properties (heat transfer capacity, heat capacity, viscosity) of the heat transfer agent. Here special dielectric organic liquids are used as the heat transfer agent;
- necessity of training of maintenance staff and keeping increased safety precautions for work with the heat transfer agent;
- necessity of more frequent cleaning of the computer room because of high fluidity of the heat transfer agent, especially in the case of leakage;
- necessity of special equipment for scheduled and emergency maintenance operations (mounting/demounting of the computational module, loading/unloading of the heat transfer liquid, etc.);
- increasing of the maintenance cost because of necessity of regular change-out of the heat transfer liquid when its service life is over and necessity of heat transfer agent management (transporting, receipt, accounting, storing, distribution, recovery of the heat transfer agent, etc.) in the corporation.

Estimating the given advantages and disadvantages of the two liquid cooling systems we can note more weighty advantages of open loop cooling systems for electronic components of computer systems. That is why for RCS computational modules designed on the base of advanced FPGA families, it is reasonable to use liquid cooling, particularly immersion of printed circuit boards of computational modules into the liquid heat transfer agent based on a mineral oil.

At present the technology of liquid cooling of servers and separate computational modules is developed by many vendors and some of them have achieved success in this direction. However, these technologies are intended for cooling computational modules which contain one or two microprocessors. All attempts of its adaptation to cooling computational modules which contain a large number of heat generating components (an FPGA field of 8 chips), have proved a number of shortcomings of liquid cooling of RCS computational modules.

The main disadvantages of existing technologies of immersion liquid cooling for computational modules which contain FPGA computational fields are:

- poor adaptation of the cooling system for placement into standard computer racks;
- inefficiency of cooling of electronic component chips with considerable (over 50 W) heat generation;
- the thermal paste between FPGA chips and radiators is washed out during long-term maintenance;
- the system of cooling liquid circulation inside the module is designed for one or two chips, but not for an FPGA field, and this fact leads to considerable thermal gradients;

- in the proposed systems, designed according to the IMMERS technology, all cooling liquid is circulating within a closed loop though the chiller, and this fact leads to some problems;
- necessity of computer complex maintenance stoppage for withdrawal separate components and devices;
- necessity of use of a power specialized pump and hydraulic equipment adapted to the cooling liquid;
- a complex system for control of cooling liquid circulation which causes periodic failures;
- high cost of the cooling liquid which is produced by the only one manufacturer.

The presented disadvantages can be considered as an inseparable part of other existing open loop liquid cooling systems because cooling of RCS computational modules which contain not less than 8 FPGA chips has some specific features in comparison with cooling of a single microprocessor. The special feature of the RCS produced in Scientific Research Centre of Supercomputers and Neurocomputers is the number of FPGAs, not less than 6–8 chips on one printed circuit board and high density of placement. This considerably increases the number of heat generating components in comparison with microprocessor modules, complicates application of the technology of direct liquid cooling IMMERS along with other final solutions of immersion systems, and requires additional technical and design solutions for effective cooling of RCS computational modules.

3   Reconfigurable Computer System Based on XILINX Ultrascale FPGAS

Since 2013 the scientific team of SRC SC and NC has actively developed the domain of creation of next-generation RCS on the base of their original liquid cooling system for printed circuit boards with high density of placement and the large number of heat generating electronic components. The basis of design criteria of the computational module (CM) of next-generation RCS with an open loop liquid cooling system are the following principles:

- the principal configuration of the computer rack is the computational module with the 3U height and the 19 width and with self-contained circulation of the cooling liquid;
- one standard 47U computer rack can contain not less than 12 computational modules with liquid cooling;
- one computational module can contain 12–16 printed circuit boards with FPGA chips;
- each printed circuit board must contain up to 8 FPGAs with dissipating heat flow of about 100 W from each FPGA;
- a standard water cooling system, based on industrial chillers, must be used for cooling the liquid.

The principal element of modular implementation of an open loop immersion liquid cooling system for electronic components of computer systems is a reconfigurable computational module of a new generation (see the design in Fig. 1-a). The CM of a new generation consists of a computational section, a heat exchange section, a casing, a pump, a heat exchanger and a fitting. In the casing, which is the base of the computational section, a hermetic container with dielectric cooling liquid and electronic components with elements that generate heat during operating, is placed. The electronic components can be as follows: computational modules (not less than 12–16), control boards, RAM, power supply blocks, storage devices, daughter boards, etc. The computational section is closed with a cover.

The computational section adjoins to the heat exchange section, which contains a pump and a heat exchanger. The pump provides circulation of the heat transfer agent in the CM through the closed loop: from the computational module the heated heat-transfer agent passes into the heat exchanger and is cooled there. From the heat exchanger the cooled heat-transfer agent again passes into the computational module and there cools the heated electronic components. As a result of heat dissipation the agent becomes heated and again passes into the heat exchanger, and so on. The heat exchanger is connected to the external heat exchange loop via fittings and is intended for cooling the heat-transfer agent with the help of the secondary cooling liquid. As a heat exchanger it is possible to use a plate heat exchanger in which the first and the second loops are separated. So, as the secondary cooling liquid it is possible to use water, cooled by an industrial chiller. The chiller can be placed outside the server room and can be connected with the reconfigurable computational modules by means of a stationary system of engineering services. The design of the computer rack with placed CMs is shown in Fig. 1-b.

The computational and the heat exchange sections are mechanically interconnected into a single reconfigurable computational module. Maintenance of the reconfigurable computational module requires its connection to the source of the secondary cooling liquid (by means of valves), to the power supply or to the hub (by means of electrical connectors).

In the casing of the computer rack the CMs are placed one over another. Their number is limited by the dimensions of the rack, by technical capabilities of the computer room and by the engineering services. Each CM of the computer rack is connected to the source of the secondary cooling liquid with the help of supply return collectors through fittings (or balanced valves) and flexible pipes; connection to the power supply and the hub is performed via electric connectors.

Supply of cold secondary cooling liquid and extraction of the heated one into the stationary system of engineering services connected to the rack, is performed via fittings (or balanced valves). A set of computer racks placed in one or several computer rooms forms a computer complex. To maintain the computer complex it is connected to the source of the secondary cooling liquid, to the power supply, and to the host computer that controls this computer complex.

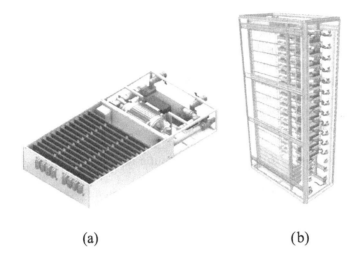

(a)   (b)

Fig. 1. The design of the computer system based on liquid cooling (a - the design of the new generation CM, b - the design of the computer rack)

Besides advantages which are typical for open loop liquid cooling systems, the considered modular implementation of the open loop liquid cooling system for electronic components of computer systems has a number of additional advantages:

– printed circuit boards of computational modules and reconfigurable computational modules are identical, relatively stand-alone and interchangeable. If one of the CMs fails or if technical diagnosis is required, then it is not needed to disconnect completely the computer rack and to stop execution of a task;
– high placement complexity of FPGAs in the CMs;
– the proposed technical solution allows, if it is necessary, increasing of performance of reconfigurable computational modules without significant increasing of dimensions (a more high-power pump and a heat exchanger can be placed into the selected dimensions). Growth of the number of printed circuit boards of the computational modules will slightly increase the dimensions (depth) of the reconfigurable computational module but the density of placement will remain unchangeable.

Owing to simplicity of design of the heat exchange section of the reconfigurable computational module, its reliability grows significantly.

The 19″ computer rack of the supercomputer (see the design in Fig. 2-a) has the following technical characteristics:

– a standard 47U computer rack with 12 3U computational modules with liquid cooling;
– each computational module contains 12 printed circuit boards with the power of 800 W each;

– each printed circuit board contains 8 Kintex UltraScale XCKU095-1FFVB2104C FPGAs, 95 million equivalent gates (134 400 logic blocks) each;
– the performance of the new generation computational module is 105 TFlops;
– the performance of the computer rack, which contains 12 CMs, is 1 PFlops;
– the power of the computer rack, which contains 12 CMs is 124 kW.

The performance of one computer rack with the liquid cooling system, which contains 12 CMs with 12 printed circuit boards each, in 6.55 times exceeds the performance of the similar rack with air cooling CMs Taygeta. Here the performance of one CM of a new generation is increased in 8.74 times in comparison with the CM Taygeta. Such qualitative increasing of the specific performance of the system is provided by the density of placement, increased more than in three times owing to original design solutions, and by increasing of the clock rate and the number of gates in one chip.

For testing technical and technological solutions, and for determination of expected technical and economical characteristics and service performance of the designed high-performance reconfigurable computer system with liquid cooling, we designed a number of models, experimental and technological prototypes. Figure 2-b shows the technological prototype of a new generation CM. For this CM new designs of printed circuit boards and computational modules with high density of placement were created.

The printed circuit board of the advanced computational module contains 8 Virtex UltraScale FPGAs of logic capacity of not less than 100 million equivalent gates each. The CM computational section contains 12–16 printed circuit boards of computational modules with the power up to 800 W each. Besides, all boards are completely immersed into an electrically neutral liquid heat-transfer agent; the heat exchange section contains pump components and the heat exchanger, which provide the flow and cooling of the heat-transfer agent. The design height of the new generation CM is 3U.

For creation of an effective immersion cooling system a dielectric heat-transfer agent was developed. This heat-transfer agent has the best electric strength, high heat transfer capacity, the maximum possible heat capacity and low viscosity. On the base of the transformer oil, a new oil called MD-4.5 with reduced viscosity was created according to the method of vacuum distillation. The oil MD-4.5 was completely tested in the heat engineering laboratory of SRC SC&NC on the technological prototype of the computational module with immersion open loop liquid cooling system (see Fig. 2-b). The performed set of laboratory and service tests proved reasonability of use of the oil MD-4.5 for cooling of electronic computer components and of use of low-power pumps for its circulation (due to its reduced viscosity).

During design of the new generation CMs we have obtained a number of break-through technical solutions, such as an immersion power supply block for the voltage of 380 V and a transducer DC/DC 380/12 V, the minimum height of the CM printed circuit board of 100 mm is provided, an original immersion control board is designed and produced. For the cooling subsystem of the new generation CMs we have determined required components of the cooling system

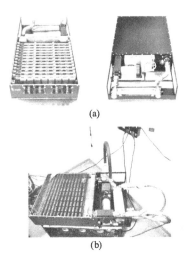

(a)

(b)

Fig. 2. The computational module with immersion open loop liquid cooling system (a - the new generation CM, b - the technological prototype of the new generation CM)

such as an original heat interface, a low-height FPGA radiator of an original design for convective heat exchange, a pump and a heat exchanger optimal for the used heat-transfer agent. Besides, we have determined the design of a volume compensator of the heat-transfer agent and control elements of the cooling subsystem such as optical level sensors and a flow sensor. The developed implementations of the cooling system design and circulation of the heat-transfer agent provide effective solution of the heat dissipation problem from the most heated components of the CM.

The complex of the developed solutions concerning the immersion liquid cooling system will provide the temperature of the heat-transfer agent not more than $30\,°C$, the power of $91\,W$ for each FPGA ($8736\,W$ for the CM) in the operating mode of the CM. At the same time, the maximum FPGA temperature does not exceed $55\,°C$. This proves that the designed immersion liquid cooling system has a reserve and can provide effective cooling for advanced families of Xilinx FPGAs (UltraScale+, UltraScale 2, etc.).

It is possible to adapt the designed cooling system to heat dissipation from other devices which contain standard chips or microprocessors, because the CM control board, which contains a microprocessor Intel Broadwell U from the families Core i3, i5 or i7 along with a loading and control unit and a DC/DC transducer, is already placed in the module. Adapting the suggested solutions of immersion liquid cooling system to other devices we must take into account, that initially it was designed for heat dissipation from a great number of tightly placed heat generators, which generate heat of about $100\,W$ each FPGA. Therefore cooling of other devices can require rather small amount of the heat transfer agent, and we can use CMs with other dimensions. So, the developed immersion liquid cooling system can be adapted to other devices, which contain electronic components, after required modification of its design.

4 Programming Tools of Computer Systems with Tight Component Arrangement

Due to increasing of the FPGA chips arrangement density in reconfigurable computer systems, requirements to their software tools grow also, because, as a rule, the chip capacity of each new Xilinx FPGA family grows twice, and the number of external pins available to the user for data exchange remains constant or even decreases. It complicates mapping of computing structures of parallel programs on to the set of interconnected FPGA chips of the RCS, and increases requirements to programming tools, particularly to the multichip synthesizer Fire!Constructor [1], which is being developed in SRC SC & NC. One of the main tasks of the synthesizer is decomposition of the information graph, which describes the computational structure of the program, into disjoint subgraphs with two conditions:

- hypothetical hardware resource of a computational structure described by a subgraph cannot exceed hardware resource of one FPGA chip;
- the number of hypothetical traced external connections of a fragment of a computational structure described by a subgraph cannot exceed the number of user available pins of the FPGA chip in which the fragment is supposed to be placed.

For existing programming tools [1] we had earlier developed and modified several methods and algorithms, based on the methods of multilevel partition scheme and recursive bisection. However, their use, when the ratio of "available hardware resource"/"FPGA pin" is continuously growing, leads to more and more limitations. That is why we have developed a new method of hierarchic graph decomposition for the modern Xilinx families such as Virtex UltraScale, Virtex UltraScale+ and advanced FPGAs of next generations. The method is based on a W-like model of multilevel partition scheme and supposes use of several selection criteria for pairs of merged vertices. Owing to such approach it is possible to obtain various results of decomposition, using effective heuristic algorithm - the multilevel scheme of graph partition. Besides, owing the new W-like model of the multilevel scheme, the suggested method, in some cases, provides evasion of local deadlock situations, in which the previously developed algorithm of information graph partition stopped.

Another direction of development of programming tools based on the high-level language COLAMO is extension of supported architectures of computer systems. According to practice, majority of real tasks implemented on modern high-performance computer systems, require both sequential and concurrent computational fragments, combined in a single computational space for effective implementation of structural and procedural fragments of calculations [1]. Many developers assume, that this problem can be solved with the help of hybrid computer systems, which contain computational nodes with different architectures, connected by channels for data transfer, and which provide implementation of structural and procedural calculations in the single computational space. Symbiosis of nodes with different architecture in one computer system theoretically allows increasing of the computer system real performance owing to opportunity of effective implementation of both structural and procedural fragments of calculations in the nodes with different architecture.

However, for effective programming of hybrid computer systems it is necessary to have language tools which allow description of fragments of calculations, which use different frequencies, data delay ratios and digital capacities of processed data. Owing to this, it is possible to scale both fragments and single circuit cores in both cases when hardware resource is increasing or decreasing, and, in addition, it is also possible to use data with variable capacity for effective use of hardware resource of the hybrid computer system. For this it is possible to use a single parallel-pipeline form of applications [1], which allows automatic adaptation of applications to the modified architecture or configuration of the hybrid computer system. Owing to this we can reasonably use resources of nodes with different architectures during hybrid computer system programming, and we have a set of necessary tools for quick development of effective resource-independent scalable parallel applications in a single language space. It simplifies hybrid computer system programming and speeds up development of parallel applications.

5 Conclusion

Use of air cooling systems for the designed supercomputers has practically reached its limit because of reduction of cooling effectiveness with growing of consumed and dissipated power, caused by growth of circuit complexity of microprocessors and other chips. That is why use of liquid cooling in modern computer systems is a priority direction of cooling systems perfection with wide perspectives of further development. Liquid cooling of RCS computational modules which contain not less than 8 FPGAs of high circuit complexity is specific in comparison with cooling of microprocessors and requires development of a specialized immersion cooling system. The designed original liquid cooling system for a new generation RCS computational module provides high maintenance characteristics such as the maximum FPGA temperature not more than $55\,°C$ and the temperature of the heat-transfer agent not more than $30\,°C$ in the operating mode. Owing to the obtained breakthrough solutions of the immersion liquid cooling system it is possible to place not less than 12 CMs of the new generation with the total performance over 1 PFlops within one 47U computer rack. Power reserve of the liquid cooling system of the new generation CMs provides effective cooling of not only existing but of the developed advanced FPGA families Xilinx UltraScale+ and UltraScale 2. Since FPGAs, as principal components of reconfigurable supercomputers, provide stable, practically linear growth of RCS performance, it is possible to get specific performance of RCS, based on Xilinx Virtex UltraScale+ FPGAs, similar to the one of the world best cluster supercomputers, and to find new perspectives of design of super-high performance supercomputers.

References

1. Kalyaev, I.A., Levin, I.I., Semernikov, E.A., Shmoilov, V.I.: Reconfigurable Multi-pipeline Computing Structures, 330 p. Nova Science Publishers, New York (2012)
2. Tripiccione, R.: Reconfigurable computing for statistical physics: the weird case of JANUS. In: IEEE 23rd International Conference on Application-specific Systems, Architectures and Processors (ASAP) (2012)
3. Baity-Jesi, M., et al.: The janus project: boosting spin-glass simulations using FPGAs. IFAC Proc. Vols., Program. Dev. Embedded Syst. **12**(1) (2013)
4. Shaw, D.E., et al.: Anton, a special-purpose machine for molecular dynamics simulation. Commun. ACM **51**(7), 91–97 (2008)
5. Levin, I.I., Dordopulo, A.I., Doronchenko, Y.I., Raskladkin, M.K.: Reconfigurable computer system on the base of Virtex UltraScale FPGAs with liquid cooling. In: Proceedings of the 10th Annual International Scientific Conference on Parallel Computing Technologies. CEUR Workshop Proceedings, vol. 1576, pp. 221–230 (2016)
6. Shah, J.M., Eiland, R., Siddarth, A., Agonafer, D.: Effects of mineral oil immersion cooling on IT equipment reliability and reliability enhancements to data center operations. In: Proceedings of the 15th InterSociety Conference on Thermal and Thermomechanical Phenomena in Electronic Systems (ITherm 2016), pp. 316–325, 20 July 2016. Article number 7517566
7. Li, L., Zheng, W., Wang, X.: Data center power minimization with placement optimization of liquid-cooled servers and free air cooling. Sustain. Comput. Inf. Syst. **11**(1), 3–15 (2016)
8. Gess, J.A., Dreher, T.A., Bhavnani, S.A., Johnson, W.B.: Effect of flow guide integration on the thermal performance of high performance liquid cooled immersion server modules. In: ASME 2015 International Technical Conference and Exhibition on Packaging and Integration of Electronic and Photonic Microsystems (InterPACK 2015), collocated with the ASME 2015 13th International Conference on Nanochannels, Microchannels and Minichannels, vol. 1 (2015)

Impact of the Investment in Supercomputers on National Innovation System and Country's Development

Yuri A. Zelenkov$^{(\boxtimes)}$ and Jibek A. Sharsheeva

Financial University under the Government of the Russian Federation,
Leningradsky, 49, 125993 Moscow, Russia
yuri.zelenkov@gmail.com, jsharsheeva.rc@gmail.com

Abstract. At all the stages of the design of new products and scientific research, numerical analysis becomes a key competitive advantage in contemporary economy. At the same time, the installation of large supercomputing facilities requires investments that in most cases are only possible with government support. Therefore, the evaluation of the influence of supercomputer technology on the national innovation system (NIS) and, ultimately, on the development of a country is a topical problem. We present in this article the results of a two-stage research. In the first stage, we built a model that relates the latent variables "country's development", "national innovation system", "investments in supercomputers", etc., and obtained quantitative estimates of these variables using the PLS-SEM method and based on data published by the World Bank, the OECD, the Human Development Index and the TOP500 ranking. Modeling results confirm that investments in supercomputers affect the NIS and are an essential component of a country's development. This is especially important for the BRIC countries. In the second stage, we investigated the cross-correlation between supercomputing installed capacity and the number of scientific publications (data were taken from scimagojr.com). The cross-correlation coefficient for most countries and scientific areas is close to 1. This finding can be deemed as a confirmation of the significant impact that investments in supercomputers have on the results of scientific activity. Russia stands out with a rather negative result, demonstrating a much smaller value of the cross-correlation coefficient, which indicates a lower effectiveness of investments in supercomputers compared with other countries. A possible explanation of this fact is the existing structure of the NIS.

Keywords: National inovation system · Country's development · Investments in supercomputers · Number of scientific publications · PLS-SEM

1 Introduction

Modern economy is increasingly acquiring features of "knowledge economy", which is characterized by the transition from mass production to mass customization. Focusing on the needs of the individual customer requires a radical shift

© Springer International Publishing AG 2017
L. Sokolinsky and M. Zymbler (Eds.): PCT 2017, CCIS 753, pp. 42–57, 2017.
DOI: 10.1007/978-3-319-67035-5_4

in the processes of design, manufacturing and logistics. Time-to-market should be reduced. This creates new requirements: rapid development of product with a high degree of customization in a completely digital form, production close to the customer, reduction of the cost of information exchange, and improvement of manufacturing and logistics operations.

Under these circumstances, knowledge is the main driving force of economic growth, social development and international competitiveness [17]. The understanding of this fact has led to the concept of a "National Innovation System" (NIS), which is defined, in its most general form, as a set of related public institutions for creation, preservation and dissemination of knowledge, skills and artifacts [15,23]. This is a system through which the government stimulates the innovation process and exerts influence on it. Each country's NIS has its own specific features and degree of effectiveness [22]. As a result, the state creates a network of actors (scientific and educational organizations, industry, government agencies) that act within the framework of a given context, and rely on various types of communication interfaces [10]. The quality level of NIS is a key factor in the economic growth and development of a country [15].

Substantially, the emergence of new contemporary business opportunities is a result of the development of modern information technology (IT). The Internet facilitates communication of participants engaged in the production and consumption process; corporate systems with different functionalities facilitate group work; personal applications radically alter the individual work; embedded systems allow to manage complex technological processes; supercomputer simulations reduce the cost of research and development [27].

These IT system classes are divided not only by functions and applications; significant differences can be noted in the methods applied to create these systems. In some cases, the creation of a new IT system is initiated by an IT supplier or a community of independent developers that carry out the initial investment, responding to specific functional needs. Some systems are developed by the users themselves, which invest their own resources. In the case of large supercomputer facilities, however, starting investments are often so large that they require the participation of the government.

The establishment of such systems always requires the development of new technologies and significant financial investments. These systems particularly attract the attention of society. Nonetheless, the degree of influence of supercomputers on both the NIS development and that of the country as a whole has not been investigated yet.

Many researchers note that each country has a particular specialization in the field of scientific research [12,13]. Therefore, the study of the relations between investments in supercomputers and the fields of national science is also of considerable interest.

This paper presents the results of a two-stage study. In the first stage, we considered the impact that investments in supercomputers have on the national innovation system and, ultimately, on the country's development. We selected for this study countries whose supercomputer installations are regularly listed in the top 500 most powerful systems. The list includes the G7 (Group of Seven) countries, the BRIC countries and South Korea. To assess the level of a country's development and its NIS, we referred to data provided by the World Bank, the Organization for Economic Cooperation and Development (OECD) and the Human Development Index (HDI) of the United Nations. Also, a structural model linking the factors of development and investments was constructed using the PLS-SEM method.

In the second stage, we investigated the correlation between the performance of computer systems and the number of scientific publications in different fields. This allows to identify the areas that are most closely related to supercomputing in each country, and evaluate investment efficiency.

2   Soft Systems and PLS-SEM Method

There are two main concepts in economics: growth due to an increase in quantitative indicators, and development, i.e. qualitative changes in a country's social and economical conditions. The term "development" usually refers to an improvement of people's living conditions. This improvement is associated, on the one hand, with the spread of knowledge and technologies and, on the other hand, with changes in social institutions [25]. According to [19], economic growth and social development are closely linked to each other. Economic growth encourages people to spend more money, not only on goods and services, but also on education, medicine, culture and science, thereby contributing to development. Development, in turn, contributes to further economic growth.

However, it should be noted that the definition of development is not sufficiently clear and does not allow to build a model for its measurement. The system under consideration (living conditions) refers to a class of so-called soft systems, which, unlike "hard" systems, do not have a fully defined structure, a fixed composition of elements and prescribed laws of behavior. There may be several simultaneous and incomplete notions of soft systems performance as a consequence of the presence of a social component in them [5]. The national innovation system also has the properties of soft systems. The measurement problem regarding NIS will be discussed below.

Key parameters of soft systems cannot be measured directly; their assessment is possible only through the use of measurable indicators. One of the most widely used approaches today is the Structural Equation Modeling (SEM) technique. It is assumed that the matrix $\mathbf{X}$ of observed values, which has size $n \times p$ (n is the number of observations, p is the number of variables), can be divided into independent units, each with an associated latent variable $LV_j, j = 1, ..., J$. Each unit contains K variables: $X_{j1}, ..., X_{jK}$. Latent variables are also often referred to as constructs or factors. The basic idea is that the observed variables are the indicators of latent variables (the so-called reflective mode), i.e. they are related by the equation

$$X_{jk} = \lambda_{0jk} + \lambda_{jk}LV_j + \epsilon_{jk}, \ k = 1, ..., K, \tag{1}$$

or they form a latent variable (the formative mode). Then the corresponding equation will appear as

$$LV_j = w_{0j} + \sum_k w_{jk} X_{jk} + \epsilon_j. \tag{2}$$

The coefficients λ_{jk} are called factor loadings; w_{jk} are weights, ϵ_{jk} and ϵ_j are measurement errors.

There are several types of problems that can be solved by SEM [20]. The most relevant in the light of the problem of our research is to build a structural regression allowing for testing the hypothesis of existence of relations among the latent variables. Mathematically, it can be written as

$$LV_j = \beta_0 + \sum_{i \to j} \beta_{ji} LV_i + \epsilon_j, \tag{3}$$

where ϵ_j is an error, β_{ji} are path coefficients, $i \to j$ indicates summation over all i except $i = j$.

Obviously, the power of connection between latent variables can be estimated by the value of β_{ji}. These links and their directions, describing the model structure, are formulated by the researcher in the form of hypotheses before model evaluation. The parameters obtained by solving the Eqs. (1)–(3) on the basis of empirical observations of indicators allow to confirm or reject a hypothesis. A large number of parameters have been proposed to evaluate the correctness of the model structure, validity of the latent variables and consistency of their indicators. The use of these parameters will be discussed below.

Thus, the steps to solve a problem of structural regression are the following:

1. Formulation of hypotheses about the model structure, i.e. about the existence of latent variables and relations among them.
2. Selection of the indicators of latent variables, data capture.
3. Numerical solution of Eqs. (1)–(3), validation of the model quality.
4. If necessary, modification of the model and return to step 3.
5. Interpretation of the results.

The system of Eqs. (1)–(3) can be solved by the general method of least squares. Note that this usually imposes restrictions on the minimum size of the observation matrix, the presence of collinearity among indicators, etc. In recent years, the partial least squares (PLS) method, which allows to relax considerably these limitations, is being increasingly used for structural regression.

The PLS method allows the number of indicators to be greater than that of observations, i.e. $p > n$. In this case, the minimal number of observations has only two restrictions [9]. The first, a technical one, is related to the convergence of the method (for example, a too small sample size can lead to a singular matrix in calculations). The second limitation is imposed by the statistical interpretation of the results. As a general rule, the larger the sample is, the narrower confidence intervals for the values of model parameters will be. Therefore, the well-known heuristic rule of regression stating that the sample size should be

10 times greater than the number of independent variables should be applied in the PLS method to the structural block (i.e. a latent variable; see Eq. (3)) with the greatest number of incoming links [6]. For a more detailed discussion of this issue, see in [16]. As for multicollinearity, the correlation of indicators is the key to developing an effective model, as follows from Eq. (2).

The method of solving the problem of structural regression using partial least squares is called PLS-SEM, and is widely used today for the empirical verification of theories in economics, management, sociology, psychology and other sciences concerned with the study of soft systems.

3   Model of the Impact Exerted by Investments in Supercomputers on a Country's Development

In accordance with the procedure formulated above, the latent variables and the relations among them must be defined first.

Since the purpose of this work is to research the investment influence on the level of a country's development, these two factors must be included in the list of latent model variables, although the relation between them is not straight but is established by means of various mediators.

Let latent variable DEV denote the level of a country's development. To identify differences in levels of development, three groups of statistical indicators (economic, demographic and quality of life) are used in economic research. The first group of parameters is generally related to economic indicators such as gross domestic product (GDP) and gross national income (GNI). To take into account the country's size, it is necessary to consider these figures as either per worker or per capita. GDP per capita would be a more correct indicator for our study purposes, since it evaluates the economic development on a territorial rather than national basis, as the GNI does. The level of goods and service exports is also an important economic indicator. It allows to make conclusions not only about the degree of development of the national economy, but also about the country's involvement in the global system of labor division.

Life expectancy, population growth rate, the proportion of economically active population, and so on, are considered as demographic indicators. The group of life quality indicators includes factors measuring the provision of social benefits (number of patients per doctor, literacy, energy consumption per capita, etc.), the level of consumption (e.g., the number of cars per 1000 inhabitants), and safety (e.g., the number of registered crimes).

Recently, a new integral indicator, Human Development Index (HDI), has been introduced. It takes into account the physical condition of people, average life expectancy, educational attainment and real income per capita. The maximum value of this index is 1. It should be noted that human capital plays a key role in the processes of economic development. Lack of human capital or its poor quality does not allow the country to create a competitive technological structure of the economy [14].

All these indicators, namely GDP per capita, volume of exports and human development index, can be used to estimate the latent variable DEV.

An important factor in a country's development is the sectoral structure of its economy, i.e. the ratio of the primary (agriculture), the secondary (industry) and the tertiary (service) sectors. It defines the sources of GDP/GNI and the employment structure of the economically active population. Most modern researchers [3,7] suppose that the development of a country is directly related to innovation and a growing role of services in the national economy. The role of the primary sector (agriculture) becomes less visible, labor productivity in this sector becomes so high that a small number of employees are able to satisfy the country's entire domestic and export needs. The role of industry in the development still remains high, although reduced due to the grow of the service sector.

It follows from this discussion that it is necessary to introduce into the structural model various latent variables: SRV—development of domestic services, IND—development of national industry, and AGR—development of national agriculture. These latent variables can be measured by quantitative indicators of volume, value added, exports in the sector, etc. To describe their relation with the variable DEV, we propose the following hypotheses:

H1: Service sector development has a positive effect on the country's development.

H2: Industrial development has a positive effect on the development of the country.

H3: The development of agriculture has no effect on the development of the country.

As already noted, the national innovation system (NIS) specifies an individual state's ability to adapt to new economic conditions. These innovations are the foundation for the development of all the sectors. Therefore, it is necessary to add in the model the latent variable NIS, which evaluates the development of the national innovation system. The following hypotheses describe its relations with other latent variables:

H4: The development of the national innovation system has a positive effect on the development of services.

H5: The development of the national innovation system has a positive effect on the development of industry.

H6: The development of the national innovation system has a positive effect on the development of agriculture.

NIS is also a weakly formalized variable, which can be measured only by means of indirect indicators [2,8,18]. Creating innovations requires more than funding of R&D. Innovations in sectors such as services depend more on intangible investments in the form of costs of market research, testing, etc. Therefore, the development level of the national innovation system is more often measured by the number of applications for patents and trademarks issued in the country [1], considering applications submitted by both residents and non-residents. This indirectly measures the flows of knowledge between countries [4].

However, research and development are the most important innovation factors; therefore, it is necessary to add in the model the latent variable *RND*, which evaluates research activities in the country. This variable can be measured by the overall national R&D spending, the number of articles in scientific and technical journals, etc. Evidently, it is positively associated with the *NIS* variable. Therefore, we can suggest the following hypothesis:

H7: The volume of research and development in a country has a positive effect on the national innovation system.

Let us introduce yet another latent variable, namely the national investment in supercomputer infrastructure, which we denote by *HPC*. This variable assesses the technical provision of both research and development, as well as the national innovation system in general, since supercomputing is widely applied in the operating activities of companies (for instance, in the analysis of large data sets and modeling of products). Thus, we assume that the place of the variable in the model is described by the following hypothesis:

H8: The volume of investments in supercomputers has a positive effect on the national innovation system.

The measurement of this variable is possible by means of indicators published in the ranking of the 500 most powerful supercomputers in the world (TOP500). Typically, systems that are included in this rating are created with the participation of the state, hence this indicators can serve as an indirect measurement of public spending.

The resulting structural model of the investigated system is shown in Fig. 1.

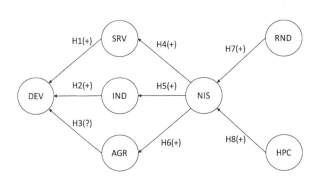

Fig. 1. Preliminary model of the system

4   Empirical Data and Results of PLS-SEM Modeling

To verify the model, we used empirical data on countries whose supercomputing systems have been constantly included in the TOP500 list during 10 years, namely from 2005 to 2014. A total of 12 countries was selected:

- G7 Countries: Canada, France, Germany, Italy, Japan, United Kingdom and United States;

– BRIC countries: Brazil, Russia, India and China;
– South Korea.

We collected data covering a 10-year period (2005–2014), making a total of 120 observations. We used as data sources the World Bank Open Data[1] (databases: WDI—World Development Indicators, and GEM—Global Economic Monitor), the Organization for Economic Cooperation and Development[2] (OECD), the Human Development Index[3] (HDI) and the TOP500 list[4]. A total of 87 different indicators was examined. After removing variables with a low level of relevance, the measurement model for each latent variable was constructed with the three indicators listed in Table 1.

Figure 2 shows the results of calculations (using the plspm software package [21]). The circles represent latent variables; the rectangles represent their indicators. The numbers near the arrows connecting the elements of the model correspond to the calculated values of the coefficients λ_{jk} and β_{ii}.

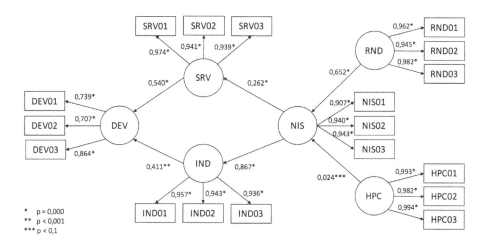

Fig. 2. Validated model

It can be noted that the hypothesis $H3$ about nonsignificant influence of the agriculture sector on the country's development was confirmed. Based on the set of empirical data described above, no statistically significant assessment of the relation between DEV and AGR could be obtained, and herein the AGR variable is excluded from the model. Hypothesis $H6$ is not confirmed on the obtained data.

---

[1] http://data.worldbank.org.
[2] http://data.oecd.org.
[3] http://hdr.yndp.org.
[4] http://www.top500.org.

Table 1. Latent variables and their indicators

Variable/Indicator	Description	Data source
DEV	*Country's Development*	
*DEV*01	GDP per worker in US dollars at the exchange rate of 2011	WDI
*DEV*02	Human development index	HDI
*DEV*03	Exports in US dollars at the exchange rate of 2005	WDI
IND	*The development of industry*	
*IND*01	Exports of goods in US dollars at current prices	WDI
*IND*02	Exports of high-tech products in US dollars at current prices	WDI
*IND*03	Industrial production in constant US dollars	GEM
SRV	*The development of services*	
*SRV*01	Sales of intellectual property rights in US dollars at current prices	WDI
*SRV*02	Exports of services in the US dollar at current prices	WDI
*SRV*03	Added value in US dollars at the rate of 2005	WDI
NIS	*The development of a national innovation system*	
*NIS*01	Patent applications, residents	WDI
*NIS*02	Applications for registration of trademarks, non-residents	WDI
*NIS*03	Applications for registration of trademarks, residents	WDI
RND	*The development of a national R&D*	
*RND*01	Patent applications, non-residents	WDI
*RND*02	The number of articles in scientific and technical journals	WDI
*RND*03	Domestic expenditure on R&D in US dollars at current prices	OECD
HPC	*Investments in supercomputers*	
*HPC*01	The total reached performance (RMax) of all country's systems included in the TOP500, GFLOPS	TOP500
*HPC*02	The number of CPU cores of all country's systems included in the TOP500	TOP500
*HPC*03	Potential peak performance (RPeak) of all country's systems included in the rating TOP500, GFLOPS	TOP500

To check the quality of the model, firstly, it is necessary to assess the internal consistency of the set of indicators describing the same latent variable. Typically, such assessment uses Cronbach's α. This coefficient can have values in the interval $[-\infty, 1]$; acceptable values are $\alpha \geq 0.7$; negative values cannot be interpreted. The values of Cronbach's α are shown in Table 2, containing also the values of Dillon–Goldstein's ρ, which is considered as a more reliable indicator of consistency [21]. Values of $\rho \geq 0.7$ are also admitted. Table 2 lists the average variance of the latent variable explained by its indicators (AVE—Average Variance Extracted).

Table 2. Model parameters

Variable	Cronbach's α	Dillon–Goldstein's ρ	AVE	R^2
DEV	0.750	0.936	0.597	0.697
SRV	0.947	0.951	0.905	0.069
IND	0.941	0.941	0.894	0.750
NIS	0.922	0.928	0.865	0.447
RND	0.961	0.965	0.927	NA
HPC	0.989	0.995	0.979	NA

The values of the factor loadings λ_{jk} also provide significant information about the model quality; this is a measure of the correlation between an indicator and the corresponding latent variable. A suitable value $\lambda_{jk} \geq 0.7$ means that at least 50% of the indicator variation $(0.7^2 = 0.49)$ is determined by the latent variable. The values of λ_{jk} are shown in Fig. 2. It is easy to see that the above requirement is satisfied.

In addition, it is necessary to be sure that the relation of an indicator with its "own" latent variable is stronger than that with "foreign" ones. This is done by comparing the cross-loadings, i.e. the correlation coefficients between indicators and latent variables. This condition is also satisfied.

The p-values of the model parameters (shown in Fig. 2) and the determination coefficients R^2 of the latent variables (see Table 2) are other parameters allowing for an assessment of the model quality. All these parameters have acceptable values confirming the quality of the model.

The results shown in Fig. 2 can be interpreted as follows. On the basis of a sample of mixed data on developed (G7) and emerging (BRIC) countries, it can be seen that in current conditions the service sector is already ahead of the industrial one regarding contribution to the development of the country. However, the national innovation system is more significant for the industrial sector. The role played by public investment in supercomputers in the development of the national innovation system is insignificant (note also that path coefficient β_{ji} from HPC to NIS has a relatively high p-value).

The G7 and BRIC groups show wide variations in economy structure. The G7 states are more uniform (although it has been noted in [24] that the European countries show a lower potential for innovation than the United States), whereas the BRIC countries are much more varied. To investigate these assumptions, we considered these groups of countries separately. The results are presented in Figs. 3 and 4. All the model parameters meet the requirements of quality indicated above.

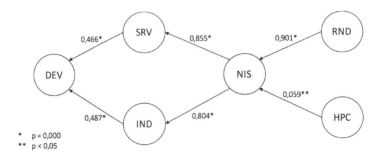

Fig. 3. Model for G7 countries

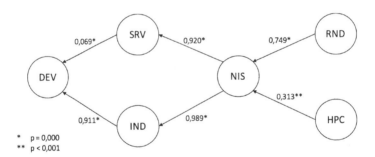

Fig. 4. Model for BRIC countries

The influence of the variables *SRV*, *IND*, *NIS*, *RND*, *HPC* on *DEV* can be estimated using the so-called total effect, which is calculated as the sum of multiplied path coefficients over all the possible paths between exogenous and endogenous variables. These data for two separate models (G7 and BRICS) are shown in Table 3.

A comparative analysis of the results obtained for the two groups of countries leads to the following conclusions. Firstly, the development of the group G7 relies much more on the service sector, whose impact on the country's development is almost equal to that of the industrial sector. In the BRIC countries, the service sector is underdeveloped, its impact on the development is inferior to that of the industrial sector by more than an order of magnitude. This finding is consistent with data available on the economy structure for these two groups of states.

Table 3. The total effect on the variable DEV

Variable	G7	BRIC
SRV	0.466	0.069
IND	0.487	0.911
NIS	0.804	0.964
RND	0.724	0.722
HPC	0.047	0.302

Secondly, the national innovation system is important for both groups of countries, its contribution is the same in both the service sector and the industrial sector in the two groups. Note also that, according to the model, the national innovation system is the most important factor in the country's development in both groups.

Thirdly, the national innovation system in the BRIC countries is much more dependent on supercomputing. Perhaps the following explanation can be proposed for this phenomenon. Developed countries (group G7) have reached the required level of saturation with supercomputers; for these states, an investment in the development of other elements of the NIS is much more important (for example, in the creation of experimental facilities, in training talents, promoting theoretical research). The BRIC countries have the so-called catching-up economical systems; their technological progress is largely based on copying the achievements of developed countries. Therefore, their share of costs in new breakthrough ideas is relatively small. An important role is played by the costs of reproduction of already-known innovative actions. It should be noted, however, that the confirmation of this assumption requires further research.

5 The Impact of Investments in Supercomputers on Scientific Specialization

As already mentioned, comparative studies of scientific publications [12, 13] show that each country has a certain scientific specialization. In this regard, one may ask in which areas of research the investments in supercomputers are most effective, and whether a national distinction of these relations exists. These relations can be rated by means of the cross-correlation coefficient between two time-series representing the number of scientific publications and the total performance ($RMax$) achieved by supercomputer systems included in the TOP500 rating.

At the same time, we should take into account two considerations. First, the preparation of a scientific publication takes a long time, suppose that in average it is two years. Assume that one year is spent on research, and the second year, on the publication of results. Second, the need for computing power in most scientific areas grows according to a law close to exponential. Therefore, we will investigate the cross-correlation between two time-series $[P_i]$ and $[L_{i-2}]$, where

P_i is the number of publications in the i-th year, and $L_{i-2} = \ln(Rmax_{i-2})$ is the natural logarithm of the total installed computer power two years before.

Scimago Journal & Country Rank[5] data were used as a source of data on scientific publications. Data on the number of publications in the 12 countries selected above for the period 2005–2014 were grouped into six categories according to information on the structure of science[6]:

- Economics, management, finance, and social sciences (EMFS). In this area, supercomputers are most often used in calculations associated with forecasting of financial markets and social systems modeling.
- Computer Science and Mathematics (CSM). In this area, publications on supercomputing are dedicated to the development of hardware and software architectures, as well as problem-solving methods and algorithms.
- Engineering (ENG). In this direction, supercomputers are used in the design of new products.
- Physics, chemistry, energy and material science (PCEM). In this area, supercomputers are used to study the properties of matter and the creation of new materials in particular.
- Earth and Planets Science (EPS). In this area, problems of geodynamics and astrodynamics are solved with the help of supercomputers.
- Genetics, molecular biology and pharmacology (GMBP). In this area, supercomputers are used to develop new drugs, study genomes and other aspects of living systems.

The values of the cross-correlation coefficient between the time series representing the number of scientific publications (for the years 2005–2014) and the natural logarithm of $RMax$ (for the years 2003–2012) are presented in Table 4.

From data in Table 4, we can make the following conclusions. The cross-correlation coefficient between the number of scientific publications and the performance of supercomputers has a value close to 1 for almost all countries and scientific disciplines. One of the exceptions concerns the publications in the "Engineering" area in developed countries (USA, Japan). Probably, this suggests that the use of supercomputers to create new products (in the case of Japan, it includes the study of new materials) in these countries is not a subject of scientific publications, but a routine technological process. At the same time, the number of publications in Brazil, China and Korea in this area has the greatest cross-correlation with supercomputer performance. In most developed countries (USA, UK), the "Genetics, Molecular Biology and Pharmacology" area has the greatest cross-correlation with supercomputer performance.

It is worth noting that these results are consistent with the priority areas of science in different groups of countries [11]. One group consists of the states where the structure of spending on science is determined by taxpayers, hence the reason that there is an interest in areas related to medicine and life science. In the second group, which is formed by developing countries, the demand for

---

[5] http://scimagojr.com.
[6] http://scimagojr.com/shapeofscience.

Table 4. Coefficients of cross-correlation between the number of scientific publications and supercomputing capacity for various countries

Country	All subjects	EMFS	CSM	ENG	PCEM	EPS	GMBP
USA	0.967	0.950	0.936	−0.331	0.947	0.976	0.977
UK	0.987	0.973	0.965	0.783	0.962	0.971	0.974
Germany	0.983	0.988	0.981	0.853	0.954	0.959	0.964
Japan	0.608	0.958	0.693	−0.844	−0.869	0.958	0.931
France	0.994	0.989	0.979	0.975	0.939	0.986	0.985
Canada	0.973	0.960	0.741	0.491	0.920	0.979	0.973
Italy	0.987	0.990	0.979	0.977	0.965	0.979	0.972
China	0.992	0.766	0.795	0.990	0.977	0.981	0.967
India	0.907	0.934	0.911	0.895	0.929	0.925	0.890
Russia	0.629	0.628	0.614	0.275	0.463	0.802	0.710
Brazil	0.944	0.942	0.950	0.950	0.934	0.947	0.944
Korea	0.946	0.928	0.881	0.952	0.931	0.922	0.940

research is based on the priorities of industrial development. The data obtained are also consistent with the results of structural modeling presented in Sect. 4. The industrial sector has the prevailing role in the BRIC countries, while the service sector is very important in the G7 countries.

It should be noted that Russia stands out, with relatively small values of the cross-correlation in all scientific areas. Obviously, this fact shows that public investment in supercomputing capacity, which is mostly readdressed to universities, has not exerted the expected effect in stimulating science. At the same time, it is impossible to talk of an insufficient volume of investments, as in terms of *RMax* Russia is surely among the top ten TOP500 leaders, despite holding only the 14–15th place in the rating by total number of scientific publications (according to scimagojr.com data). Perhaps the features of the national innovation system play the most important role in this disparity. In the case of Russia, these are: resource constraints, low levels of scientific capacity on the part of industry, problems of communication between science and industry, and the availability of competitive foreign products [26]. In addition, a detailed analysis of particular scientific fields (e.g., research in the field of "enterprise information systems" [28]) shows that the focus of scientists that publish results in the Russian language is significantly shifted relative to the priorities of international science, and published works are of an inferior quality. This is because publications on the Russian language fall out of the scope of international expertise, and Russian researchers are isolated from the international scientific community, which is due to poor knowledge of the English language.

6 Conclusions

The results presented here demonstrate that investments in supercomputing technology are an essential element of the national innovation system. In turn, NIS is a key factor contributing to a country's development. The impact of investments in supercomputers that is observed in the BRIC countries is particularly significant, although the mechanisms of this effect are not completely clear.

Russia stands out compared with other countries: it has a small cross-correlation between the achieved performance of supercomputers and the number of scientific publications, which means that investments in supercomputing infrastructure are used less efficiently.

References

1. Acs, Z.J., Anselin, L., Varga, A.: Patents and innovation counts as measures of regional production of new knowledge. Res. Policy **31**(7), 1069–1085 (2002)
2. Adam, F.: Measuring National Innovation Performance. Springer, Heidelberg (2014)
3. Anderson, C.: The Long Tail: Why the Future of Business is Selling Less of More. Hyperion, New York (2006)
4. Castellacci, F., Natera, J.M.: The dynamics of national innovation systems: a panel cointegration analysis of the coevolution between innovative capability and absorptive capacity. Res. Policy **42**(3), 579–594 (2013)
5. Checkland, P.B.: Soft systems methodology. In: Rosenhead, J., Mingers, J. (eds.) Rational Analysis for a Problematic World Revisited, pp. 61–90. Wiley, Chicchester (2001)
6. Chin, W.: The partial least square approach for structural equation modeling. In: Marcoulides, G. (ed.) Modern Methods for Business Research, pp. 295–336. Lawrence Erlbaum, Mahwah (1998)
7. Drucker, P.: Post-Capitalist Society. Harper Business, New York (1993)
8. Guan, J., Chen, K.: Modeling the relative efficiency of national innovation systems. Res. Policy **41**(1), 102–115 (2012)
9. Hensler, J., Hubona, G., Ash Ray, P.: Using PLS path modeling in new technology research: updated guidelines. Ind. Manag. Data Syst. **116**(1), 2–20 (2016)
10. Howells, J.: Intermediation and the role of intermediaries in innovation. Res. Policy **35**(5), 715–728 (2006)
11. King, D.A.: The scientific impact of nations. Nature **430**(6997), 311–316 (2004)
12. Kotsemir, M.: Dynamics of Russian and World Science through the prism of international publications. Foresight Russ. **6**(1), 38–58 (2012)
13. Kotsemir, M., Kuznetsova, T., Nasybulina, E., Pikalova, A.: Identifying directions for Russia's science and technology cooperation. Foresight Russ. **9**(4), 54–72 (2015)
14. Kuznets, S.: Economic Development, the Family, and Income Distribution: Selected Essays. Cambridge University Press, Cambridge (2002)
15. Lundvall, B.-A.: National innovation systems - Analytical concept and development tool. Ind. Innov. **14**(1), 95–119 (2007)
16. Marcoulides, G.A., Chin, W.W.: You write, but others read. In: Abdi, H., et al. (eds.) New Perspectives in Partial Least Squares and Related Methods, pp. 31–64. Springer, New York (2013)

17. OECD: Managing National Innovation Systems (1999)
18. OECD: Measuring Innovation: A New Perspective (2010)
19. Ranis, G., Stewart, F., Ramirez, A.: Economic growth and human development. World Dev. **28**(2), 197–219 (2000)
20. Raykov, T., Marcoulides, G.: A First Course in Structural Equation Modeling, 2nd edn. Lawrence Erlbaum Associates, Mahwah (2006)
21. Sanchez, G.: PLS path modeling with R (2013). http://www.gastonsanchez.com/PLSPathModelingwithR.pdf
22. Schmoch, U., Rammer, C., Legler, H.: National Systems of Innovation in Comparison. Structure and Performance Indicators for Knowledge Societies. Springer, Heidelberg (2006)
23. Sharif, N.: Emergence and development of the national innovation system concept. Res. Policy **35**(5), 745–766 (2006)
24. Schibany, A., Reiner, C.: Can basic research provide a way out of economic stagnation? Foresight Russ. **8**(4), 54–63 (2014)
25. Todaro, M.R., Smith, S.C.: Economic Development, 12th edn. Pearson, New York (2015)
26. Zaichenko, S., Kuznetsova, T., Roud, V.: Features of interaction between Russian Enterprises and Research Organisations in the field of innovation. Foresight Russ. **8**(1), 6–23 (2014)
27. Zelenkov, Y.: Supercomputing in the context of the knowledge economy. In: CEUR Workshop Proceedings, vol. 1482, pp. 465–475 (2015)
28. Zelenkov, Y.: Enterprise information systems through the prism of scientific publications. Open Syst. DBMS **2**, 44–46 (2016)

Heuristic Anticipation Scheduling in Grid with Non-dedicated Resources

Victor V. Toporkov[ID], Dmitry M. Yemelyanov$^{(\boxtimes)}$, and Petr A. Potekhin

National Research University "Moscow Power Engineering Institute", Moscow, Russia
{ToporkovVV,YemelyanovDM,PotekhinPA}@mpei.ru

Abstract. A heuristic user job-flow scheduling approach to grid virtual organizations with non-dedicated resources is discussed in this article. Users' and resource providers' preferences, virtual organization's internal policies, resources geographical distribution along with local private utilization impose specific requirements for efficient scheduling according to different, usually contradictive, criteria. The available resources set and the corresponding decision space decrease as resources utilization increases. This introduces further complications into the task of efficient scheduling. We propose a heuristic anticipation scheduling approach to improve the overall scheduling efficiency. Initially, it generates a near optimal but infeasible scheduling solution which is then used as a reference for efficient allocation of resources.

Keywords: Scheduling · Grid · Resources · Utilization · Heuristic · Job batch · Virtual organization · Anticipation

1 Introduction and Related Works

In distributed environments with non-dedicated resources, such as utility grids, the computational nodes are usually partly utilized by local high-priority jobs coming from resource owners. Thus, the resources available for use are represented with a set of slots, i.e. time intervals during which the individual computational nodes are capable of executing parts of independent users' parallel jobs. These slots generally have different start and finish times and present a difference in performance. The presence of a set of slots deprives the problem of a coordinated selection of the resources that are necessary to execute the job flow coming from computational environment users. Resource fragmentation also results in a decrease of the total computing environment utilization level [1,2].

Two established trends may be outlined among diverse approaches to distributed computing. The first one is based on the available resources utilization

This work was partially supported by the Council on Grants of the President of the Russian Federation for State Support of Young Scientists and Leading Scientific Schools (grants YPhD-2297.2017.9 and SS-6577.2016.9), RFBR (grants 15-07-02259 and 15-07-03401), the Ministry of Education and Science of the Russian Federation (project no. 2.9606.2017/BCh).

L. Sokolinsky and M. Zymbler (Eds.): PCT 2017, CCIS 753, pp. 58–70, 2017.
DOI: 10.1007/978-3-319-67035-5_5

and application level scheduling [3]. As a rule, this approach does not imply any global resource sharing or allocation policy. Another trend is related to the formation of user's virtual organizations (VO) and a job flow scheduling [4,5]. In this case, a metascheduler is an intermediate chain between the users, and local resource management and job batch processing systems.

Uniform rules of resource sharing and consumption, in particular based on economic models, make it possible to improve the job-flow level scheduling and resource distribution efficiency. VO policy may offer optimized scheduling to satisfy both users' and VO common preferences. The VO scheduling problems may be formulated as follows: to optimize users' criteria or utility function for selected jobs [6,7], to keep resource overall load balance [8,9], to have job run in strict order or maintain job priorities [10], to optimize overall scheduling performance by some custom criteria [11,12], and so on.

VO formation and performance largely depends on mutually beneficial collaboration between all the related stakeholders. However, users' preferences and VO common preferences (owners' and administrators' combined) may conflict with each other. Users are likely to be interested in the fastest possible running time for their jobs with least possible costs, whereas VO preferences are usually directed to available resources load balancing or node owners' profit boosting. Thus, VO policies in general should respect all members, and the most important aspect of the rules suggested by VO is their fairness.

A number of works understand fairness as it is defined in the theory of cooperative games, such as fair job flow distribution [8], fair quotas [13,14], fair user jobs prioritization [10], non-monetary distribution [15]. The cyclic scheduling scheme (CSS) [16] implements a fair scheduling optimization mechanism that ensures stakeholders interests to some predefined extent.

The downside of a majority of centralized metascheduling approaches is that they loose their efficiency and optimization features in distributed environments with a limited resource supply. For example, in [2], a traditional backfilling algorithm provided better scheduling outcome when compared to different optimization approaches in resource domain with a minimal performance configuration. The common root cause is that, in fact, the same scarce set of resources (being efficient or not) has to be used for a job-flow execution, otherwise some jobs might hang in the queue. And under such conditions, user jobs priority and ordering greatly influence the scheduling results. At the same time, application-level brokers are still able to ensure user preferences and optimize the job's performance under free-market mechanisms.

A main contribution of this paper is a heuristic CSS-based job-flow scheduling approach that retains optimization features and efficiency even in distributed computing environments with limited resources. The rest of the paper is organized as follows. Section 2 presents a general CSS fair scheduling concept. The proposed heuristic-based scheduling technique is presented in Sect. 3. Section 4 contains a simulation experiment setup and results for the proposed scheduling approach. Finally, Sect. 5 summarizes the paper.

2   Cyclic Alternative-Based Fair Scheduling Model and Limited Resources

Scheduling of a job flow using the CSS is performed in time cycles known as scheduling intervals, by job batches [16]. The actual scheduling procedure consists of two main steps. The first step involves a search for alternative scenarios of each job execution or simply alternatives [17]. During the second step, dynamic programming methods [16] are used to choose an optimal alternatives combination (one alternative is selected for each job) with respect to the given VO and user criteria. This combination represents the final schedule based on current data regarding resources load and possible alternative executions.

An example of a user scheduling criterion may be a minimization of overall job running time, a minimization of overall running cost, etc. This criterion describes user's preferences for that specific job execution, and expresses a type of additional optimization to perform while searching for alternatives. Alongside with time (T) and cost (C) properties, each job execution alternative has a user utility (U) value: a user evaluation against the scheduling criterion. A common VO optimization problem may be stated as either minimization or maximization of one of the properties, having other fixed or limited, or involve a Pareto-optimal strategy search involving both kinds of properties [4,16,18].

We consider the following relative approach to represent a user utility U. A job alternative with the minimum (the best) user-defined criterion value $Z_{\min}$ corresponds to the left interval boundary $(U = 0\%)$ of all possible job scheduling outcomes. An alternative with the worst possible criterion value $Z_{\max}$ corresponds to the right interval boundary $(U = 100\%)$. In the general case, for each alternative with value Z of the user criterion, U is defined, depending on its position in the interval $[Z_{\min}; Z_{\max}]$, according to the following formula:

$$U = \frac{Z - Z_{\min}}{Z_{\max} - Z_{\min}} \cdot 100\%. \tag{1}$$

Thus, each alternative gets its utility in relation to the "best" and the "worst" optimization criterion values that a user could expect according to the job's priority. And the more some alternative corresponds to user's preferences, the smaller is the value of U. Examples of user utility functions for a job with four alternatives and a cost minimization criterion are presented in Table 1.

For a fair scheduling model, the second step of the VO optimization problem could be expressed in the form: $C \rightarrow \max$, $\lim U$ (maximize total job-flow execution cost while respecting user's preferences to some extent); $U \rightarrow \min$, $\lim T$ (meet user's best interests while ensuring some acceptable job-flow execution time), and so on.

The launch of any job requires a co-allocation of a specified number of slots, in the same manner as in the classic backfilling variation. A single slot is a time span that can be assigned to run a part of a multiprocessor job. The target is to scan a list of N_s available slots, and to select a window of m parallel slots with the length of the required resource reservation time (see Fig. 1). The user job

Table 1. User utility examples for a job with execution cost minimization

Job execution alternatives	Execution cost	Utility
First alternative	5	0%
Second alternative	7	20%
Third alternative	11	60%
Fourth alternative	15	100%

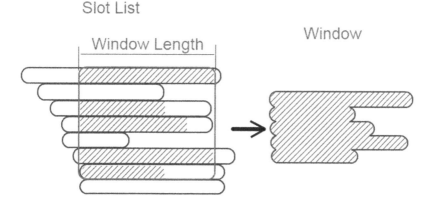

Fig. 1. An example of a window allocation procedure

requirements are arranged into a resource request containing a resource reservation time, characteristics of computational nodes (clock speed, RAM amount, disk space, operating system, etc.), and limitations on the selected window maximum cost. ALP, AMP and AEP window search algorithms were discussed in [17].

The job batch scheduling requires allocation of a multiple *nonintersecting* (in terms of slots) alternatives for each job. Otherwise irresolvable collisions for resources may occur, if different jobs will share the same time slots. Sequential alternatives search and resources reservation procedures help to prevent such scenario. However, in an extreme case, when resources are limited or overutilized, only at most one alternative execution could be reserved for each job. In this case, alternatives-based scheduling result will be no different from First Fit resources allocation procedure [2]. First Fit resource selection algorithms [19] assign any job to the first set of slots matching the resource request conditions, without any optimization.

3   Heuristic Anticipation Scheduling

3.1   General Anticipation Scheduling Scheme

In order to address this problem, the following heuristic job batch scheduling scheme, consisting of three main steps, is proposed.

1. First, a set of all possible execution alternatives is found for each job, without considering time slots intersections or any resource reservation. The resulting intersecting alternatives found for each job reflect a full range of different job execution possibilities that a user may expect on the current scheduling interval. It may be noticed that this set is guaranteed to include the best and the worst alternatives according to any scheduling criterion, including user and VO criteria.
2. Second, the CSS procedure is performed to select alternatives combination (one alternative for each job of the batch) optimal according to the VO policy. The resulting alternatives combination most likely corresponds to an infeasible scheduling solution, since possible time-slots intersections will cause collisions in the resources allocation stage.

   The main idea of this step is that the obtained infeasible solution will provide some heuristic insights on how each job should be handled during the scheduling. For example, whether time-biased or cost-biased execution is preferred, how it should correspond to user criterion and VO administration policy, and so on.
3. Third, a feasible resources allocation is performed by replicating alternatives selected in step 2. The base for this replication step is an Algorithm searching for Extreme Performance (AEP) described in details in [17]. In the current step, AEP helps to find and reserve feasible execution alternatives most similar to those selected in the near-optimal infeasible solution.

After these three steps are performed, the resulting solution is both feasible and efficient, as it reflects a scheduling pattern obtained from a near-optimal reference solution from step 2.

The following subsections will discuss these scheduling steps in more details.

3.2 Finding a Near Optimal Infeasible Scheduling Solution

CSS results strongly depend on the diversity of alternatives sets obtained for batch jobs. The task of finding all possible execution alternatives for each job of the batch may become impractical, since the number of different resources combinations may reach $C(p, m)$, where p is the total number of different resource types available, and m is the number of resources requested by the user. Moreover, if we consider non-dedicated resources, then this task will be additionally complicated by local resources utilization. In this case, not all the resources combinations may be available during the scheduling cycle.

However, as we need to find alternatives for an *a priori* infeasible reference solution, a reasonable diverse set of possible execution alternatives will do. An important feature of this set is that it should contain extreme execution alternatives according to different criteria, e.g. the most expensive, the least time-consuming alternative, and so on.

Further, this set of possible alternatives may be used to evaluate actual user job execution against the job execution possibilities according to Eq. (1). We assume that such a set may represent a fair uniform basis for a user utility expectations. This *uniform user utility* U_U may be used to compare different scheduling algorithms from the user's point of view.

We used a modification of the AEP to allocate a diverse set of execution alternatives for each job. Originally, the AEP is able to find only one alternative execution that satisfies the user resource request and is optimal according to the user custom criterion. The main idea of the current modification is to save all intermediate AEP search results to a dedicated list as shown in the following algorithm description:

Data: *slotList* — a list of available slots; *job* — a job for which the search is
          performed
Result: *alternativesSet* — a set of possible alternatives
slotList = orderSystemSlotsByStartTime();
for *each slot in slotList* **do**

    if *not(properHardwareAndSoftware(slot.node))* **then**
       | continue;
    end
    windowSlotList.add(slot);
    windowStartTime = slot.startTime;
    for *each wSlot in windowSlotList* **do**
       minLength = wSlot.node.getWorkingTimeEstimate();
       if *(wSlot.endTime - windowStartTime) < minLength* **then**
          | windowSlotList.remove(wSlot);
       end
    end
    if *windowSlotList.size() ≥ job.nodesNeed* **then**
       minCostWindow = getMinCostWindow(windowSlotList);
       maxCostWindow = getMacCostWindow(windowSlotList);
       minRuntimeWindow =
         getMinRuntimeWindow(windowSlotList);
       alternativesSet.add(minCostWindow);
       alternativesSet.add(maxCostWindow);
       alternativesSet.add(minRuntimeWindow);
    end
end

Algorithm 1. AEP modification to allocate a set of possible job execution alternatives

In this algorithm, an expanded window *windowSlotList* of size M *moves* through a list *slotList* of all available slots sorted by their start time in ascending order. At each step, any combination of m slots inside *windowSlotList* (in the case when $m \leq M$) can form a window that meets all the requirements to run the job. The main difference from the original AEP algorithm is indicated in bold. Instead of searching for a single window with a maximum criterion value, we allocate several windows with extreme criteria values from every instance of *windowSlotList*, and save to *alternativesSet*. By the end of *slotList*, *alterna-tivesSet* will contain a diverse set of possible job execution alternatives. And since every possible *windowSlotList* instance is processed by the AEP, *alternativesSet* is guaranteed to contain alternatives with extreme criteria values (maxCost/minCost/minTime), as well as a variety of alternatives with some intermediate criteria values.

After sets of possible intersecting execution alternatives are allocated for each job, a CSS scheduling optimization procedure selects an optimal alternatives combination according to VO and users criteria [16].

3.3   Replication Scheduling and Resources Allocation

The resulting near-optimal scheduling solution in most cases is infeasible, since the selected alternatives may share the same time slots, thereby causing resource collisions. However, we suggest to use it as a reference solution, and replicate it into a feasible resources allocation.

For the purpose of replication, a new *Execution Similarity* criterion was introduced, which assists AEP in finding a window with minimum *distance* to a reference alternative. Generally, we define a *distance* between two different alternatives (windows) as a relative difference or *error* between their significant criteria values. For example, if the reference alternative has total cost C_{ref}, and some candidate alternative cost is C_{can}, then the relative cost error E_C is calculated as

$$E_C = \frac{|C_{\mathrm{ref}} - C_{\mathrm{can}}|}{C_{\mathrm{ref}}}.$$

If one needs to consider several criteria, then the *distance* D between two alternatives may be calculated as a linear sum of criteria errors,

$$D_m = E_C + E_T + .. + E_U,$$

or as a geometric distance in a parameters space,

$$D_g = \sqrt{E_C^2 + E_T^2 + ... + E_U^2}.$$

For a feasible job batch resources allocation, the AEP consequentially allocates for each job a single execution window with a minimum *distance* to a reference alternative. Time slots allocated to the i-th job are reserved and excluded from the slot list when the AEP search algorithm is performed for the following jobs $i + 1, i + 2, ... N$. Thus, this procedure prevents any conflicts between resources and provides a scheduling solution that in some sense reflects a near-optimal reference solution.

4   Simulation Study

4.1   Simulation Environment Setup

An experiment was prepared as follows, using a custom distributed environment simulator [20]. Virtual organization and computing environment properties:

- The resource pool includes 80 heterogeneous computational nodes.
- The specific cost of a node is an exponential function of its performance value (base cost) with an added variable margin distributed normally as ± 0.6 of the base cost.
- The scheduling interval length is 800 time quanta. The initial resource load with owner jobs is distributed hyper-geometrically resulting in 5 to 10% of time quanta excluded in total.

Job batch properties:

- The number of jobs in a batch is 125.
- The number of nodes needed for a job is a whole number distributed evenly in $[2; 6]$.
- The node reservation time is a whole number distributed evenly in $[100; 500]$.
- The job budget varies in a way that some of the jobs can pay as much as 160% of the base cost, whereas some may require a discount.
- Every request contains a specification of a custom user criterion, namely one of the following: job execution runtime or overall execution cost.

During each experiment, a VO domain and a job batch were generated, and the following scheduling schemes were simulated and studied.

First, a general *CSS* solved the optimization problems $T \rightarrow \min$, $\lim U$ with different limits $U_a \in \{0\%, 1\%, 4\%, 10\%, 16\%, 32\%, 100\%\}$. U_a stands for the average user utility for one job, e.g. $\lim U_a = 10\%$ means that at average the resulting deviation from the best possible outcome for each user did not exceed 10%.

Second, a near-optimal but infeasible reference solution *REF* (see Sect. 3.2) was obtained for the same problems.

Third, a replication procedure CSS_{rep} was performed based on the *CSS* solution to demonstrate the replication process accuracy.

For the heuristic anticipation scheduling *ANT*, the same replication procedure was performed based on the *REF* solution.

Finally, two independent job batch scheduling procedures were performed to find the scheduling solutions most suitable for VO users ($USER_{\text{opt}}$) and VO administrators (VO_{opt}). $USER_{\text{opt}}$ was obtained applying only user criteria to allocate resources for jobs without taking into account VO preferences. VO_{opt} was obtained by using one VO optimization criterion (the runtime minimization $T \rightarrow \min$ in our example) for each job scheduling without taking into account user preferences.

4.2   Simulation Results

1000 single scheduling experiments were simulated. The average number of alternatives found for a job in *CSS* was 2.6. This result shows that usually a few alternative executions were found for relatively *small* jobs, whereas *large* jobs usually had at most one possible execution option (remember that according to the simulation settings the difference between jobs execution time could be up to 15-fold). At the same time, the *REF* algorithm at average considered more than 100 alternative executions for each job. *CSS* failed to find any alternative execution at least for one job of the batch in 209 experiments; *ANT* did the same in 155 experiments.

These results show that the simulation settings provided a quite diverse job batch and, at the same time, a limited set of resources not allowing to execute all the jobs during every experiment.

Figure 2 shows the average job execution time (VO criterion) in a $T \rightarrow$ min, $\lim U$ optimization problem. Different limits $U_a \in \{0\%, 1\%, 4\%, 10\%, 16\%, 32\%, 100\%\}$ specify to what extent user preferences were taken into account. The two horizontal lines $USER_{opt}$ and VO_{opt} indicate, respectively, the practical T values when only user or VO administration criteria are optimized.

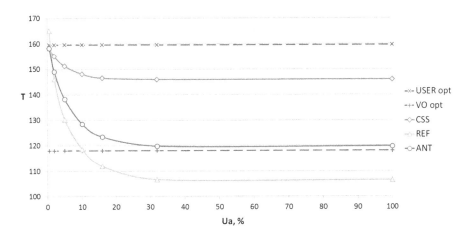

Fig. 2. Average job execution time in $T \rightarrow$ min, $\lim U$ problem

The first thing that attracts our attention in Fig. 2 is that *REF* provides for $U > 10\%$ a better (smaller) job execution time value than those of VO_{opt}. This behavior is, nonetheless, expected, as *REF* generates an infeasible solution and may use time-slots from more suitable (according to VO preferences) resources several times for different jobs.

On the other hand, *ANT* provided a better VO criterion value than *CSS* for all $U > 0\%$. The relative advantage reaches 20% when $U > 20\%$.

Interestingly, the *ANT* algorithm graph gradually changes from the $USER_{opt}$ value at $U = 0\%$ to almost the VO_{opt} value at $U = 100\%$ just as the average user utility limit changes. Therefore, *ANT* represents a general scheduling approach allowing to balance between VO stakeholders criteria according to a specified scenario, including VO or user criteria optimization.

A similar pattern can be observed in Fig. 3, where the $C \rightarrow$ max, $\lim U$ scheduling problem is represented. In this scenario, however, the *ANT* advantage over *CSS* amounts to 10% against the VO criterion.

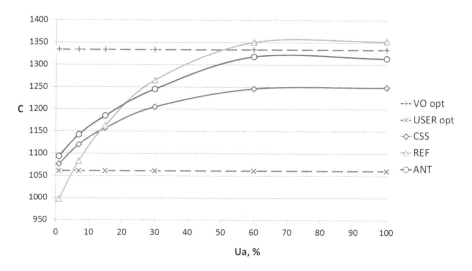

Fig. 3. Average job execution cost in the $C \to \max$, $\lim U$ problem

The advantage of *ANT* over *CSS* can be explained by a scarce set of alternatives found for user jobs by the latter algorithm. To compare *ANT* and *CSS* scheduling results against user criteria we used U_U uniform user utility metric (see Sect. 3.2). Figure 4 shows average uniform user utility U_U for *ANT* and *CSS* recalculated based on reference alternatives from *REF* using Eq. 1.

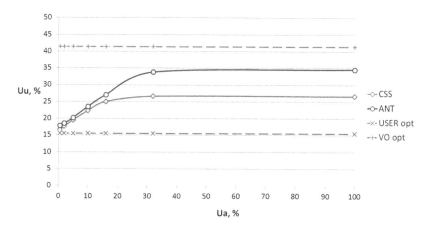

Fig. 4. Uniform user utility value in the $T \to \min$, $\lim U$ problem

As it can be seen from Fig. 4, *ANT* provides a higher uniform user utility in each experiment compared with *CSS*. Even more, *ANT* generally operates in a wider range of possible user utilities: from 17% when $U = 0\%$ (only user criteria are optimized) to 35% when $U = 100\%$ (only the VO criterion is optimized). At the same time, CSS is able to change uniform user utility in the interval $(16\%; 26\%)$. Wider U_U interval represents greater decision space for ANT which implies optimization advantage.

It may seem that both algorithms operate in a rather small uniform user utilities interval and are not efficient enough. However, uniform user utilities are based on an infeasible set of alternatives and cannot be entirely replicated in a feasible solution. In Fig. 4, the VO_{opt} and $USER_{opt}$ horizontal lines roughly represent a feasible interval for uniform user utility values. In this case, ANT covers more than half of this feasible U_U interval.

Finally, Fig. 5 shows the average replication error (or *distance*) for each job of the batch provided by replicating a feasible CSS solution for $U_a = 30\%$ ($CSS_{ref}30$) and infeasible REF solutions with lim $U_a = 0\%$ (ANT) and lim $U_a = 30\%$ ($ANT30$). In this experiment, we used time and cost errors to calculate a geometric distance Dg between reference and allocated alternatives. Figure 5 shows that the CSS_{ref} error is practically independent from ordinal job number, reaching 0.05 (or 5%) for the last job of the batch. Thus, we can conclude that a feasible solution generally may be replicated with a good accuracy even when resources are limited.

Completely different results are provided by ANT. Depending on U_a, the ANT error may reach 0.35 (or 35%) for the last jobs of the batch. Unlike CSS, an infeasible REF solution may require, for example, the allocation of the nodes with the highest possible performance for each job. In this case, the replication process will not be able to reserve the required amount of high performance nodes, and the error for the last jobs may increase greatly.

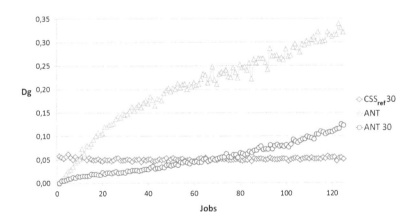

Fig. 5. Average replication error for user jobs

5   Conclusions and Future Work

In this paper, we have studied the problem of a fair job batch scheduling with a relatively limited resources supply. The main problem arising is a scarce set of job execution alternatives, which eliminates the scheduling optimization efficiency. We proposed an algorithm to obtain a diverse set of possible execution alternatives for each job. This set may be used as a uniform basis for a fair *uniform user utility* calculation to rate a scheduling solution. Then we proposed a heuristic scheduling scheme that generates a near-optimal but infeasible reference solution and, after that, replicates it to allocate a feasible accessible solution.

A computer simulation was performed to study these algorithms and evaluate their efficiency. The obtained results show that the new heuristic approach provides flexible and efficient solutions for different fair scheduling scenarios. The advantage over the general CSS against VO preferences (for example, when minimizing the total job batch execution time) reaches 25%. The above-mentioned replication procedure showed a relatively high accuracy providing less than 5% error when replicating a batch of 125 user jobs.

Future work will focus on the study of the replication algorithm and its possible application to fulfill complex user preferences expressed in a resource request.

References

1. Dimitriadou, S.K., Karatza, H.D.: Job scheduling in a distributed system using backfilling with inaccurate runtime computations. In: Proceedings of 2010 International Conference on Complex, Intelligent and Software Intensive Systems, pp. 329–336 (2010). doi:10.1109/CISIS.2010.65
2. Toporkov, V., Toporkova, A., Tselishchev, A., Yemelyanov, D., Potekhin, P.: Heuristic strategies for preference-based scheduling in virtual organizations of utility grids. J. Ambient Intell. Humanized Comput. **6**(6), 733–740 (2015). doi:10. 1007/s12652-015-0274-y
3. Buyya, R., Abramson, D., Giddy, J.: Economic models for resource management and scheduling in grid computing. J. Concurr. Comput. **14**(5), 1507–1542 (2002). doi:10.1002/cpe.690
4. Kurowski, K., Nabrzyski, J., Oleksiak, A., Weglarz, J.: Multicriteria aspects of grid resource management. In: Nabrzyski, J., Schopf, J.M., Weglarz, J. (eds.) Grid Resource Management. State of the Art and Future Trends, vol. 64, pp. 271–293. Springer, Boston (2003). doi:10.1007/978-1-4615-0509-9_18
5. Rodero, I., Villegas, D., Bobro, N., Liu, Y., Fong, L., Sadjadi, S.M.: Enabling interoperability among grid meta-schedulers. J. Grid Comput. **11**(2), 311–336 (2013). doi:10.1007/s10723-013-9252-9
6. Ernemann, C., Hamscher, V., Yahyapour, R.: Economic scheduling in grid computing. In: Feitelson, D.G., Rudolph, L., Schwiegelshohn, U. (eds.) JSSPP 2002. LNCS, vol. 2537, pp. 128–152. Springer, Heidelberg (2002). doi:10.1007/3-540-36180-4_8
7. Rzadca, K., Trystram, D., Wierzbicki, A.: Fair game-theoretic resource management in dedicated Grids. In: IEEE International Symposium on Cluster Computing and the Grid (CCGRID 2007), pp. 343–350 (2007). doi:10.1109/ccgrid.2007.52

8. Penmatsa, S., Chronopoulos, A.T.: Cost minimization in utility computing systems. Concurr. Comput. Pract. Exp. **16**(1), 287–307 (2014). doi:10.1002/cpe.2984. Wiley
9. Vasile, M., Pop, F., Tutueanu, R., Cristea, V., Kolodziej, J.: Resource-aware hybrid scheduling algorithm in heterogeneous distributed computing. J. Future Gener. Comput. Syst. **51**, 61–71 (2015). doi:10.1016/j.future.2014.11.019
10. Mutz, A., Wolski, R., Brevik, J.: Eliciting honest value information in a batch-queue environment. In: 8th IEEE/ACM International Conference on Grid Computing, pp. 291–297. IEEE Computer Society (2007). doi:10.1109/grid.2007.4354145
11. Blanco, H., Guirado, F., Lérida, J.L., Albornoz, V.M.: MIP model scheduling for multi-clusters. In: Caragiannis, I., et al. (eds.) Euro-Par 2012. LNCS, vol. 7640, pp. 196–206. Springer, Heidelberg (2013). doi:10.1007/978-3-642-36949-0_22
12. Takefusa, A., Nakada, H., Kudoh, T., Tanaka, Y.: An advance reservation-based co-allocation algorithm for distributed computers and network bandwidth on QoS-guaranteed Grids. In: 15th International Workshop JSSP 2010, vol. 6253, pp. 16–34 (2010). doi:10.1007/978-3-642-16505-4_2
13. Carroll, T., Grosu, D.: Divisible load scheduling: an approach using coalitional games. In: Proceedings of the Sixth International Symposium on Parallel and Distributed Computing (ISPDC 2007), p. 36 (2007). doi:10.1109/ispdc.2007.16
14. Kim, K., Buyya, R.: Fair resource sharing in hierarchical virtual organizations for global Grids. In: Proceedings of the 8th IEEE/ACM International Conference on Grid Computing, pp. 50–57 (2007). doi:10.1109/grid.2007.4354115
15. Skowron, P., Rzadca, K.: Non-monetary fair scheduling cooperative game theory approach. In: Proceedings of the Twenty-Fifth Annual ACM Symposium on Parallelism in Algorithms and Architectures (SPAA 2013), pp. 288–297 (2013). doi:10.1145/2486159.2486169
16. Toporkov, V., Yemelyanov, D., Bobchenkov, A., Tselishchev, A.: Scheduling in grid based on VO stakeholders preferences and criteria. In: Zamojski, W., Mazurkiewicz, J., Sugier, J., Walkowiak, T., Kacprzyk, J. (eds.) Dependability Engineering and Complex Systems. AISC, vol. 470, pp. 505–515. Springer, Cham (2016). doi:10.1007/978-3-319-39639-2_44
17. Toporkov, V., Toporkova, A., Tselishchev, A., Yemelyanov, D.: Slot selection algorithms in distributed computing. J. Supercomput. **69**(1), 53–60 (2014). doi:10.1007/s11227-014-1210-1
18. Farahabady, M.H., Lee, Y.C., Zomaya, A.Y.: Pareto-optimal cloud bursting. IEEE Trans. Parallel Distrib. Syst. **25**, 2670–2682 (2014). doi:10.1109/tpds.2013.218
19. Cafaro, M., Mirto, M., Aloisio, G.: Preference-based matchmaking of grid resources with CP-Nets. J. Grid Comput. **11**(2), 211–237 (2013). doi:10.1007/s10723-012-9235-2
20. Toporkov, V., Tselishchev, A., Yemelyanov, D., Bobchenkov, A.: Composite scheduling strategies in distributed computing with non-dedicated resources. Procedia Comput. Sci. **9**, 176–185 (2012). doi:10.1016/j.procs.2012.04.019

Parallel Numerical Algorithms

On the Parallel Strategies in Mathematical Modeling

Valery Il'in[(✉)]

Institute of Computational Mathematics and Mathematical Geophysics RAS,
Novosibirsk State University, 6 Pr. Lavrentieva, 630090 Novosibirsk, Russia
ilin@sscc.ru

Abstract. The article considers parallel strategies and tactics at different stages of mathematical modeling. These technological steps include geometrical and functional modeling, discretization and approximation, algebraic solvers and optimization methods for inverse problems, post-processing and visualization of numerical results, as well as decision-making systems. Scalable parallelism can be provided by combined application of MPI tools, multi-thread computing, vectorization, and the use of graphics accelerators. The general method to achieve high-performance computing consists in minimizing data communications, which are the most time and energy consuming. The construction of efficient parallel algorithms and code optimization is based on various approaches at different levels of computational schemes. The implementation of the biggest interdisciplinary direct and inverse problems in cloud computing technologies is considered. The corresponding applied software with a long life cycle is represented as integrated environment oriented to large groups of end users.

Keywords: Scalable parallelism · Domain decomposition · Runtime · Communications · Multi-thread computing · Vectorization · Exchange buffers · Hierarchical memory · Speedup · Accelerators

1 Introduction

The idea of parallelization is very old, and consists in the simultaneous operation of different hardware units. Modern heterogeneous supercomputer multi-processor systems (MPS) have a rich architecture: a large net of nodes with distributed memory, sets of multi-core CPUs with shared hierarchical memory and very fast registers, and several graphics accelerators (at each node) of GPGPU or Intel Phi type, with a complicated internal structure. We can, accordingly, consider four-level hybrid programming tools: Message Passage Interface (MPI system), multi-thread computing (OpenMP), CUDA system, and vectorization of machine operations by applying AVX instructions inside the CPU or the Phi

The work was supported by the RFBR grant N 16-29-15122 and the RSF grant N 15-11-10024.

L. Sokolinsky and M. Zymbler (Eds.): PCT 2017, CCIS 753, pp. 73–85, 2017.
DOI: 10.1007/978-3-319-67035-5_6

unit. Moreover, programmers can use different hints for code optimization, taking into account detailed peculiarities of the memory access. We should emphasize that the evolution of computer platforms is dramatically fast, and applied software must be flexibly adapted to hardware changes in order to provide a long numerical life cycle of the environment.

Scalable parallelism is a challenging problem when solving interdisciplinary direct and inverse super-tasks concerned with mathematical modeling (MM), which now constitute the main tool for obtaining new fundamental knowledge and optimizing industrial production. One of the main trends of development of computational and informational technologies (CIT) to study various processes and phenomena consists in the convergence of MM approaches and CAD-CAE-CAM applications. Other important feature of the current situation is associated with a rapid upsurge of new results in theoretical and computational mathematics. It is well-known that the most "clever" and efficient algorithms are difficult to parallelize. Thus, the most urgent issue concerning MM support is to strike the right balance between the mathematical efficiency of an algorithm and the computer performance of its program implementation.

The bottleneck is programmer's labor productivity, which lags far behind the growth rates of supercomputer capacity. Overcoming the world-wide crisis requires a new paradigm of development. The existing long-term practice is the implementation of applied software packages (ASPs), either commercial or publicaccess, for concrete classes of problems. Examples of such products are ANSYS [1] and FeniCS [2]. Developments of other types are program libraries that implement a totality of algorithms for a certain type of computational tasks. For instance, Netgen [3] is responsible for mesh generation, PETSc [4] is a suite of algebraic solvers, and so on. Another versions that are becoming increasingly popular nowadays are instrumental computational systems Open-FOAM [5], DUNE (Distributed Unified Numerical Environment) [6] and Basic System of Modeling (BSM) [7]. Some general issues that arise when creating a program environment for mathematical modeling are considered in [8]. It is worth mentioning an interesting project devoted to algorithms and their parallel implementations: the Open Encyclopedia of Properties of Algorithms [9].

This paper is organized as follows. Section 2 contains the algorithmic description of the main technological stages of mathematical modeling. In Sect. 3, we discuss specific features of parallel tactics at different steps of a large-scale numerical experiment. In the Conclusions, we make some remarks on parallelization strategies for cloud computing and Data Center frameworks.

2   Technological Stages of Large-Scale Numerical Experiments

Regardless of the subject orientation of applied software, a computational experiment goes through similar technological stages. We can implement these steps almost independently if we define the internal interfaces correctly, in accordance with Virt rule: "Program = Algorithms + Data Structure". Some performance and intellectuality issues of supercomputer modeling are discussed in [10].

Geometric and functional modeling. At the first stage, the user formulates a computational task, which may include a description of a complex geometric configuration consisting of subdomains with different material properties. While geometric objects and operations have long been assimilated in numerous CAD products (CAE, CAM, PLM) and graphics systems, functional modeling requires operating with formalisms such as equations in subdomains, boundary conditions on border segments, various coefficients, etc.

The formal description of an initial boundary value problem for partial differential equations (PDEs) can be presented, for example, as follows:

$$
\begin{aligned}
&L\boldsymbol{u} = f(\boldsymbol{x}, t), \quad \boldsymbol{x} \in \bar{\Omega}, \ 0 < t \leq T < \infty, \\
&l\boldsymbol{u} = \boldsymbol{g}(\boldsymbol{x}, t), \quad \boldsymbol{x} \in \Gamma, \ \boldsymbol{u}(\boldsymbol{x}, 0) = \boldsymbol{u}^0(\boldsymbol{x}), \\
&L = A\frac{\partial}{\partial t} + \nabla B \nabla + C \nabla + D, \ \Gamma = \Gamma_D \cup \Gamma_N, \\
&\boldsymbol{u}\big|_{\Gamma_D} = \boldsymbol{g}_D, \ (D_N \boldsymbol{u} + A_N \nabla_n \boldsymbol{u})\big|_{\Gamma_N} = \boldsymbol{g}_N.
\end{aligned}
\tag{1}
$$

Here $\boldsymbol{x}$ and t are spatial and time variables, L and ℓ are differential operators, $\boldsymbol{u}$ is the solution sought (in general, a vector), A, B, C, D are some matrices, and Γ_D, Γ_N are border segments with different types of boundary conditions.

In addition to input data, we should specify what we want to obtain and in which form. Methods to be applied or even detailed computational schemes, which unambiguously determine the process of mathematical modeling in a concrete environment, may also be prescribed. Emphasizing the above aspects, we have come, in fact, to the automation of the model and algorithm construction. Some questions on these topics, including geometric and functional data structures (GDS and FDS), are discussed in [11].

Problem discretization. The solution of non-trivial mathematical equations essentially always begins with constructing a grid. To show the diversity of questions that arise in this respect, it suffices to mention the most popular types of grids, such as adaptive, structured, unstructured and quasi-structured, matching, non-matching and mortar, regular and irregular, static and dynamic, and so on. Modern real super-tasks require a quite large number of nodes (10^9 and more). Important questions on the performance of this stage, as well as a review of algorithms and numerical software are presented in [12].

The most effective approaches to discretization are associated with sufficiently complex discrete objects and their transformations, including sequences of hierarchical grids and their local refinement, decomposition of grid domains into subdomains, dynamic reconfiguration of grids, and an *a posteriori* and/or *a priori* account of the properties of the desired solutions. Although there are quite a few indicators of the quality of grids, the determination of the optimal grid remains a very complicated problem, which practical studies do not even formulate. The most frequently used principle of choice can be reduced to an empirical approach: the use of distribution densities of mesh nodes according to general qualitative considerations. Individual methodological recommendations relate to particular cases and are mere exceptions to the rule. The global applied-software market offers both very expensive and free mesh generators, which use a certain number of mesh data structures (MDSs) recognized by the computational community. The effective use of this colossal materialized intellectual potential appears to be a very important task.

Discretization is a technological stage that is important for both resource intensity as a whole and computational resolution, which largely determines the success in application of the modeling. This is especially true for problems with a complex spatial and temporal behavior of the solution, including actual situations with strong multi-scale characteristics. Therefore, mesh generation is a highly intelligent methodology; the lack of substantial theoretical and algorithmic results causes us to orient not toward automatic but toward automated mesh construction accompanied by immediate visualization and participation of an expert in controlling the computational process.

Approximation of equations. When the previous stages have been executed and an MDS has been formed, which, together with the geometric and functional data structures (GDS and FDS), reflects the whole information about the initial problem at a discrete level, its approximation becomes possible. The result is a system of finite-dimensional algebraic relations, i.e. an algebraic data structure (ADS) that can effectively use widespread matrix representations for sparse algebraic systems. As an example, let us mention the Compressed Sparse Row format (CSR).

The operations performed in this case are the most science-based and are represented by diverse theoretical approaches: finite-difference, finite-volume and finite-element methods (FDM, FVM and FEM); different spectral algorithms; integral equation methods, etc. The logical complexity of the "approximators" particularly increases when methods of a high order of accuracy are used, especially on unstructured grids, formulas for which would eventually extend over several pages. This circumstance hinders their wide dissemination despite their significant advantages. A cardinal solution to this situation is the use of artificial intelligence potentialities, namely the means of automating analytic symbolic transformations. In principle, such tools are present in large specialized systems of the Reduce or Maple types, and are successfully used, for example, in the FEniCS package [2]. In the above cases, the problem is simplified by the FEM- and FVM-based unique element-by-element technology of independent and easily parallelized computation of local matrices with the subsequent assembly of a global matrix. Some general questions of approximation techniques are described in [13].

Solving algebraic problems. At this stage, various matrix-vector operations are performed that require the largest computer resources, since the volume of both arithmetic operations and necessary memory often grows nonlinearly as the number of degrees of freedom (d.o.f.) of the problem grows. The performed computations may require implementing recurrent sequences, solving systems of algebraic equations (linear, SLAEs, and nonlinear, SNAEs), solving eigenvalue problems, and optimizing algorithms for mathematical programming. These tasks constitute vast fields of computational algebra, characterized by a colossal diversity of conceptual approaches, concrete versions of methods, and particular ways of their application. It is here where the issues of parallelization of algorithms and their implementation on MPS architectures, particularly on cluster systems containing heterogeneous nodes with classical and specialized processors, arise.

The international market offers a big amount of algebraic software, which is continuously updated and expanded, owing to adaptive modifications for new computer platforms and architectures, and the rapid development of new algorithms. The rapid growth and regular updating create the problem of coordinated re-use of existing products. Let us note that there are serious achievements in this area: standard universal data structures and libraries with the basic set of matrix-vector operations (BLAS, SPARSE BLAS) [14].

The diversity of algebraic methods is associated, first of all, with a variety of types of considered matrices: Hermitian and non-Hermitian, real and complex, symmetrical and non-symmetrical, degenerate and non-degenerate, positive-definite and indefinite, and so on. Matrices of all types are divided into dense and sparse, and approaches to their processing are significantly different. Moreover, the choice of optimal algorithms largely depends on structural properties of matrices (band, triangular, etc.), as well as on their dimension (the notion of "large" matrices continuously changes depending on the capacity of the current generation of computers, being in the post-petaflops era 10^9 to 10^{12} orders of magnitude). Ill-posed problems with a strong instability of numerical solutions relative to inherent or computer rounding errors are particularly complicated.

The most efficient modern algorithms are characterized by high logical complexity: algebraic multigrid approaches, domain decomposition methods, variable ordering optimization, matrix scaling techniques, and so on. We can affirm that the most resource-intensive algebraic methods also require an active use of artificial intelligence. The conception of an integrated numerical environment for computational algebra is presented in [15].

Optimization approaches to solving inverse problems. The solution of direct problems of mathematical modeling (1), which require to find the desired functions, given the coefficients of the equations and the initial and boundary conditions, may have a high computational complexity. But this is usually only a part of the difficulties associated with the solution of an inverse problem. The latter is characterized by the fact that some of its initial data depend on unknown parameters, which should be found by minimizing the described objective functional under certain additional restrictions on the problem properties. For example, when computing technical devices or instruments, the engineer usually aims not only at studying their properties but also at the computer-aided design of optimal configurations that would ensure the required characteristics. In addition, almost always there are additional restrictions associated with the size, weight or other functional conditions. Another characteristic example of an inverse problem is the identification of the parameters of a mathematical model based on comparison of estimated results with data from natural measurements.

The optimization statement of an inverse problem can be written in the form

$$
\begin{aligned}
&\Phi_0(\boldsymbol{u}(\boldsymbol{x},t,\boldsymbol{p}_{opt})) = \min_{\boldsymbol{p}} \Phi_0(\boldsymbol{u}(\boldsymbol{x},t,\boldsymbol{p})), \quad \boldsymbol{p} = \{p_k\}, \\
&L\boldsymbol{u}(\boldsymbol{p}) = \boldsymbol{f}, \quad p_k^{min} \le p_k \le p_k^{max}, \quad k = 1, ..., m_1, \\
&\Phi_l(\bar{u}(\boldsymbol{x},t,\boldsymbol{p})) \le \delta_l, \quad l = 1, ..., m_2,
\end{aligned}
\tag{2}
$$

where Φ_0 is the goal functional, $\boldsymbol{p}$ is an unknown vector parameter, p_k^{min} and p_k^{max} define linear constraints, Φ_ℓ and δ_ℓ are non-linear constraints, and $L u(\boldsymbol{p}) = \boldsymbol{f}$ denotes a constitutive equation that is defined, in fact, by the whole direct problem (1).

The main universal approaches to solving inverse problems rely on the use of constrained minimization methods, which imply a directed sequential search for a local or a global minimum and the intermediate values of the objective functional being computed at each step, which is nothing but the solution of a direct problem. Consequently, in the general case, the solution of an inverse problem requires repeated solutions of direct problems.

In recent decades, optimization methods have been actively developed, giving rise to new trends, such as algorithms of interior points, sequential quadratic programming and trust regions. Note, however, that the minimization of functionals with complex geometric characteristics, especially those of the ravine type, is something at the interface between science and art. That is why a fully automated computational process is possible only in the simplest situations. In fact, even in this case, highly intelligent technologies are necessary, entailing a step-by-step implementation of the entire problem in dialogue with the user, who, based upon his experience, should control the behavior of sequential approximations and govern the parameters of the algorithms to achieve the ultimate goal as soon as possible.

Post-processing and visualization of results. Computational process control and decision-making tools. The results of algebraic computations lack any physical meaning and obviousness, primarily owing to their large volumes. For example, the FEM makes it possible to obtain the coefficients of the expansion of the required solutions with respect to the basic functions used in grid cells, whereas the user needs a compact and illustrative picture of multidimensional vector fields. Hence the reason why applied software should have a developed set of instruments to construct typical representations, such as isosurfaces, force lines, cross sections, various graphs, and so on. This is the first requirement. The second one is associated with the fact that one cannot foresee everything, and an intelligent modeling system should contain the means for automating the programming of various possible characteristics of final data. Finally, the third factor is that end users may come from different professions, and all of them want to obtain a comfortable representation of the results of using a computer, determining its production effect.

It is important that even ideal applied software does not preclude the fact that computer-aided modeling of complicated processes or phenomena is a multifold creative activity. For example, to study some applications systemically, one should first make sure that the models and methods applied meet the specifications; for this, it is necessary, first, to perform test computations and then to analyze whether the data obtained are adequate. Then it appears possible to start the study itself, which can be a large-scale machine experiment preceded by planning and method-selection procedures. The latter are unattainable without providing for the flexible to compile computational schemes, which implies the creation of the corresponding languages (declarative or imperative) to control the computational processes. Finally, modeling is not an end in itself but a tool for cognitive or production activities, therefore, to ensure the adoption of a decision with computational results, applied software should contain either some cognitive principles or means for connecting to CAD infrastructures or technologies that support and optimize the operation regimes of concrete processes. However, these issues are beyond the frame of mathematical modeling.

3   Scalable Parallelism: Problems and Solutions

Modern mathematical modeling offers a huge set of real applications, models and numerical methods, as well as an essential diversity of hardware and computing platforms. This provides a great possibility to choose tactics and strategies for the optimization of computational processes.

Some general issues of parallelization. A universal requirement on applied software is the absence of software restrictions on both the number of d.o.f. of the problem to be solved and the quantity of processors and/or cores used. At the same time, it should be recalled that there are important algorithm parallelization characteristics such as weak and strong scaling. Weak scaling means that computation time remains practically the same as the number of d.o.f. and the number of computing devices grow. Strong scaling means a proportional time decrease for a fixed problem as the number of computers grows.

Ideally, a solution to the problem of automation and optimization of algorithm parallelization should be sought through simulation of a computer system as a whole. However, this is too complicated: that is why one has to employ semi-empirical techniques or the simplest models of computer calculations. Two values can be mentioned as examples of parallelization characteristics, namely the coefficients of computational speedup and the processor utilization efficiency:

$$S_p = T_1/T_p, \quad E_p = S_p/P,$$

where T_p is the time required for the execution of a task or algorithm on P processors. An ideal situation is that where the value of S_p is directly proportional to P and $E_p = 1$. In practice, however, we often have to content ourselves with efficiency factors of only several percent.

Let us note that if the portion of consecutive operations is equal to θ, then the maximum speedup is defined by Amdahl's law [16]:

$$S_p = T_1/(\theta T_1 + (1 - \theta T_1))/P = P/[1 + \theta(P - 1)].$$

The development of supercomputer technologies occurs in two main directions: high-performance computing (HPC) and operations with Big Data. Note that the convergence of these two trends (intensive data computing) has recently been observed. On the whole, the evolution of MPS generations and extremum modeling problems is accompanied by a similar growth of RAM speed and capacity (the number of teraflops or petaflops is quite similar to the number of terabytes or petabytes of the computer).

The main objective of programming parallel algorithms is to minimize the information exchange, since the total problem time T equals the sum of two terms:

$$T = T_a + T_c, \ \ T_a = N_a \tau_a, \ \ \ T_c = M(\tau_0 + N_c \tau_c),$$

where τ_a and τ_c are the average times required, respectively, for one arithmetic operation and for one transfer, N_a is the number of arithmetic operations, M is the number of memory accesses, τ_0 is the exchange operation delay (setting) time, and N_c is the average volume of one transferred array. We should bear in mind the characteristic relation $\tau_0 \gg \tau_c \gg \tau_a$. The requirement to reduce data transfer is explained not only by a need to increase speed, but also by the energy consumption of communications.

It is evident from all the above that the notion of the quality of algorithms changes for large tasks: from any two methods being compared, the best is not the one that requires fewer computations but the one that is executed faster on MPSs of the type under consideration. In other words, there appears a new concept of computation optimization based on the search for approaches that would significantly reduce the volume of data transferred between processors, even if they increase the number of arithmetic operations.

An important point of interest is the necessity to overcome the uzer mental inconvenience when we have no supercomputer "within reach". With modern cloud technologies, it is sufficient to have Internet access to a computing center for collective users (CCCU or Data Center). Of course, to intellectualize the user interface, a workstation should be equipped with specialized means; however, this is beyond the scope of this paper.

Characteristic features of the parallelization of technological studies. Parallelization tactics at each computation stage are determined by the volume of data and the number of operations. The stage of geometric and functional modeling, substantial in intellectual loads and crucial for the user input interface, deals with macro-objects, which should not be too many (tens, hundreds or, at worst, thousands). Therefore, it seems desirable to manage without exchanges, copying the computations in all MPI processes and storing in them the geometric and functional data structures obtained.

The mesh generation may formally be represented as a data transformation: GDS + FDS → MDS. Note that the mesh data structure for the entire computational space may have a large volume. For this reason, it is natural to create an MDS by each MPI process for "its" mesh subdomain (with a certain overlapping). The formation of distributed data at the initial stage is reasonable, so much so that the decomposition of domains is the main instrument of parallelization. However, since the estimated mesh domain should also be identified as an integral object, all its nodes and other elementary objects (edges, faces, cells) should be numbered twice, namely locally by a subdomain and globally. Decomposition problems can employ two tactics: subdomain construction, which precedes mesh generation (for example, it is natural to separate media with contrasting material properties), or direct formation of mesh subdomains. We should also bear in mind that many efficient algorithms are based on special rearrangements of components (one may speak of such tasks in terms of graphs as well), and all the respective procedures should be accessible to all MPI processes or subdomains, which will, on the whole, substantially reduce data exchange. The popular software packages METIS, parMETIS and other tools for graph partition are effective re-arrangement instruments (see, for example, the review in Algowiki [9]).

Moving boundary problems are the most computationally complex, since they imply that adaptive grids are dynamically reconfigurable as well. Many efficient methods are based on a local refinement and multigrid approaches, whose instrumental support should also be distributed.

Upon obtaining the distributed data arrays, mapping onto MDS, GDS and FDS, one can approximate the original problem in parallel. For this purpose, FEM and FVM have a unique technology for computing local matrices and assembling a global matrix. Since the "approximator" works in parallel by subdomain or MPI processes, with already distributed necessary data, the obtained matrix-vector structures must be in their subdomain. Therefore, this stage can be perfectly implemented without exchanges. The principal operations performed by the mesh cells independently of each other can be effectively parallelized using multi-thread computing. In non-stationary problems and also in nonlinear or optimization computations, approximations are repeated. However, from the point of view of adaptation to computing devices, this usually changes nothing.

Linear systems are the most important intermediate elements when solving algebraic problems. Owing to this, we will focus on them. Special attention should be given to very large SLAEs with sparse matrices, which emerge after the FEM- or FVM-assisted approximation of differential or respective variational multi-dimensional problems on unstructured grids. From the point of view of the classification of algorithms, the SLAEs to be solved can be divided into two major classes: special and general ones. For the former, which comprise systems occurring in boundary value problems with separable variables, there are superfast direct and/or iterative problem solvers, as the fast Fourier transform or alternating direction implicit (ADI) methods with optimal sets of iterative parameters. These approaches have been in demand over the last decades, because of practical requirements to solve the actual Lyapunov and Sylvester matrix equations.

Direct methods for large sparse SLAEs of the general type are actively improving; however, in the most advanced versions of the popular PARDISO [14] and MUMPS programs, their applicability is limited, mainly because of their requirements on RAM volume. Iterative additive domain decomposition methods (DDMs) constitute the main tool for a parallel highly productive solution of SLAEs of this type. DDMs are covered in a considerable body of specialized literature (see, for example, the review in [17]), and have been discussed at 23 major international conferences devoted to this topic. The essence of these decomposition methods consists in dividing a computational mesh domain into subdomains with parametrized overlapping (in a particular case, without intersections) at the internal boundaries of which certain interface boundary conditions are set to determine informational interrelations between neighboring subdomains. In the simplest case, iterations are formed according to the block Jacobi method, which leads to solving auxiliary SLAEs in subdomains simultaneously with data exchange between them. To accelerate this process, optimal algorithms in Krylov subspaces are primarily used. For further increase in speed, various two- or multi-level approaches are employed, such as deflation, aggregation, coarse-grid correction, low-rank matrix approximations, which are implemented in the library KRYLOV [15]. A systematic analysis of modern approaches in algebraic DDMs is presented in [18]. As examples of well-known libraries for parallel solving SLAEs, we can mention PETSc [4] and pARMs [19].

Parallel time integration methods for solving evolution problems are a special topic of interest. An overview of the exciting and rapidly developing area of parallel time algorithms is given in [20].

To attain scalable parallelization, hybrid programming technologies are used: MPI processes are formed over the memory distributed by computational nodes, one per subdomain, inside which multi-threaded computations are performed using OpenMP in common memory. Note that a substantial acceleration is achievable if inter-processor exchanges are matched with synchronous performance of arithmetic operations in subdomains. A separate problem is how to effectively use universal graphics accelerator cards with a great number of computer cores but relatively slow communications (General Purpose Graphic Processor Units, GPGPU), as well as Intel Xeon Phi units and advanced Field Programmable Gate Arrays (FPGAs).

The adaptation of modern decomposition methods to existing computer platforms is, in terms of philosophy and methodology, a problem of mapping algorithms onto the MPS architecture. This basic (in terms of significance) scientific trend is largely experimental, and only numerous comparisons of real performance measurements can be the foundation for elaborating practical recommendations on solving classes of problems.

A special topic of parallelization analysis is that of optimization approaches to solving inverse problems. Some computational issues of this important area are described in [21]. Usually, the solution of an inverse problem requires solving successively a set of direct tasks. In this case, the speedup of parallel computing does not change. Exceptions must be made for the search for several minima of the goal functional and the solution of a global minimization problem. In these cases, we can decompose the domain in the spaces of parameters that should be determined in the original problem, and find an auxiliary inverse constrained subproblem independently in each subdomain.

The post-processing and visualization of computational results is the most favorable field for parallelization. Despite its apparent mathematical simplicity, this technological stage is the key to the success of a large modeling project. High-quality color graphics, especially with dynamic scenarios and regular control of intermediate data, requires significant computer resources, and in a large-scale computational experiment, it can take the lion's share of machine time. Since one of the main requirements on the quality of visualization is high speed of image generation, a natural technical solution is the use of a high-speed graphics processor. An important feature of visualization is that the resultant multi-dimensional vector fields, which should be graphically presented to the user, are distributed over hierarchical memory units of various processors. Another circumstance is related to the presence of a large number of professional graphic products (Visual Studio, OpenGL, and so on), and one of the main problems for the developers of a modeling system is their effective re-use.

From the point of view of large-scale parallelization, optimization methods and computational experimentation control are a superstructure over data-intensive computing stages, and we can expect no special problems here, although the decisions made at the upper block level play a significant role in reaching the final high performance.

4   Conclusions

From the previous analysis of computational models, algorithms and technologies, we can conclude that the infrastructure of large-scale mathematical modeling constitutes a sufficiently large and complicated system. Also, the optimal control of parallel computing requires a careful analysis of the peculiarities of every technological stage. Tactics and strategies of scalable parallelism can be different in terms of both the algorithm and the total task to be solved. We want to make two more comments. First, creating a high-performance integrated numerical environment for solving a wide class of applications on heterogeneous supercomputers with distributed and hierarchical shared memory is a big management problem, which can be solved on the base of common component architecture (CCA) principles (see a discussion in [22]). And second, the implementation of large mathematical experiments should be actually done in a cloud computing framework, with the task-flow technologies at Data Center, and here we have another view to parallelism problems. It is possible to analyze the optimization statement for the runtime, the performance or the speedup in terms of one algorithm, a particular technological stage or a concrete applied mathematical problem. In a sense, we obtain at different levels various local constrained minimization or global multi-variable minimization problems, and the final strategy solutions will be different. However, the main objective of this paper has just been to outline some issues, while the real solutions of these subjects are topics for a special additional research.

References

1. ANSYS - Simulation Driven Product Development. http://www.ansys.com
2. Logg, F., Mardal, K.-A., Wells, G.N. (eds.): Automated Solution of Partial Differential Equations by Finite Element Method: The FEniCS Book. Springer, Heidelberg (2011)
3. Schoberl, J.: Netgen - an advancing front 2D/3D mesh generator based on abstract rules. Comput. Vis. Sci. **1**, 41–52 (1997). doi:10.1007/s007910050004
4. PETSc. http://www.mcs.anl.gov/petsc
5. Open FOAM - The Open Source Computational Fluid Dynamics (CFD) Toolbox. http://www.open-foam.com
6. DUNE Numerics. Distributed Unified Numerical Environment. http://www.dune-project.org
7. Il'in, V.P., Skopin, I.N.: Computational programming technologies. Programm. Comput. Softw. **37**(4), 210–222 (2011). doi:10.1134/s0361768811040037
8. Ilin, V.P.: Fundamental Issues of mathematical modeling. Herald Rus. Acad. Sci. **86**, 118–126 (2016). doi:10.1134/s101933161602009x
9. AlgoWiki: Open encyclopedia of algorithm properties. http://algowiki-project.org
10. Il'in, V.P., Skopin, I.N.: About performance and intellectuality of supercomputer modeling. Program. Comput. Softw. **42**, 5–16 (2016). doi:10.1134/s0361768816010047
11. Golubeva, L.A., Il'in, V.P., Kozyrev, A.N.: Programming technologies in geometric aspects of mathematical modeling. Vestnik NGU. Seria: Inf. Tekhnol. **10**, 25–33 (2012). (in Russian)
12. Il'in, V.P.: DELAUNAY: A technological environment for mesh generation. Siberian Zhurnal Indus. Math. **16**, 83–97 (2013). (in Russian)
13. Butygin, D.S., Il'in, V.P.: Chebyshev: Principles of automation of algorithm construction in an integrated environment for mesh approximations of initial boundary value problems. In: Proceedings of International Conference on Parallel Computational Technologies 2014, Izd. YuUrGU, Chelyabinsk, pp. 42–50 (2014). (in Russian)
14. Intel Math. Kernel Library. http://software.intel.com/en-us/intel-mkl
15. Butyugin, D.S., Gurieva, Y.L., Il'in, V.P., et al.: Functionality and technologies of algebraic solvers in Krylov's library. Mathematical Modeling, Programming & Computer software, pp. 76–86. Bulletin of south Ural state university, Russain Federation (2013). (in Russian)
16. Konshin, I.N.: Parallel Computational Models to Estimate an Actual Speedup of Analyzed Algorithm. Russian Supercomputing Days: Proceedings of the International Conference, pp. 269–280. MSU Publ. (2016) (in Russian). doi:10.1007/978-3-319-55669-7_24
17. Dolean, V., Jolivet, P., Nataf, F.: An Introduction to Domain Decomposition Methods: algorithms, theory and parallel implementation. doi:10.1137/1.9781611974065.. https://archives-ouvertes.fr/cel-01100932
18. Il'in, V.P.: Problems of parallel solution of large systems of linear algebraic equations. J. Math. Sci. **216**, 795–804 (2016). doi:10.1007/s10958-016-2945-4
19. Saad, Y., Sosonkina, M.: pARMS: A package for the parallel iterative solution of general large sparse linear systems user's guide. Report UMSI 2004–8, Minnesota Supercomp. Inst., Univer. of Minnesota, MN (2004)
20. Gander, M.J.: 50 Years of Time Parallel Time Integration. doi:10.1007/978-3-319-23321-5_3. http://www.unige.ch/~gander/Preprints/50YearsTimeParallel.pdf

21. Il'in, V.P.: Numeric solving of direct and inverse problems of electromagnetic prospecting. Sibirian Zhurnal of Vychislitelnoi Mathematiki **6**, 381–394 (2003). (in Russian)
22. Il'in, V.P.: Component technologies of high-performance mathematical modeling. In: Proceedings of the International Conference on Parallel Computational Technologies - 2015 (UFU, IMM UrO RAN, Yekaterinburg), pp. 166–171 (2015). (in Russian)

On the Solution of Linear Programming Problems in the Age of Big Data

Irina Sokolinskaya and Leonid B. Sokolinsky$^{(\boxtimes)}$

South Ural State University, 76 Lenin Prospekt, Chelyabinsk 454080, Russia
{Irina.Sokolinskaya,Leonid.Sokolinsky}@susu.ru

Abstract. The Big Data phenomenon has spawned large-scale linear programming problems. In many cases, these problems are non-stationary. In this paper, we describe a new scalable algorithm called *NSLP* for solving high-dimensional, non-stationary linear programming problems on modern cluster computing systems. The algorithm consists of two phases: *Quest* and *Targeting*. The *Quest* phase calculates a solution of the system of inequalities defining the constraint system of the linear programming problem under the condition of dynamic changes in input data. To this end, the apparatus of Fejer mappings is used. The *Targeting* phase forms a special system of points having the shape of an n-dimensional axisymmetric cross. The cross moves in the n-dimensional space in such a way that the solution of the linear programming problem is located all the time in an ε-vicinity of the central point of the cross.

Keywords: *NSLP* algorithm · Non-stationary linear programming problem · Large-scale linear programming · Fejer mapping

1 Introduction

The Big Data phenomenon has spawned large-scale linear programming (LP) problems [1]. Such problems arise in many different fields. In [2], the following large-scale industrial optimization problems are presented within the context of big data:

- schedule crews for 3400 daily flights in 40 countries;
- buy ads in 10–15 local publications across 40 000 zip codes;
- pick one of 742 trillion choices in creating the US National Football League schedule;
- select 5 offers out of 1000 for each of 25 000 000 customers of an online store;
- place 1000 stock keeping units on dozens of shelves in 2000 stores;
- decide among 200 000 000 maintenance routing options.

The reported study has been partially supported by the RFBR according to research project No. 17-07-00352-a, by the Government of the Russian Federation according to Act 211 (contract No. 02.A03.21.0011) and by the Ministry of Education and Science of the Russian Federation (government order 2.7905.2017/8.9).

© Springer International Publishing AG 2017
L. Sokolinsky and M. Zymbler (Eds.): PCT 2017, CCIS 753, pp. 86–100, 2017.
DOI: 10.1007/978-3-319-67035-5_7

Each of these problems uses Big Data from the subject field. Such a problem is formalized as a linear programming problem involving up to tens of millions of constraints and up to hundreds of millions of decision variables.

Gondzio [3] presents a certain class of large-scale optimization problems arising in quantum information science and related to Bell's theorem. These problems are two-level optimization problems. The higher-level problem is a non-convex non-linear optimization task. It requires solving hundreds of linear programming problems, each of which can contain millions of constraints and millions of variables.

Mathematical modeling in economics is another source of large-scale LP problems. In many cases, LP problems arising in mathematical economy are non-stationary (dynamic). For example, Sodhi [4] describes a dynamic LP task for asset-liability management. This task involves 1.7 billion constraints and 5.1 billion variables. Algorithmic trading is another area that generates large-scale non-stationary linear programming problems [5–7]. In such problems, the number of variables and inequalities in the constraint system formed by using Big Data can reach tens and even hundreds of thousands, and the period of input data change is within the range of hundredths of a second.

Until now, one of the most popular methods for solving LP problems is the class of algorithms proposed and designed by Dantzig on the basis of the simplex method [8]. The simplex method has proved to be effective in solving a large class of LP problems. However, Klee and Minty [9] gave an example showing that the worst-case complexity of the simplex method is exponential time. Nevertheless, Khaciyan [10] proved that the LP problem can be solved in polynomial time by a variant of an iterative ellipsoidal algorithm developed by Shor [11]. Attempts to apply the ellipsoidal algorithm in practice have been unsuccessful so far. In most cases, this algorithm demonstrated much worse efficiency than the simplex method did. Karmarkar [12] proposed the *interior-point method*, which runs in polynomial time and is also very efficient in practice.

The simplex method and the interior-point method remain today the main methods for solving the LP problem. However, these methods may prove ineffective in the case of large-scale LP problems with rapidly changing and (or) incomplete input data. The authors described in [13] a parallel algorithm for solving LP problems with non-formalized constraints. The main idea of the proposed approach is to combine linear programming and discriminant analysis methods. Discriminant analysis requires two sets of patterns M and N. The first set must satisfy the non-formalized constraints, while the second must not. To obtain representative patterns, methods of data mining [14] and time series analysis can be used [15]. To overcome the problem of non-stationary input data, the authors proposed in [16,17] the pursuit algorithm for solving non-stationary LP problems on cluster computing systems. The pursuit algorithm uses Fejer mappings (see [18]) to build a pseudo-projection onto a convex bounded set. The pseudo-projection operator is similar to a projection, but in contrast to the last, it is stable to dynamic changes in input data. In [19], the authors investigated the efficiency of using Intel Xeon Phi multi-core processors to calculate the pseudo-projections.

In this paper, we describe the new *NSLP* (Non-Stationary Linear Programming) algorithm for solving large-scale non-stationary LP problems on cluster computing systems. The *NSLP* algorithm is more efficient than the pursuit algorithm, since it uses a compute-intensive pseudo-projection operation only once (the pursuit algorithm computes pseudo-projections K times at each iteration, K being the number of processor nodes). The rest of the paper is organized as follows. Section 2 gives a formal statement of an LP problem and presents the definitions of the Fejer process and the pseudo-projection onto a polytope. Section 3 describes the new *NSLP* algorithm. Section 4 summarizes the obtained results and proposes directions for future research.

2   Problem Statement

Let a non-stationary LP problem be given in the vector space $\mathbb{R}^n$:

$$\max \left\{ \langle c_t, x \rangle \mid A_t x \leq b_t, \ x \geq 0 \right\}, \tag{1}$$

where the matrix A_t has m rows. The non-stationarity of the problem means that the values of the elements of the matrix A_t and the vectors b_t, c_t depend on time $t \in \mathbb{R}_{\geq 0}$. We assume that the value of $t = 0$ corresponds to the initial time:

$$A_0 = A, b_0 = b, c_0 = c. \tag{2}$$

Let us define the map $\varphi_t \colon \mathbb{R}^n \to \mathbb{R}^n$ as follows:

$$\varphi_t(x) = x - \frac{\lambda}{m} \sum_{i=1}^{m} \frac{\max \left\{ \langle a_{ti}, x \rangle - b_{ti}, 0 \right\}}{\|a_{ti}\|^2} \cdot a_{ti}, \tag{3}$$

where a_{ti} is the i-th row of the matrix A_t, and $b_{t1}, \ldots, b_{tm}$ are the elements of the column b_t. Let us denote

$$\varphi(x) = \varphi_0(x) = x - \frac{\lambda}{m} \sum_{i=1}^{m} \frac{\max \left\{ \langle a_i, x \rangle - b_i, 0 \right\}}{\|a_i\|^2} \cdot a_i. \tag{4}$$

Let M_t be the polytope defined by the constraints of the non-stationary LP problem (1). Such a polytope is always convex. It is known (see [18]) that φ_t is a continuous single-valued M_t-fejerian[1] map for the relaxation factor $0 < \lambda < 2$.
   By definition, put

$$\varphi_t^s(x) = \underbrace{\varphi_t \ldots \varphi_t(x)}_{s}. \tag{5}$$

---

[1] A single-valued map $\varphi \colon \mathbb{R}^n \to \mathbb{R}^n$ is said to be *fejerian* relatively to a set M (or briefly, M-*fejerian*) if

$$\varphi(y) = y, \forall y \in M;$$
$$\|\varphi(x) - y\| < \|x - y\|, \forall x \notin M, \forall y \in M.$$

The *Fejer process* generated by the map φ_t for an arbitrary initial approximation $x_0 \in \mathbb{R}^n$ is the sequence $\{\varphi_t^s(x_0)\}_{s=0}^{+\infty}$. It is known (see Lemma 39.1 in [20]) that the Fejer process for a fixed t converges to a point belonging to the polytope M_t:

$$\{\varphi_t^s(x_0)\}_{s=0}^{+\infty} \to \bar{x} \in M_t. \tag{6}$$

Let us consider the simplest non-stationary case, which is a translation of the polytope $M = M_0$ by the fixed vector $d \in \mathbb{R}^n$ in one unit of time. In this case, $A_t = A, c_t = c$, and the non-stationary problem (1) takes the form

$$\max\left\{\langle c, x \rangle \,|\, A(x - td) \le b,\ x \ge 0\right\}, \tag{7}$$

which is equivalent to

$$\max\left\{\langle c, x \rangle \,|\, Ax \le b + Atd,\ x \ge 0\right\}.$$

Comparing this with (1), we obtain $b_t = b + Atd$. In this case, the M_t-fejerian map (3) is converted to the following:

$$\varphi_t(x) = x - \frac{\lambda}{m} \sum_{i=1}^{m} \frac{\max\left\{\langle a_i, x \rangle - (b_i + \langle a_i, td \rangle), 0\right\}}{\|a_i\|^2} \cdot a_i,$$

which is equivalent to

$$\varphi_t(x) = x - \frac{\lambda}{m} \sum_{i=1}^{m} \frac{\max\left\{\langle a_i, x - td \rangle - b_i, 0\right\}}{\|a_i\|^2} \cdot a_i \tag{8}$$

The *φ-projection* (*pseudo-projection*) of the point $x \in \mathbb{R}^n$ on the polytope M is the map $\pi_M^\varphi(x) = \lim_{s \to \infty} \varphi^s(x)$.

3   The *NSLP* Algorithm

The *NSLP* (*Non-Stationary Linear Programming*) algorithm is designed to solve large-scale non-stationary LP problems on cluster computing systems. It consists of two phases: *Quest* and *Targeting*. The *Quest* phase calculates a solution of the system of inequalities defining the constraint system of the linear programming problem under the condition of dynamic changes in input data. To this end, the apparatus of Fejer mappings is used. The *Targeting* phase forms a special system of points having the shape of an n-dimensional axisymmetric cross. The cross moves in the n-dimensional space in such a way that the solution of the LP problem remains permanently in an ε-vicinity of the central point of the cross. Let us describe both phases of the algorithm in more detail.

3.1   The *Quest* Phase

Without loss of generality, we can assume that all the calculations are performed in the region of positive coordinates. At the beginning, we choose an arbitrary point $z_0 \in \mathbb{R}_{\geq 0}^n$ with non-negative coordinates. This point plays the role of initial approximation for the problem (1). Then we organize an iterative Fejer process of the form (6). During this process, the Fejer approximations are consecutively calculated by using the Fejer mapping (3). This process converges to a point located on the polytope M_t. Owing to the non-stationary nature of the problem (1), the polytope M_t can change its position and shape during the calculation of the pseudo-projection. An input data update is performed every L iterations, L being some fixed positive integer that is a parameter of the algorithm. Let us denote by $t_0, t_1, \ldots, t_k, \ldots$ sequential time points corresponding to the instants of input data update. Without loss of generality, we can assume that

$$t_0 = 0, t_1 = L, t_2 = 2L, \ldots, t_k = kL, \ldots. \tag{9}$$

This corresponds to the case when one unit of time is equal to the time spent by the computer to calculate one value of the Fejer mapping using Eq. (3).

Let the polytope M_t take shapes and locations

$$M_0, M_1, \ldots, M_k, \ldots$$

at time points (9). Let

$$\varphi_0, \varphi_1, \ldots, \varphi_k, \ldots$$

be the Fejer mappings determined by Eq. (3) taking into account the changes in input data of problem (1) at time points (9). In the *Quest* phase, the iterative process calculates the following sequence of points (see Fig. 1):

$$\{z_1 = \varphi_0^L(z_0), z_2 = \varphi_1^L(z_1), \ldots, z_k = \varphi_{k-1}^L(z_{k-1}), \ldots\}.$$

Let us briefly denote this iterative process as

$$\left\{\varphi_k^L(z_0)\right\}_{k=0}^{+\infty}. \tag{10}$$

It terminates when[2]

$$\mathrm{dist}\left(\varphi_k^L(z_{k-1}), M_k\right) < \varepsilon,$$

where $\varepsilon > 0$ is a positive real number being a parameter of the algorithm. One of the most important issues is the convergence of the iterative process (10). In the general case, this issue remains open. However, the following theorem holds for the non-stationary problem (7).

---

[2] Here $\mathrm{dist}(z, M) = \inf \{\|z - x\| : x \in M\}$.

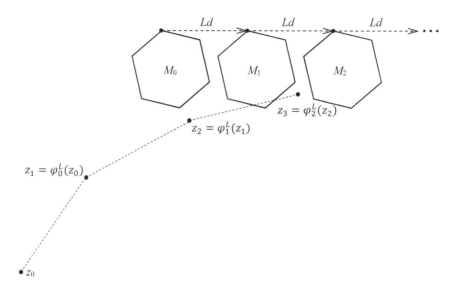

Fig. 1. The iterative process in the *Quest* phase for problem (7)

Theorem 1. *Let a non-stationary LP problem be given by (7). Let the Fejer mappings* $\varphi_0, \varphi_1, \ldots, \varphi_k, \ldots$ *be defined by the equation*

$$\varphi_k(x) = x - \frac{\lambda}{m} \sum_{i=1}^{m} \frac{\max\{\langle a_i, x - kLd\rangle - b_i, 0\}}{\|a_i\|^2} \cdot a_i. \tag{11}$$

This equation is derived using (8) and (9). By definition, put

$$z_k = \varphi_{k-1}^{L}(z_{k-1}) \tag{12}$$

where $k = 1, 2, \ldots$ *. Then*

$$\lim_{k \to \infty} \text{dist}(z_k, M_k) = 0 \tag{13}$$

under the following condition:

$$\forall x \in \mathbb{R}^n \backslash M \left(\|Ld\| < \text{dist}(x, M) - \text{dist}(\varphi^L(x), M) \right). \tag{14}$$

The Theorem 1 gives a sufficient condition for the convergence of the iterative process shown in Fig. 1. To prove this theorem, we will need the following auxiliary lemma.

Lemma 1. *Under the conditions of Theorem 1, we have*

$$v - u = pLd \Rightarrow \varphi_p^l(v) - \varphi^l(u) = pLd \tag{15}$$

for any $p = 0, 1, 2, \ldots,$ $l = 1, 2, 3, \ldots$ *and* $u, v \in \mathbb{R}^n$.

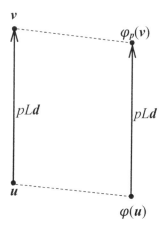

Fig. 2. Illustration to the proof of Lemma 1

Proof. The proof is by induction on l.

Induction base. Let $l = 1$, then the following condition holds:

$$v - u = pLd. \tag{16}$$

Then using (4), (11) and (16), we get

$$\varphi_p(v) - \varphi(u) = \varphi_p(u + pLd) - \varphi(u)$$

$$= u + pLd - \frac{\lambda}{m} \sum_{i=1}^{m} \frac{\max\left\{\langle a_i, u \rangle - b_i, 0\right\}}{\|a_i\|^2} \cdot a_i - \varphi(u)$$

$$= u + pLd - \frac{\lambda}{m} \sum_{i=1}^{m} \frac{\max\left\{\langle a_i, u \rangle - b_i, 0\right\}}{\|a_i\|^2} \cdot a_i$$

$$- u + \frac{\lambda}{m} \sum_{i=1}^{m} \frac{\max\left\{\langle a_i, u \rangle - b_i, 0\right\}}{\|a_i\|^2} \cdot a_i = pLd.$$

Thus, (15) holds if $l = 1$ (see Fig. 2).

Inductive step. Assume that condition (16) is true. Using the induction hypothesis, we get

$$\varphi_p^{l-1}(v) - \varphi^{l-1}(u) = pLd. \tag{17}$$

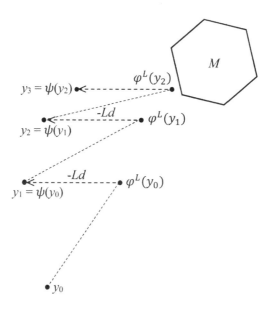

Fig. 3. The process defined by (18)

Then, combining (4), (5), (17) and (11), we obtain

$$\varphi_p^l(v) - \varphi^l(u) = \varphi_p(\varphi_p^{l-1}(v)) - \varphi(\varphi^{l-1}(u))$$

$$= \varphi_p(\varphi^{l-1}(u) + pLd) - \varphi(\varphi^{l-1}(u))$$

$$= \varphi^{l-1}(u) + pLd - \frac{\lambda}{m} \sum_{i=1}^{m} \frac{\max\left\{\langle a_i, \varphi^{l-1}(u)\rangle - b_i, 0\right\}}{\|a_i\|^2} \cdot a_i - \varphi(\varphi^{l-1}(u))$$

$$= \varphi^{l-1}(u) + pLd - \frac{\lambda}{m} \sum_{i=1}^{m} \frac{\max\left\{\langle a_i, \varphi^{l-1}(u)\rangle - b_i, 0\right\}}{\|a_i\|^2} \cdot a_i$$

$$- \varphi^{l-1}(u) + \frac{\lambda}{m} \sum_{i=1}^{m} \frac{\max\left\{\langle a_i, \varphi^{l-1}(u)\rangle - b_i, 0\right\}}{\|a_i\|^2} \cdot a_i = pLd.$$

This completes the proof of Lemma 1.

Proof (of Theorem 1). Let us fix an arbitrary point $z_0 \in \mathbb{R}^n \backslash M$. Let the map $\psi \colon \mathbb{R}^n \to \mathbb{R}^n$ be given by

$$\psi(x) = \varphi^L(x) - Ld, \forall x \notin M;$$
$$\psi(x) = x, \forall x \in M.$$
$$(18)$$

By definition, put

$$y_0 = z_0 \tag{19}$$

and

$$y_k = \psi(y_{k-1}) \tag{20}$$

for $k = 1, 2, \ldots$ (see Fig. 3).

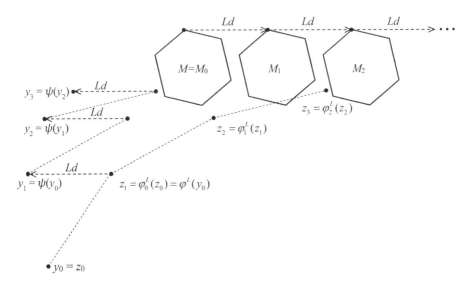

Fig. 4. Illustration to Eq. (21)

Now let us show by induction on k that

$$z_k - y_k = kLd \tag{21}$$

for $k = 0, 1, 2, \ldots$ (see Fig. 4).

Induction base. Equation (21) holds for $k = 0$. Taking into account (19), we see that the equation

$$z_0 - y_0 = 0 \cdot Ld$$

holds.

Inductive step. Suppose that

$$z_{k-1} - y_{k-1} = (k-1)Ld \tag{22}$$

for $k > 0$. Substituting $u = y_{k-1}, v = z_{k-1}, l = L, p = k - 1$ in Lemma 1, and using (15), we obtain

$$z_{k-1} - y_{k-1} = (k-1)Ld \Rightarrow \varphi^L_{k-1}(z_{k-1}) - \varphi^L(y_{k-1}) = (k-1)Ld.$$

Comparing this with (22), we have

$$\varphi^L_{k-1}(z_{k-1}) - \varphi^L(y_{k-1}) = (k-1)Ld. \tag{23}$$

Combining (12), (18), (20) and (23), we get

$$z_k - y_k = z_k - \psi(y_{k-1}) = z_k - \varphi^L(y_{k-1}) + Ld$$
$$= \varphi^L_{k-1}(z_{k-1}) - \varphi^L(y_{k-1}) + Ld = (k-1)Ld + Ld = kLd.$$

Thus, Eq. (21) holds.

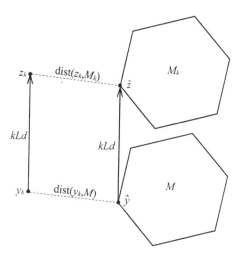

Fig. 5. Illustration to Eq. (24)

Now we show that
$$\text{dist}(z_k, M_k) = \text{dist}(y_k, M) \tag{24}$$
for all $k = 0, 1, 2, \ldots$. Let us choose a point $\hat{y} \in M$ that satisfies the following condition:
$$\|\hat{y} - y_k\| = \text{dist}(y_k, M). \tag{25}$$
Such a point exists and is unique since the polytope M is a bounded, closed and convex set. The polytope M_k is the result of translating the polytope M by the vector kLd (see Fig. 5). Since $\hat{y} \in M$, it follows that the point $\hat{z} = \hat{y} + kLd$ belongs to the polytope M_k. Taking into account (21), we conclude that the points $\{y_k, z_k, \hat{z}, \hat{y}\}$ are the vertices of a parallelogram. Therefore,
$$\|\hat{z} - z_k\| = \|\hat{y} - y_k\|. \tag{26}$$
Let us show that
$$\|\hat{z} - z_k\| = \text{dist}(z_k, M_k). \tag{27}$$
Assume for a contradiction that $\exists z' \in M_k$ such that
$$\|z' - z_k\| < \|\hat{z} - z_k\|. \tag{28}$$
Since $z' \in M_k$, it follows that the point $y' = z' - kLd$ belongs to the polytope M. Now, if we recall that the points $\{y_k, z_k, \hat{z}, \hat{y}\}$ are the vertices of a parallelogram, we get
$$\|y' - y_k\| = \|z' - z_k\|.$$
Combining this with (25), (26) and (28), we obtain
$$\|y' - y_k\| = \|z' - z_k\| < \|\hat{z} - z_k\| = \|\hat{y} - y_k\| = \text{dist}(y_k, M).$$

It follows that
$$\exists y' \in M \left(\|y' - y_k\| < \mathrm{dist}(y_k, M) \right).$$

This contradicts the definition of the distance between a point and a set. Therefore, Eq. (27) holds. Combining (25), (26) and (27), we get that Eq. (24) also holds.

Further, the map ψ defined by Eq. (18) is single-valued and continuous (this follows from the fact that φ is a single-valued and continuous map). Let us show that the map ψ is M-fejerian. Let $x \in \mathbb{R}^n \backslash M$ be an arbitrary point not belonging to the polytope M. Let us choose a point $\hat{x} \in M$ that satisfies the following condition

$$\left\| \varphi^L(x) - \hat{x} \right\| = \mathrm{dist}(\varphi^L(x), M). \tag{29}$$

Such a point exists and is unique because the polytope M is a bounded, closed and convex set. Combining the *dist* definition, Eq. (18), the triangle inequality for the norm and Eqs. (14) and (29), we get

$$
\begin{aligned}
\mathrm{dist}(\psi(x), M) &\le \|\psi(x) - \hat{x}\| = \left\| \varphi^L(x) - Ld - \hat{x} \right\| \\
&\le \|Ld\| + \left\| \varphi^L(x) - \hat{x} \right\| = \|Ld\| + \mathrm{dist}(\varphi^L(x), M) < \mathrm{dist}(x, M).
\end{aligned}
$$

It follows that ψ is M-fejerian. Therefore,

$$\left\{ \psi^k(y_0) \right\}_{k=0}^{+\infty} \to \bar{y} \in M.$$

This means that $\lim\limits_{k \to \infty} \mathrm{dist}(y_k, M) = 0$. Taking into account (24), we conclude that $\lim\limits_{k \to \infty} \mathrm{dist}(z_k, M_k) = 0$. This completes the proof of the theorem.

From a non-formal point of view, Theorem 1 states that the Fejer process must converge faster than the polytope M "runs away". Manycore processors can be used to increase the Fejer mapping calculation speed. In [19], the authors investigated this issue on Intel Xeon Phi multi-core coprocessors with MIC architecture [21]. It was shown that the Intel Xeon Phi may be used efficiently for solving large-scale problems.

3.2   The *Targeting* Phase

The *Targeting* phase begins after the *Quest* phase. At the *Targeting* phase, an n-dimensional axisymmetric cross is formed. An *n-dimensional axisymmetric cross* is a finite set $G = \{g_0, \dots, g_P\} \subset \mathbb{R}^n$. The cardinality of G equals $P + 1$, where P is a multiple of $n \ge 2$. The point g_0 is singled out from the point set G. This point is called the *central point* of the cross. Initially, the central point is assigned the coordinates of the point z_k calculated in the *Quest* phase by using the iterative process (10).

The set $G \backslash \{g_0\}$ is divided into n disjoint subsets C_i ($i = 0, \dots, n - 1$) called the *cohorts*:

$$G \backslash \{g_0\} = \bigcup_{i=0}^{n-1} C_i,$$

where n is the dimension of the space. Each cohort C_i consists of

$$K = \frac{P}{n} \tag{30}$$

points lying on the straight line that is parallel to the i-th coordinate axis and passes through the central point g_0. By itself, the central point does not belong to any cohort. The distance between any two neighbor points of the set $C_i \cup \{g_0\}$ is equal to a constant s. An example of a two-dimensional cross is shown in Fig. 6. The number of points in one dimension, excluding the central point, is equal to K. The symmetry of the cross supposes that K takes only even values greater than or equal to 2. Using Eq. (30), we obtain the following equation for the total number of points contained in the cross:

$$P + 1 = nK + 1. \tag{31}$$

Since K can take only even values greater than or equal to 2 and $n \geq 2$, it follows from Eq. (31) that P can also take only even values, and $P \geq 4$. In Fig. 6, we have $n = 2$, $K = 6$, $P = 12$.

Each point of the cross G is uniquely identified by a marker being a pair of integer numbers (χ, η) such that $0 \leq \chi < n$, $|\eta| \leq K/2$. Informally, χ specifies the number of the cohort, and η specifies the sequential number of the point in the cohort C_χ counted from the central point. The corresponding marking of points in the two-dimensional case is given in Fig. 6(a). The coordinates of the point $x_{(\chi,\eta)}$ having the marker (χ, η) can be reconstructed as follows:

$$x_{(\chi,\eta)} = g_0 + (0, \ldots, 0, \underbrace{\eta \cdot s}_{\chi}, 0, \ldots, 0). \tag{32}$$

The vector being added to g_0 in the right part of Eq. (32) has a single non-zero coordinate in position χ. This coordinate equals $\eta \cdot s$, where s is the distance between neighbor points in a cohort.

The *Targeting* phase includes the following steps.

1. Build the n-dimensional axisymmetric cross G that has K points in each cohort, the distance between neighbor points equaling s, and the center at point $g_0 = z_k$, where z_k is obtained in the *Quest* phase.
2. Calculate $G' = G \cap M_k$.
3. Calculate $C'_\chi = C_\chi \cap G'$ for $\chi = 0, \ldots, n - 1$.
4. Calculate $Q = \bigcup_{\chi=0}^{n-1} \{\arg \max \{\langle c_k, g \rangle \mid g \in C'_\chi, C'_\chi \neq \emptyset\}\}$.
5. If $g_0 \in M_k$ and $\langle c_k, g_0 \rangle \geq \max_{q \in Q} \langle c_k, q \rangle$, then $k := k + 1$, and go to step 2.
6. $g_0 := \frac{\sum_{q \in Q} q}{|Q|}$.
7. $k := k + 1$.
8. Go to step 2.

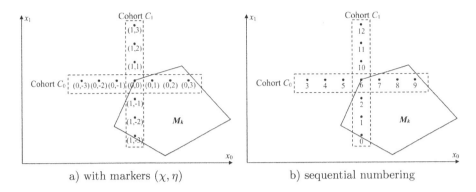

a) with markers (χ, η) b) sequential numbering

Fig. 6. A two-dimensional cross

Thus, in the *Targeting* phase, the steps 2–7 form a perpetual loop in which the approximate solution of the non-stationary LP problem is permanently recalculated. From a non-formal point of view, in Step 2, we determine which points of the cross G belong to the polytope M_k. To do this, we check the condition $A_k g \leq b_k$ for each point $g \in G$. Such checks can be executed in parallel by different processor nodes of a cluster computing system. For this goal to be achieved, P MPI-processes can be exploited, where P is defined by Eq. (31). We use sequential numbering for distributing the cross points among the MPI-processes. Each point of the cross is assigned a unique number $\alpha \in \{0, \dots, P-1\}$. The sequential number α can be converted to a marker (χ, η) by means of the following equations[3]:

$$\chi = ||\alpha - K| - 1| \div (K/2);$$
$$\eta = \mathrm{sgn}\,(\alpha - K) \cdot (((|\alpha - K| - 1) \bmod (K/2)) + 1).$$

The backward conversion can be performed by means of the equation

$$\alpha = \eta + \mathrm{sgn}(\eta)\frac{\chi}{2}K + K.$$

Figure 6(b) demonstrates the sequential numbering of points that corresponds to the marking in Fig. 6(a).

4   Conclusion

In this paper, a new *NSLP* algorithm for solving non-stationary linear programming problems of large dimension has been described. This algorithm is oriented to cluster computing systems with manycore processors. The algorithm consists of two phases: *Quest* and *Targeting*. The *Quest* phase calculates a solution of the system of inequalities defining the constraint system of the linear programming

---

[3] The symbol $\div$ denotes integer division.

problem under the condition of input data dynamic changes. To do this, we organize a Fejer process that computes a pseudo-projection onto the polytope M defined by the constraints of the LP problem. In this case, input data changes occur during calculation of the pseudo-projection. A convergence theorem for the described iterative process is proved in the case of translation of the polytope M. The *Targeting* phase forms a special system of points having the shape of an n-dimensional axisymmetric cross. The cross moves in the n-dimensional space in such a way that the solution of the linear programming problem is located all the time in an ε-vicinity of the central point of the cross. A formal description of the *Targeting* phase is presented in the form of a sequence of steps. Our future goal is a parallel implementation of the *NSLP* algorithm in the C++ language using the MPI library, as well as the development of computational experiments on a cluster computing system using synthetic and real LP problems.

References

1. Chung, W.: Applying large-scale linear programming in business analytics. In: Proceedings of the 2015 IEEE International Conference on Industrial Engineering and Engineering Management (IEEM), pp. 1860–1864. IEEE (2015)
2. Tipi, H.: Solving super-size problems with optimization. Presentation at the meeting of the 2010 INFORMS Conference on O.R. Practice. Orlando, Florida, April 2010. http://nymetro.chapter.informs.org/prac_cor_pubs/06-10%20Horia%20Tipi%20SolvingLargeScaleXpress.pdf. Accessed 7 May 2017
3. Gondzio, J., et al.: Solving large-scale optimization problems related to Bells Theorem. J. Comput. Appl. Math. **263**, 392–404 (2014)
4. Sodhi, M.S.: LP modeling for asset-liability management: a survey of choices and simplifications. Oper. Res. **53**(2), 181–196 (2005)
5. Dyshaev, M.M., Sokolinskaya, I.M.: Predstavlenie torgovykh signalov na osnove adaptivnoy skol'zyashchey sredney Kaufmana v vide sistemy lineynykh neravenstv [Representation of trading signals based Kaufman adaptive moving average as a system of linear inequalities]. Vestnik Yuzhno-Ural'skogo gosudarstvennogo universiteta. Seriya: Vychislitel'naya matematika i informatika [Bull. South Ural State Univ. Ser. Comput. Math. Softw. Eng.] **2**(4), 103–108 (2013)
6. Ananchenko, I.V., Musaev, A.A.: Torgovye roboty i upravlenie v khaoticheskikh sredakh: obzor i kriticheskiy analiz [Trading robots and management in chaotic environments: an overview and critical analysis]. In: Trudy SPIIRAN [SPIIRAS Proceedings], vol. 3, no. 34, pp. 178–203 (2014)
7. Radenkov, S.P., Gavryushin, S.S., Sokolyanskiy, V.V.: Avtomatizirovannyye torgovyye sistemy i ikh installyatsiya v rynochnuyu sredu (chast' 1) [Automated trading systems and their installation in the market environment (Part 1)]. Voprosy ekonomicheskikh nauk [Probl. Econ.] **6**(76), 70–74 (2015)
8. Dantzig, G.: Linear Programming and Extensions. Princeton University Press, Princeton (1998). 656 pp.
9. Klee, V., Minty, G.J.: How good is the simplex algorithm? In: Proceedings of the Third Symposium on Inequalities, University of California, Los Angeles, CA, pp. 159–175. Academic Press, New York-London, 1–9 September 1969. Dedicated to the Memory of Theodore S. Motzkin

10. Khachiyan, L.G.: Polynomial algorithms in linear programming. USSR Comput. Math. Math. Phys. **20**(1), 53–72 (1980)
11. Shor, N.Z.: Cut-off method with space extension in convex programming problems. Cybern. Syst. Anal. **13**(1), 94–96 (1977)
12. Karmarkar, N.: A new polynomial-time algorithm for linear programming. In: Proceedings of the Sixteenth Annual ACM Symposium on Theory of Computing, pp. 302–311. ACM (1984)
13. Sokolinskaya, I.M., Sokolinskii, L.B.: Parallel algorithm for solving linear programming problem under conditions of incomplete data. Autom. Remote Control **71**(7), 1452–1460 (2010)
14. Rechkalov, T.V., Zymbler, M.L.: Accelerating medoids-based clustering with the Intel many integrated core architecture. In: Proceedings of the 9th International Conference on Application of Information and Communication Technologies, Rostov-on-Don, Russia, pp. 413–417. IEEE, 14–16 October 2015
15. Zymbler, M.: Best-match time series subsequence search on the intel many integrated core architecture. In: Morzy, T., Valduriez, P., Bellatreche, L. (eds.) ADBIS 2015. LNCS, vol. 9282, pp. 275–286. Springer, Cham (2015). doi:10.1007/978-3-319-23135-8_19
16. Sokolinskaya, I.M., Sokolinsky, L.B.: Implementation of parallel pursuit algorithm for solving unstable linear programming problems. Bull. South Ural State Univ. Ser. Comput. Math. Softw. Eng. **5**(2), 15–29 (2016). doi:10.14529/cmse160202. (in Russian)
17. Sokolinskaya, I., Sokolinsky, L.: Solving unstable linear programming problems of high dimension on cluster computing systems. In: Proceedings of the 1st Russian Conference on Supercomputing - Supercomputing Days 2015, Moscow, Russian Federation. CEUR Workshop Proceedings, vol. 1482, pp. 420–427. CEUR-WS.org, 28–29 September 2015
18. Eremin, I.I.: Fejerovskie metody dlya zadach linejnoj i vypukloj optimizatsii [Fejer Methods for Problems of Convex and Linear Optimization]. Publishing of the South Ural State University, Chelyabinsk (2009). 200 pp.
19. Sokolinskaya, I., Sokolinsky, L.: Revised pursuit algorithm for solving non-stationary linear programming problems on modern computing clusters with many-core accelerators. In: Voevodin, V., Sobolev, S. (eds.) RuSCDays 2016. CCIS, vol. 687, pp. 212–223. Springer, Cham (2016). doi:10.1007/978-3-319-55669-7_17
20. Eremin, I.I.: Teoriya lineynoy optimizatsii [The theory of linear optimization]. Publishing House of the "Yekaterinburg", Ekaterinburg (1999). 312 pp.
21. Thiagarajan, S.U., Congdon, C., Naik, S., Nguyen, L.Q.: Intel Xeon Phi coprocessor developers quick start guide. White Paper. Intel (2013). https://software.intel.com/sites/default/files/managed/ee/4e/intel-xeon-phi-coprocessor-quick-start-developers-guide.pdf. Accessed 7 May 2017

A Distributed Parallel Algorithm for the Minimum Spanning Tree Problem

Artem Mazeev, Alexander Semenov$^{(\boxtimes)}$, and Alexey Simonov

JSC NICEVT, Moscow, Russia
{a.mazeev,semenov,simonov}@nicevt.ru

Abstract. In this paper, we present and evaluate a parallel algorithm for solving the minimum spanning tree (MST) problem on supercomputers with distributed memory. The algorithm relies on the relaxation of the message processing order requirement for one specific message type compared to the original GHS (Gallager, Humblet, Spira) algorithm. Our algorithm adopts hashing and message compression optimization techniques as well. To the best of our knowledge, this is the first parallel implementation of the GHS algorithm that linearly scales to more than 32 nodes (256 cores) of an InfiniBand cluster.

Keywords: Large graphs · MST · GHS · Supercomputers · MPI

1   Introduction

Given a connected, weighted undirected graph $G = (V, E)$, a spanning tree is a tree in this graph that contains all its vertices. A Minimum Spanning Tree (MST) [1] is a spanning tree having the minimum possible weight, if the weight of the tree is defined as the sum of the weights of all the edges contained in it.

The paper considers the minimum spanning tree problem in large graphs. By large graphs, we mean graphs that will not fit in the memory of the typical node of a distributed memory system.

The MST problem is encountered in many areas, for example, in bioinformatics, computer vision, and also when designing various networks. Requirements to the size of the processed graphs in real problems are constantly increasing. For example, in bioinformatics, when considering clustering problems [2] that can be solved by constructing a MST, graphs may take up to one petabyte or even more memory.

There are many algorithms [3] that solve the MST problem; the best known are Prim's [4], Kruskal's [5] and Boruvka's algorithms [6]. Some algorithms are suitable for shared memory parallelization, there are a lot of such implementations (see, for example, [7–10]).

Some of the mentioned algorithms are adapted for implementation on distributed memory systems [13–17]. Among the parallel implementations mentioned above, there is no implementation scalable to at least one hundred parallel processes. There are algorithms specially designed for distributed systems,

© Springer International Publishing AG 2017
L. Sokolinsky and M. Zymbler (Eds.): PCT 2017, CCIS 753, pp. 101–113, 2017.
DOI: 10.1007/978-3-319-67035-5_8

for example, the GHS (Gallager, Humblet, Spira) algorithm [11] and Awerbuch algorithm [12]. To the best of our knowledge, there is only one paper [16] that describes the implementation of the GHS algorithm, but it does not present any good experimental results.

In this paper, we present a parallel algorithm for solving the minimum spanning tree problem on distributed memory systems. The algorithm has been developed on the basis of the GHS algorithm, it allows processing of large-scale graphs and linearly scales to more than two hundred parallel processes.

2   GHS Algorithm

The GHS algorithm has been chosen for the research as a fundamental distributed parallel MST algorithm. This algorithm is based on a vertex-centric programming model [18]. The idea of the algorithm is as follows. All vertices perform the same procedure, which consists of sending, receiving and processing the messages from adjacent vertices. The messages can be transmitted independently in both directions of an edge, the order of messages must be preserved along the edge direction.

At any time, the set of graph vertices is represented as a union of a certain number of fragments, i.e. disjoint sets of vertices.

Initially, each vertex is a fragment. Each fragment finds an edge of minimum weight among the edges outgoing from this fragment to other fragments. The fragments are then combined over these edges. The edges that are used to combine the fragments will compose a minimum spanning tree when there is only one fragment comprising all the remaining vertices.

Let us consider the algorithm in detail. There are three possible vertex states: *Sleeping*, *Find* and *Found*, where *Sleeping* is the initial state of all vertices. The vertex will be in the state *Find* when is participating in a fragment's search for the minimum-weight outgoing edge, and is in the state *Found* in other cases. Each fragment has an L variable characterizing its level. Initially the level of each fragment is 0. Two fragments of the same level L can be combined into a level $L + 1$ fragment. A fragment cannot join to another fragment of a lower level.

The following is the detailed description of the searching process of the minimum-weight outgoing edge of a fragment. In the trivial case where the fragment consists of a single vertex and its level is 0, the vertex locally chooses its minimum-weight outgoing edge, marks this edge as a branch of the minimum spanning tree, sends a message called *Connect* over this edge, and goes into the *Found* state.

Now consider the case where a fragment level is greater than 0. Suppose that a new fragment at level L has just been formed by the combination of two level $L - 1$ fragments with the same outgoing edge, which becomes the core of the new fragment. The weight of this core edge is used as the identity of the fragment. Then an *Initiate* message is broadcast all over the fragment starting from the vertices adjacent to the core, so that all vertices receive a new fragment level and identity, and are placed in the *Find* state. When a vertex receives the *Initiate* message, it starts finding the minimum-weight outgoing edge.

Each edge of the graph can be in one of three states: *Branch*, if the edge belongs to the minimum spanning tree; *Rejected*, if the edge is not part of the minimum spanning tree; and *Basic*, if it is not yet known whether the edge is part of the minimum spanning tree. In order to find the minimum-weight outgoing edge, for each vertex v, all edges in the *Basic* state are sorted starting from the most light-weight edge. Each edge is probed by sending *Test* messages along that edge. The *Test* message contains fragment level and identity as arguments. When a vertex u receives the *Test* message, it compares its own fragment identity with the one received in the message. If the identities are equal, then the vertex u sends the *Reject* message back, and then both vertices put the edge in the *Reject* state. In this case, the vertex v that has sent the *Test* message continues the search, analyzing the next best edge, and so on. If the fragment identity in the received *Test* message is different from that of the receiving vertex u, and if the receiving vertex fragment level is greater or equal to the one in the *Test* message, then an *Accept* message is sent back. In this case, the state of the edge incident to the vertex v is changed to *Branch*. However, if the fragment level of the vertex u is smaller than the one in the received message, then the message is postponed until the fragment level of the vertex u increases to the necessary value.

Finally, each vertex finds a minimum-weight outgoing edge, if any. Now the vertices are sending *Report* messages (see Fig. 1.a) to find the minimum-weight outgoing edge of the whole fragment. If none of the graph vertices have outgoing edges in the *Basic* state, then the algorithm terminates, and the edges in the *Branch* state are the minimum spanning tree.

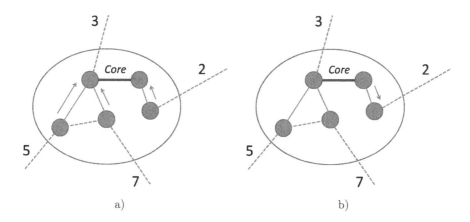

Fig. 1. Scheme of the GHS algorithm execution. In (a), arrows denote the sending of the *Report* messages. In (b), the arrow denotes the sending of the *Change core* message towards the minimum-weight outgoing edge of the fragment. The edges shown by solid lines are in the *Branch* state. Numbers on edges are weight values.

Report messages are sent by the following rules. Each leaf vertex of the fragment sends *Report(w)* along the only incident edge in the *Branch* state (*w* is either the weight of the minimum outgoing edge from the vertex, or infinity if there are no outgoing edges). Each internal vertex finds its own minimum-weight outgoing edge, and waits for all messages from all the subtrees. Then the vertex chooses the minimum weight from all the weight values. If the minimum is achieved with a value that came from the subtrees, then the number of the outgoing branch is placed into the vertex variable *best_edge*, otherwise the number of the minimum-weight outgoing edge is placed into this variable. This is done with the purpose of easily restoring the path by moving to where *best_edge* is pointing.

Further, there is a sending of the *Report* message up the tree of the fragment with an argument equal to the minimum value found among all the weight values. When the vertex sends the *Report* message, it also goes into the *Found* state. Finally, two vertices that are incident to the core edge send *Report* messages along the core and determine the weight of the minimum outgoing edge and the direction to this edge.

To try to connect one fragment to another over the found minimum-weight outgoing edge of the fragment, it is possible to use the *best_edge* variable in every vertex to trace the path from the core to the minimum-weight outgoing edge. For this purpose, a *Change core* message is sent from one of the core vertices that is closer to the minimum-weight outgoing edge (see Fig. 1.b). A vertex that has received this message sends it further in accordance with its own *best_edge* value, and so on. When the message reaches the vertex having the minimum-weight outgoing edge, then this vertex becomes the root of the tree formed by the fragment. This vertex sends the *Connect(L)* message over the minimum-weight outgoing edge, where L is the fragment level. If two level L fragments have the same minimum-weight outgoing edge, then each of them sends the *Connect(L)* message over this edge, and this edge becomes the core of the new level $L + 1$ fragment, which immediately starts to send the *Initiate* message with a new level number and identity all over the fragment.

When a level L fragment with identity F sends the *Connect* message into a level $L' > L$ fragment with identity F', the larger fragment will send the *Initiate* message with L' and F' into the smaller fragment.

The time complexity of the GHS algorithm is $O(N \log N)$, and the number of communication messages is $5N \log N + 2M$, where N is the number of vertices, and M is the number of edges in the graph. Not all the occurring cases are considered in this algorithm description but only the basic ones.

3   MST Algorithm

The GHS algorithm considered in [11] is only a description and analysis of the necessary high-level steps that must be performed at each vertex. As far as we know, there is no paper describing details of an implementation of the algorithm that scales well.

It is necessary to reasonably choose and develop a set of techniques, and solve a number of problems for the development of a parallel algorithm that finds an MST using the GHS algorithm. The implementation of the proposed algorithm has been made using the C++ language with an MPI library. When running on a supercomputer, the number of vertices in the graph is significantly larger than the number of MPI processes, so a large number of vertices and all related information are typically stored in the memory allocated to each process. All graph vertices are sequentially distributed in blocks among the processes. The local part of the graph in each process is stored in the CRS (Compressed Row Storage) format.

3.1   Preprocessing of the Original Graph

A preprocessing of the graph is conducted before searching for a minimum spanning tree in the graph, namely loops and multiple edges are removed from the graph. The removal of multiple edges is performed to fulfil the GHS algorithm condition that requires all the edges to be unique. The time spent on preprocessing is negligible and is not included in the total time of algorithm execution.

3.2   Base Version

The base version of the algorithm was developed at the beginning of the work.

Every MPI process supports a queue where vertices can postpone a message when necessary. The aggregation of messages is implemented to speed up the algorithm; a separate buffer is created in every process for each possible receiving process.

Implementation scheme of the base version of the parallel algorithm for construction of an MST using an MPI library; executed in parallel in every MPI process.

```
Input: local_G - local part of the graph
Output: local part of the MST

While (True) {
    /* read messages and push them to the queue */
    read_msgs ();
    /* queue processing, sending messages (write to the buffer) */
    If (time_to_process_queue) {
        process_queue ();
    }
    If (time_to_send) {
        /* send all aggregated messages */
        send_all_bufs ();
    }
    /* checking for algorithm completion using MPI_Allreduce */
    check_finish ();
}
```

Besides information that is necessary for the algorithm execution, messages also contain service information: the number of the sending vertex and the number of the receiving vertex, as well as the message type.

It is important to note that the GHS algorithm requires the original graph to be connected. This is not necessary for the proposed algorithm because it will work until the interconnect is in the "silence" state, when all queues are empty, all messages are processed and there are no undelivered messages in the network. Thus, the proposed algorithm allows finding not only an MST in a connected graph, but also a minimum spanning forest in a graph with any number of connected components.

Since the GHS algorithm requires the weights of graph edges to be different, a special identity *special_id* is added to the usual weight of the edge. The *special_id* identity is calculated for each graph edge e as follows: let u and v be the vertices that are incident to the edge e, then *special_id* in binary representation is equal to the consecutively recorded binary representations of $min(u, v)$ and $max(u, v)$. Such an arrangement enables the algorithm to work correctly even if the input graph has two different edges with the same weights.

3.3 Searching Local Edges

When an MPI process receives an incoming message, it is necessary to find the edge (an edge index in the list of local edges) over which the message came, i.e. to find the index of the edge formed by the two vertices (the sending and the receiving) in the list of local edges. The search is necessary since a change of local data related to that edge may be required.

The base version uses a linear search for this operation. During the linear search, all edges that are incident to the receiving vertex are sorted. If the vertex on the other end of the edge is equal to the sending vertex, then the right edge is found.

The first possible way to optimize this operation is to sort all incident edges in each vertex of the original graph in ascending order of the vertex numbers on the opposite ends of the edges. In this approach, it is necessary to spend some time on sorting at the beginning of the algorithm execution but later, during the algorithm execution, a binary search can be used instead of a linear one. This approach provides a small gain in performance.

The second optimization that was considered is hashing. It is possible to create a hash table in every process instead of sorting and binary search. Let u be the vertex sending a message, v is the receiving vertex, a vertex identifier is a 32-bit machine word. Let us define the hash function $get\_hash(u, v)$ as

$$((u \ll 32) \mid v) \textbf{ mod } hash\_table\_size, \tag{1}$$

where $\ll$ is left bit-shift, $\mid$ is bitwise OR, **mod** is the remainder of the division, *hash_table_size* is the hash table size (several times larger than the number of local edges).

The hashing method used in the proposed algorithm is called *linear search and insertion* [19]. Thus, the identity of the local edge can be determined by two adjacent vertices in time $O(1)$. But firstly, it is necessary to create and populate the hash table. This procedure is a part of the algorithm initialization and takes very little time, hence the reason it is not included in the total time of algorithm execution.

3.4   Separate Processing of the *Test* Messages

It is not always possible to process immediately certain types of messages (*Connect*, *Test* and *Report*), since several conditions must be satisfied to perform the processing. The condition is satisfied when some data are changed, and to change a specific piece of data, it is necessary to wait for a specific message. Thus, there are situations when a message should be postponed, and then an attempt should be made again to process it. It is not known when it will be processed.

The original GHS algorithm requires that the order of the messages be preserved. The study of the algorithm execution showed, however, that *Test* messages constitute a significant part of all messages. It was determined that it would be more effective to organize a separate queue for *Test* messages, and process this queue much less frequently than the main one.

3.5   Messages Length Optimization

To achieve the maximum possible performance of the algorithm implementation on a distributed memory system, it is necessary to minimize the size of the communication messages. It is, therefore, important that the structure that stores a message takes as little memory as possible.

At first messages were grouped into "short" (*Connect, Accept, Reject, Change core*) and "long" messages (*Initiate, Test, Report*). The main difference is that "long" messages contain a weight, which takes a significant amount of memory (64 bit).

In the beginning of each structure, for both "long" and "short" messages, a packed bit field of 16 bits is stored (actually, only 9 bits are necessary: 3 bits for message type, 5 bits for fragment level, and 1 bit for vertex state).

Further, the structure stores the identifiers of both the sending vertex and the receiving vertex (a vertex identifier is a 32-bit machine word).

Additionally, long messages store the extended weight of the edge (*special_id*) and the weight itself.

Finally, the following optimization is implemented. Instead of storing *special_id* in the extended weight (concatenation of the identifiers of two vertices, 64 bits in total), it is possible to store the minimum number of all the MPI processes that store this edge, after verifying that the weights of all the edges in each process are different. Indeed, if the weights of all the edges in every process are different, then two different edges with the same weight can be only in different processes, thus, the numbers of the relevant processes are enough to understand that such edges are different.

As a result, short and long messages are 80 and 152 bits in size, respectively.

3.6 Parameters of the Proposed Algorithm

There are relevant implementation parameters:

- MAX_MSG_SIZE is the maximum size of aggregated messages (by default, 10 000 bytes);
- $SENDING\_FREQUENCY$ is the frequency of flushing the aggregated messages (by default, every 5 iterations of the *while* loop);
- $CHECK\_FREQUENCY$ is the frequency of processing the queue of *Test* messages (by default, every 5 iterations of the *while* loop);
- $EMPTY\_ITER\_CNT\_TO\_BREAK$ is the frequency of checking for completion (by default, every 100 000 iterations of the *while* loop);
- $HASH\_TABLE\_SIZE$ is the size of the hash table, i.e. the number of elements contained in it. The default value is $local\_actual\_m * 5 * 11 / 13$, where $local\_actual\_m$ is the number of local edges in the MPI process after removing multiple edges and self-loops.

4   Experimental Results

RMAT, SSCA2 and Uniformly Random graphs are used for the performance evaluation of the algorithm.

- RMAT [20] graphs represent real-world large-scale graphs from social networks and the Internet, and are complex enough to analyze, thereby being often used to evaluate performance of graph processing algorithms.
- SSCA2 [21] graphs represent a set of randomly connected cliques.
- In Uniformly Random [22] graphs, the neighbors of each vertex are chosen randomly.

In the paper we examine graphs with the following parameters: the number of vertices is a pow of 2, the average vertex degree is 32. The weights of the edges are real numbers from the $(0, 1)$ interval. The $SCALE$ parameter specifies the number of vertices in the graph. If n is the $SCALE$ parameter, then 2^n is the number of vertices in the graph. For example, a graph with $SCALE = n$ is hereinafter referred to as RMAT-n.

The default values of the algorithm parameters listed in Subsect. 3.6 are used for performance evaluation.

We focus our design and experimental evaluation on the MVS-10P cluster system.

Table 1 provides an architecture overview of the system.

4.1 Impact of Optimizations

In this subsection, an RMAT graph with scale 23 (RMAT-23) is used for testing. The number of MPI processes per node of the MVS-10P cluster is 8.

If binary search is used instead of linear search when finding a local edge, then the execution time on a cluster node is reduced by 2%. If hashing is used

Table 1. MVS-10P cluster system configuration

	MVS-10P
Nodes	2x Xeon E5-2690 (8 cores, 2.9 GHz)
Number of nodes	207
Memory	64 GB
Interconnect	InfiniBand 4xFDR
MPI	Intel MPI 4.1

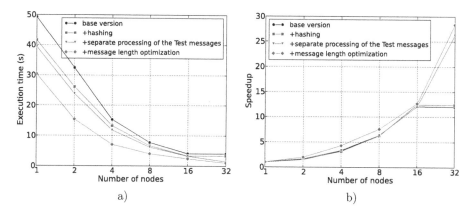

a)   b)

Fig. 2. Impact of optimizations: from the base version to the final version (with all the optimizations). MVS-10P cluster, RMAT-23 graph, 8 MPI processes per node.

instead of linear search, then the execution time on a node is approximately 18% less (MVS-10P cluster, RMAT-23 graph, 8 MPI processes per node). Thus, hashing was chosen for the final version.

Figure 2.a shows how the runtime changes (in seconds) as the optimizations described in Subsects. 3.3, 3.4 and 3.5 are added. Figure 2.b shows the scalability of the same runs, i.e. the ratio of the problem solution time on one node to that on a given number of nodes.

Figure 3.a shows the profiling results of the algorithm version with one optimization of the local edge search, while Fig. 3.b shows the profiling results of the final version of the algorithm.

The profiling shows that most of the time is spent on processing queues. Some messages are processed repeatedly, including *Test* messages. Therefore, in the final version of the algorithm, in which *Test* messages are processed less frequently, the part of queue processing in the total execution time is less than in the version with hashing optimization only. Precisely this optimization provided a two-fold increase in algorithm scalability (see Fig. 2.b).

Also, the message length optimization made a considerable contribution to the algorithm performance. This optimization reduced the execution time of the final version of the algorithm on any number of nodes by approximately 50%.

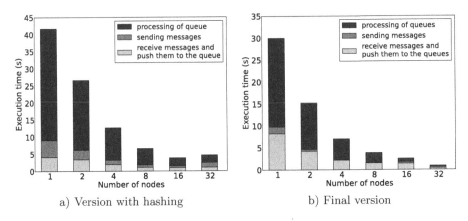

a) Version with hashing   b) Final version

Fig. 3. Profiling results. MVS-10P cluster, RMAT-23 graph, 8 MPI processes per node.

4.2   Scaling

Table 2 shows performance evaluation results of the final algorithm version on the MVS-10P cluster for the RMAT-24, SSCA2-24 and Random-24 graphs. The number of MPI processes per node is 8.

Table 2. Performance of the proposed algorithm on the MVS-10P cluster system (all used graphs have scale 24).

Number of nodes			1	2	4	8	16	32	64
MVS-10P	RMAT-24	Time (s)	63.27	36.12	17.98	8.47	5.41	2.04	1.45
		Scaling	1.00	1.75	3.52	7.47	11.7	31.01	43.63
	SSCA2-24	Time (s)	54.69	32.37	11.90	6.02	3.63	1.72	n/a
		Scaling	1.00	1.69	4.60	9.08	15.07	31.62	n/a
	Random-24	Time (s)	88.61	51.65	21.47	10.27	6.68	3.23	n/a
		Scaling	1.00	1.72	4.13	8.63	13.26	27.43	n/a

$SCALE$ 24 is the largest graph scale that fits into a single node memory of the MVS-10P cluster. The size of such graphs is approximately 6.5 GB. The rest of the memory node is needed for the algorithm implementation. In particular, a large amount of memory is required to organize the hash table.

The scalable mode in Intel MPI 4.1 on the MVS-10P cluster provides linear scaling on 32 nodes. On 64 nodes (512 cores) of the MVS-10P, the scaling is 43.6.

In Fig. 4 we show the dependence between the average size of communication messages and the execution time of the final algorithm version. Here message size refers to an aggregated message sent over the interconnect. The value of the MAX_MSG_SIZE aggregation parameter is 20 000 bytes. The figure shows that an increasing in the number of nodes yields a decrease in message size. On 32 nodes, messages are short; their size does not exceed 2 KB. It is also clear that the size of the messages depends on the algorithm execution time.

We suppose that the main limitation factor of the algorithm performance can be latency or injection rate of short messages.

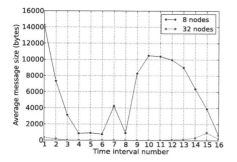 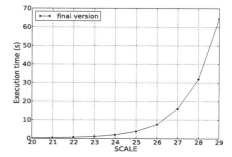

Fig. 4. Average size (over all MPI processes) of communication messages in bytes, depending on interval number (the total execution time of the algorithm is divided into equal intervals). MVS-10P cluster, RMAT-23 graph, 8 MPI processes per node.

Fig. 5. Execution time of the final algorithm version for graphs of different sizes. 32 nodes of the MVS-10P cluster, 8 MPI processes per node.

Figure 5 shows a weak scaling for RMAT graphs on 32 nodes of the MVS-10P cluster. RMAT-29 is the largest graph that fits into the memory of 32 nodes; it takes a total of 205 GB. It should be noted that the implementation of the algorithm for the solution of the MST problem is scalable in-memory, i.e. it is possible to increase the size of the graph by increasing the number of nodes.

5   Conclusion

The paper presents a parallel algorithm for finding a minimum spanning tree (forest) in a graph on distributed memory systems, and the corresponding algorithm implementation using MPI.

Compared with the original GHS algorithm, the proposed parallel algorithm has the following key features:

- the requirement of message processing order has been relaxed for *Test* messages, which doubled the scaling of the algorithm;
- the algorithm is generalized for the case of disconnected graphs. It builds a minimum spanning forest in this case (the original algorithm is only applicable to connected graphs).

The proposed algorithm adopts some optimization techniques, namely hashing for local edge search and compression of communication messages. The algorithm implementation linearly scales on 32 nodes of the MVS-10P InfiniBand cluster.

In the future, we intend to improve the algorithm scaling and develop its hybrid MPI+OpenMP implementation.

The work was supported by the grant No. 17-07-01592A of the Russian Foundation for Basic Research (RFBR).

References

1. Cormen, T., Leiserson, C., Rivest, R., Stein, C.: Minimum spanning trees. Introduction to Algorithms, 2nd edn, pp. 561–579. MIT Press and McGraw-Hill, Cambridge (2001). Chap. 23

2. Rubanov, L.I., Seliverstov, A.V., Zverkov, O.A.: Ultraconservative elements in the simplest of subtype Alveolata. Modern Inf. Technol. IT Educ. **2**, 581–585 (2015)

3. Eisner, J.: State-of-the-art algorithms for minimum spanning trees. A Tutorial Discussion, University of Pennsylvania (1997)

4. Prim, R.C.: Shortest connection networks and some generalizations. Bell Syst. Tech. J. **36**, 1389–1401 (1957)

5. Kruskal, J.B.: On the shortest spanning subtree of a graph and the traveling salesman problem. AMS **7**, 48–50 (1956)

6. Boruvka, O.: O jistem problemu minimalnim (about a certain minimal problem). Prace Mor. Prirodoved. Spol. v Brne, III **3**, 37–58 (1926)

7. Kolganov, A.S.: Parallel implementation of minimum spanning tree algorithm on CPU and GPU. Parallel computational technologies (2016)

8. Mariano, A., Lee, D., Gerstlauer, A., Chiou, D.: Hardware and software implementations of Prim's algorithm for efficient minimum spanning tree computation. In: Schirner, G., Götz, M., Rettberg, A., Zanella, M.C., Rammig, F.J. (eds.) IESS 2013. IAICT, vol. 403, pp. 151–158. Springer, Heidelberg (2013). doi:10.1007/978-3-642-38853-8_14

9. Wang, W., Huang, Y., Guo, S.: Design and implementation of GPU-based Prim's algorithm. Int. J. Modern Educ. Comput. Sci. **3**(4), 55–62 (2011)

10. Katsigiannis, A., Anastopoulos, N., Nikas, K.: An approach to parallelize Kruskal's algorithm using helper threads. In: IEEE 26th International Parallel and Distributed Processing Symposium Workshops and PhD Forum, pp. 1601–1610 (2012)

11. Gallager, R.G., Humblet, P.A., Spira, P.M.: A distributed algorithm for minimum-weight spanning trees. ACM Trans. Program. Lang. Syst. **5**, 66–77 (1983)

12. Awerbuch, B.: Optimal distributed algorithms for minimum weight spanning tree, Counting, Leader Election, and Related Problems. In: 19th ACM Symposium on Theory of Computing (STOC), New York, pp. 230–240 (1987)

13. Gregor, D., Lumsdaine, A.: The parallel BGL: a generic library for distributed graph computations. In: Parallel Object-Oriented Scientific Computing (2005)

14. Loncar, V., Skrbic, S.: Parallel implementation of minimum spanning tree algorithms using MPI. In: IEEE 13th International Symposium on Computational Intelligence and Informatics (CINTI), pp. 35–38 (2012)

15. Loncar, V., Skrbic, S., Balaz, A.: Parallelization of minimum spanning tree algorithms using distributed memory architectures (2014)

16. Sireta, A.: Comparison of parallel and distributed implementation of the MST algorithm (2016). http://delaat.net/rp/2015-2016/p41/report.pdf
17. Ramaswamy, S.I., Patki, R.: Distributed minimum spanning trees (2015). http://stanford.edu/~rezab/classes/cme323/S15/projects/distributed_minimum_spanning_trees_report.pdf
18. McCune, R.R., Weninger, T., Madey, G.: Thinking like a vertex: a survey of vertex-centric frameworks for distributed graph processing. ACM Comput. Surv. **48** (2015)
19. Knuth, D.: The Art of Computer Programming, 2nd edn., vol. 3, pp. 513–558. Addison-Wesley (1998)
20. Chakrabarti, D., Zhan, Y., Faloutsos, C.: R-MAT: a recursive model for graph mining. In: Proceedings of the Fourth SIAM International Conference on Data Mining (2004). http://repository.cmu.edu/cgi/viewcontent.cgi?article=1541&context=compsci
21. Bader, D.A., Madduri, K.: Design and implementation of the HPCS graph analysis benchmark on symmetric multiprocessors. In: Bader, D.A., Parashar, M., Sridhar, V., Prasanna, V.K. (eds.) HiPC 2005. LNCS, vol. 3769, pp. 465–476. Springer, Heidelberg (2005). doi:10.1007/11602569_48
22. Erdoos, P., Reyni, A.: On random graphs. I. Publicationes Math. **6**, 290–297 (1959)

Applying Volunteer and Parallel Computing for Enumerating Diagonal Latin Squares of Order 9

Eduard I. Vatutin[1], Stepan E. Kochemazov[2], and Oleg S. Zaikin[2(✉)]

[1] Southwest State University, Kursk, Russia
evatutin@rambler.ru
[2] Matrosov Institute for System Dynamics and Control Theory SB RAS,
Irkutsk, Russia
veinamond@gmail.com, zaikin.icc@gmail.com

Abstract. In this paper we design the algorithm aimed at fast enumeration of diagonal Latin squares of small order. This brute force based algorithm uses straightforward representation of a Latin square and greatly depends on the order in which we fill in Latin square cells. Moreover, we greatly improved its effectiveness by careful optimization using bit arithmetic. By applying this algorithm we have enumerated diagonal Latin squares of order 9. This problem was previously unsolved. In order to ensure the accuracy of the obtained result, two separate large-scale experiments were carried out. In the first one a computing cluster was employed. The second one was performed in a BOINC-based volunteer computing project. Each experiment took about 3 months. As a result we obtained two similar numbers.

Keywords: Combinatorics · Latin square · Enumeration · Bit arithmetic · Volunteer computing · BOINC · Parallel computing

1 Introduction

Latin square of order N is a square table with N rows and N columns, filled with elements from the set $\{0, \ldots, N - 1\}$ in such a way that each element appears exactly once in each row and each column [1]. A Latin square is called diagonal, if both its main diagonal and main antidiagonal contain all elements from 0 to $N-1$.

Stepan Kochemazov and Oleg Zaikin are funded by Russian Science Foundation (project No. 16-11-10046). Stepan Kochemazov and Oleg Zaikin are additionally supported by Council for Grants of the President of the Russian Federation (stipends SP-1829.2016.5 and SP-1184.2015.5 respectively). Eduard Vatutin is partially supported by Council for Grants of the President of the Russian Federation (grant MK-9445.2016.8) and by Russian Foundation for Basic Research (grant 17-07-00317-a). We thank all Gerasim@home volunteers, whose computers took part in the experiment. Authors contribution: E. Vatutin: general scheme of algorithm proposed in Sect. 2, the experiment described in Subsect. 4.2; S. Kochemazov: bit arithmetic implementation of algorithm outlined in Sect. 3; O. Zaikin: The experiment described in Subsect. 4.3.

© Springer International Publishing AG 2017
L. Sokolinsky and M. Zymbler (Eds.): PCT 2017, CCIS 753, pp. 114–129, 2017.
DOI: 10.1007/978-3-319-67035-5_9

Latin squares represent an interesting example of combinatorial design that is easy to understand and experiment with, but at the same time there remain exceptionally hard related open problems. The most well known one is to answer the question whether there exist three mutually orthogonal Latin squares of order 10. One possible way to obtain an answer to such a problem is to generate all possible Latin squares of order 10 and for each one check if it participates in said triple. Due to combinatorial nature of Latin squares, the number of species representatives grows very fast with the increase of N. Therefore it is usually considered unrealistic to enumerate or generate all Latin squares of a specific order (even taking into account equivalence classes). However, the rapid growth of supercomputing and computational combinatorics in the last decades makes it possible to change this point of view.

In this paper we present the fast algorithm for enumerating diagonal Latin squares of small order. In fact, the proposed algorithm has a very simple structure: essentially, it represents Latin square as an integer array and uses $\leq N^2$ nested loops to traverse all possible variants how it can be filled. However, here the details play a very important role. In particular, the order in which we fill the cells with numbers greatly influences the algorithm performance. The same goes with the implementation details: how we choose the value to put in the next cell, etc. The current version of our algorithm makes it possible to enumerate up to 7 millions of diagonal Latin squares of order 9 per second on one CPU core. Nevertheless, according to our estimations, a sequential implementation of this algorithm (aimed at launching on a PC) can't enumerate diagonal Latin squares of order 9 in reasonable time. That is why we made two implementations of this algorithm: one for a volunteer computing project and one for a computing cluster. They made it possible to solve the considered problem.

Let us give a brief outline of the paper. In the next section we first discuss possible approaches to generation of Latin squares. After this we describe the basic structure of our algorithm that serves as a basis for further modifications. In Sect. 3 we show that in practice the performance of the algorithm can be greatly increased by using bit arithmetic in quite non-trivial ways and perform experimental evaluation of different algorithm versions. In Sect. 4 we describe two large-scale computational experiments aimed at enumerating diagonal Latin squares of order 9. Both experiments were performed employing the improved bit-arithmetic-based version of the algorithm (which is the fastest one). After this we discuss related works and draw conclusions.

2   Algorithm Description

First, we would like to assume that for the purposes of our research, *enumeration* and *generation* mean the same and therefore these terms are interchangeable. Indeed, as it will be seen later, when we enumerate, we each time construct a specific Latin square, so the difference between enumeration and generation consists in whether we store the constructed Latin square or not.

Second, we are interested in algorithms that are deterministic and complete, i.e. they generate all possible representatives of the species satisfying specific constraints. The latter condition is justified by the fact that since we already know that the amount of Latin squares is very large and we do not intend to store them all, then the generation should proceed in a fixed order, so that we can be sure that we do not process some instances more than once, and at the same time we traverse the entire search space. It means that the algorithm from [2] that allows to generate Latin squares according to the uniform distribution does not suite our purposes.

Taking into account the outlined conditions, let us next consider possible approaches to generating Latin squares.

2.1   Approaches to Generating Latin Squares

It is easy to see that each row and each column of a Latin square represents a permutation of N elements. Therefore for small N ($N \leq 10$) it is possible to generate all permutations and construct Latin squares from these permutations: if we fill the square row-by-row, then we only need to ensure that rows do not "intersect" (do not have equal elements in the same positions), because within any row the elements satisfy the uniqueness constraint by definition. However, once we filled several rows, the number of possible variants for the next row significantly decreases, and we need to apply some additional techniques to take this into account. One way of doing it is to represent the problem as *exact cover* instance [3] and apply state-of-the-art algorithms, such as DLX [4], to solve it. Another alternative is to represent permutations as bit vectors, and apply bit arithmetic to make fast checks and comparisons. We can also split the space of permutations into disjoint subsets separated according to values of several first elements to speed up the deeper parts of the search. Additionally if we are interested only in diagonal Latin squares we need to tune the algorithms specifically for this purpose. In our empirical evaluation DLX and bit arithmetic algorithms based on the outlined approaches make it possible to generate about 3×10^5 diagonal Latin squares of order 9 per second.

Another general approach to generating Latin squares consists in representing a Latin square of order N as an array of N integer values. It means that in the most basic variant we implement generation procedure as N^2 nested **for** loops with checks within each iteration. On the first glance it may seem that this approach is much worse than the one where we employ permutations. Indeed, the most simple implementation gives us a generation speed of about 6×10^3 diagonal Latin squares of order 9 per second on one CPU core. However, there is a number of optimizations which make it possible to greatly improve the algorithm performance. We consider them in detail in the following subsections.

We would like to note, that on the preliminary stage we experimented with many different approaches to Latin squares generation, including recurrent implementations, using such meta-heuristic methods as simulated annealing, etc., but all these versions showed worse performance than the ones with nested cycles.

2.2   Algorithm Design

Assume that we consider enumeration of diagonal Latin squares of order N. For this purpose our algorithm employs the following auxiliary constructs:

1. Integer array $LS[N][N]$ which contains a Latin square;
2. Two-dimensional integer arrays $Rows[N][N]$ and $Columns[N][N]$, that contain data regarding which elements are already "occupied" in each row/column.
3. Two integer arrays $MD[N]$ and $AD[N]$ that contain data regarding which elements are "occupied" on main diagonal and main antidiagonal.
4. Integer value $SquaresCnt$ which contains the number of squares.

```
Data: LS[N][N], Rows[N][N], Columns[N][N], MD[N], AD[N], SquaresCnt
/* Assume that all variables are initialized by 0.              */
/* Iterate over all possible values of cell [0][0]              */
for LS[0][0] = 0; LS[0][0] < N; LS[0][0] = LS[0][0] + 1 do
    /* if the value is ''occupied'' within the row, column or main
       diagonals - continue to the next value of loop variable   */
    if Rows[0][LS[0][0]]||Columns[0][LS[0][0]]||MD[LS[0][0]] then
      | continue
    /* Otherwise mark the value as ''occupied'' and proceed       */
    Rows[0][LS[0][0]] = 1;
    Columns[0][LS[0][0]] = 1;
    MD[LS[0][0]] = 1;
    for LS[0][1] = 0; LS[0][1] < N; LS[0][1] = LS[0][1] + 1 do
        if Rows[0][LS[1][1]]||Columns[1][LS[0][1]] then continue
        Rows[0][LS[0][1]] = 1;
        Columns[1][LS[0][1]] = 1;
        ...
        /* Increment the counter if reached the last element     */
        for LS[N − 1][N − 1] = 0; LS[N − 1][N − 1] < N;
        LS[N − 1][N − 1] = LS[N − 1][N − 1] + 1 do
            | if Rows[N − 1][LS[N − 1][N − 1]]||Columns[N − 1][LS[N − 1][N −
            | 1]]||MD[LS[N − 1][N − 1]] then continue;
            | SquaresCnt = SquaresCnt + 1;
        end
        ...
        Rows[0][LS[0][1]] = 0;
        Columns[1][LS[0][1]] = 0;
    end
    /* On exit mark value as 'free'                              */
    Rows[0][LS[0][0]] = 0;
    Columns[0][LS[0][0]] = 0;
    MD[LS[0][0]] = 0;
end
```

Algorithm 1. General outline of the algorithm

As we already mentioned, the order in which we traverse cells plays a crucial role in the effectiveness of the algorithm. Without the loss of generality, let us first introduce the general outline of the algorithm for enumerating diagonal Latin squares with simple order when we fill the square from the first (topmost leftmost) element to the last. Its pseudocode is presented as Algorithm 1.

Note that one property of Latin squares allows us to make simple optimization right from the start. Any ordinary Latin square (i.e. non-diagonal) can be effectively transformed (via row and column permutations) in such a way that its first row and first column appear in ascending order $0, 1, \ldots, N - 1$. It means that we can safely fix the values of corresponding variables in the array $LS[N][N]$ and modify the initialization stage. It results in $(N - 1)^2$ inner loops instead of N^2. In the case of diagonal Latin squares we can only fix either the first row to $0, 1, \ldots, N - 1$ or the first column (because if they become equal we violate the constraints on diagonals), thus having $N^2 - N$ inner loops.

It is easy to see that essentially, the algorithm consists of simple blocks corresponding to loops over values of specific cells. The order in which we fill cells is reflected by the order of loops. The structure of each particular loop is presented as Algorithm 2.

Data: $LS[N][N]$, $Rows[N][N]$, $Columns[N][N]$, $MD[N]$, $AD[N]$, $SquaresCnt$
```
/* Iterate over all possible values of cell [i][j]                    */
for LS[i][j] = 0; LS[i][j] < N; LS[i][j] = LS[i][j] + 1 do
    /* We check if the value is ''occupied'' in the current row,
       column and diagonals (if applicable)                          */
    bool Condition_{i,j} = Rows[i][LS[i][j]]||Columns[j][LS[i][j]];
    if i = j then Condition_{i,j} = Condition_{i,j}||MD[LS[i][j]];
    if i + j = N-1 then Condition_{i,j} = Condition_{i,j}||AD[LS[i][j]];
    if Condition_{i,j} then continue;
    /* Otherwise mark the value as ''occupied'' and proceed          */
    Rows[i][LS[i][j]] = 1;
    Columns[j][LS[i][j]] = 1;
    if i = j then MD[LS[i][j]] = 1;
    if i + j = N - 1 then AD[LS[i][j]] = 1;
    BODY OF INNER LOOP FOR NEXT CELL LS[i'][j'];
    Rows[i][LS[i][j]] = 0;
    Columns[j][LS[i][j]] = 0;
    if i = j then  MD[LS[i][j]] = 0;
    if i + j = N - 1 then  AD[LS[i][j]] = 0;
end
```
Algorithm 2. Inner loop structure

The body of the loop for the last cell according to specified order contains only instruction to increment the counter $SquaresCnt$ (if the corresponding value of Latin square cell does not violate any conditions). Note that all statements in each particular inner loop are hard-coded, i.e. there are no **if** statements which check if $i = j$, etc. Now let us discuss the best ways to traverse the Latin square cells.

2.3   On the Optimal Order of Cells

In our preliminary experiments we noticed a peculiar pattern. The generation speed for diagonal Latin squares greatly depended on the order in which we fill cells. We finally figured out the following simple strategy that works best. It should be noted, that this strategy is quite similar to the one suggested in [5].

Let us consider the generation of diagonal Latin squares of order 9. Assume that we fill the cells of a Latin square $A = \{a_{i,j}\}$. In accordance with the previous section, we fix the first row of A to $0, 1, \ldots, 8$. After this we use the iterative process to choose what cell to assign next. Note that we do not want to think about specific values that we assign, we need a general strategy. Each cell with coordinates i, j participates in two to four 'uniqueness' constraints: one for i-th row, one for $j - th$ column, one for main diagonal (if $i = j$) and one for main antidiagonal (if $i = 9 - j - 1$). It means that for each cell we can consider the 4-component vector $V_{i,j}^k(r_{i,j}, c_{i,j}, md_{i,j}, ad_{i,j})$ in which $r_{i,j}$ is the number of assigned cells in the i-th row, $c_{i,j}$ – the number of assigned cells in the j-th column, etc. For elements that do not lie on main diagonal or main antidiagonal we assume that $md_{i,j}$ and $ad_{i,j}$ are 0. The index k here corresponds to the step number since the contents of $V_{i,j}$ clearly depend on currently assigned cells. We can choose vector norm and calculate norms for each element and then choose the one with maximal value of norm as the next cell to be assigned.

In our experiments we obtained the best results with the simplest norm $|V_{i,j}| = |(r_{i,j}, c_{i,j}, md_{i,j}, ad_{i,j})| = r_{i,j} + c_{i,j} + md_{i,j} + ad_{i,j}$. We also apply additional heuristics here: if in some row, column, main diagonal or main antidiagonal after assigning current element there are $N - 1$ assigned elements, then the remaining element is automatically assigned next. Thus, returning to the generation of diagonal Latin squares of order 9, the first cell to be assigned is the one that is situated on the intersection of main diagonal and main antidiagonal (cell $4, 4$) because $V_{44}^1 = (0, 1, 1, 1)$, so the sum of components is 3 and for all the other cells the sum is at most 2. On the next step we actually have several alternatives with the same value, in this case we choose at random, etc. For the considered enumeration problem the order of cells looks as follows (see Fig. 1).

```
 -  -  -  -  -  -  -  -  -
20  2 16 17 21 18 19  3 22
25 26  6 23 27 24  7 28 29
55 56 57 10 53 11 58 59 60
61 63 65 45  1 47 67 69 70
62 64 66 12 54 14 68 71 72
32 33  8 30 34 31  9 35 36
39  4 42 37 40 38 43  5 41
13 49 50 44 48 46 51 52 15
```

Fig. 1. Order of cells for generation of diagonal Latin squares of order 9 (the first row is omitted because the values of the corresponding elements are fixed)

It is easy to see that for diagonal Latin squares it is best to start with diagonal elements first and then fill the rest. The corresponding enumeration algorithm that employs this order makes it possible to generate about 1.2 million squares per second on one core of Intel Core i7-6700 CPU. Note, that for ordinary Latin squares the trivial order of cells (row by row, column by columns) works best, so we do not need special heuristics to construct it. For both cases, the performance of enumeration algorithm can be significantly improved by the following optimizations.

2.4   Optimizations

Here we introduce two basic optimizations that in conjunction make it possible to improve the performance of the algorithm to about 1.8 million squares per second (on the aforementioned platform).

Use Formula to Compute the Last Element in a Row/Column/Diagonal. It is easy to see that in certain points of our algorithm there appear situations when in some particular row, column, main diagonal or main antidiagonal exactly $N - 1$ element out of N are assigned. In this case the value of remaining element can be computed directly, thus we do not need to use the **for** loop at all. Without the loss of generality assume that we have already assigned first $N - 1$ out of N elements of the j-th row. Then the formula for the remaining element can be written as follows:

$$a_{j,N-1} = N \times (N - 1)/2 - \sum_{l=0}^{N-2} a_{l,j}. \tag{1}$$

If the corresponding value does not violate the other uniqueness constraints then we can safely proceed deeper into the search space.

Lookahead Heuristic. The next optimization comes from the general area of combinatorial search and represents a variant of lookahead heuristics. Its basic idea consists in the fact, that on certain levels of search when the number of assigned cells is relatively large, we can look ahead before branching and spend a little more computational resources in order to sometimes avoid spending much more.

   Here we would like to once again use the vectors of constraints that we applied to produce the optimal order of cells. Using the same (sum-of-elements) norm, it is easy to see that at some point for many cells the total number of constraints will exceed N. It means that in a *lucky* situation there will be no possible assignment for the corresponding Latin square element. For the non-diagonal element $a_{i,j}$ it corresponds to the situation when the following equation holds (remind that $Rows[i][l] = 1$ iff there already exists an element with value l within the i-th row).

$$\sum_{l=0}^{N-1} (Rows[i][l] || Columns[j][l]) = N. \tag{2}$$

An important notice here is that excessive use of this heuristic may clutter the search procedure and slow it down. Therefore there is a tradeoff achieved by choosing the levels of depth on which to apply it. We have empirically found that for generating diagonal Latin squares of order 9 the best results are achieved when we apply this lookahead heuristic within inner loops from number 51 to 60.

It turned out that we can in a way fuse both optimizations and improve the general performance of the algorithm by using bit arithmetic.

3   Bit-arithmetic-based Version of the Algorithm

Since we consider Latin squares of small order, and the algorithm employs lots of checks whether some element is "occupied" or not, it is natural to employ bit arithmetic to speed up its implementation. Hereinafter, without the loss of generality, we assume that all integer values contain at least 16 bits.

Let us describe the modifications we introduce to the algorithm outlined above. First, we drop one dimension from arrays $Rows$, $Columns$, MD, AD since we fuse it within one integer value with $\geq N$ bits. Second, we represent the values of Latin square cells in a different manner: when $a_{i,j} = k$ instead of $LS[i][j] = k$ we now write $LS[i][j] = 1 \ll k$, where $\ll k$ means left bit shift for k positions.

We also introduce array of auxiliary variables $CR[N][N]$ to keep track of the number of current constraints on each Latin square element. Then the inner loop structure looks as follows (we omit the entries for main diagonal and main antidiagonal for simplicity).

Data: $LS[N][N]$, $CR[N][N]$, $Rows[N]$, $Columns[N]$, MD, AD, $SquaresCnt$
/* Compute vector of possible values for cell [i][j]   */
$CR[i][j] = Rows[i]|Columns[j]$;
/* Iterate over all possible values of cell [i][j]   */
for $LS[i][j] = 1$; $LS[i][j] < (1 \ll N)$; $LS[i][j] = LS[i][j] \ll 1$ **do**
  | /* Check if the value is ''occupied''   */
  | **if** $(CR[i][j]\&LS[i][j])! = 0$ **then** continue;
  | /* Otherwise mark the value as ''occupied'' and proceed   */
  | $Rows[i] = Rows[i]|LS[i][j]$;
  | $Columns[j] = Columns[j]|LS[i][j]$;
  | **BODY OF INNER LOOP FOR NEXT CELL** $LS[i'][j']$;
  | $Rows[i] = Rows[i] \oplus LS[i][j]$;
  | $Columns[j] = Columns[j] \oplus LS[i][j]$;
end

Algorithm 3. Inner loop structure in the bit-arithmetic-based version of the algorithm

---

Even without any optimizations considered in the end of the previous section this version of the algorithm makes it possible to generate about 2.6×10^6 diagonal Latin squares of order 9 per second. Nevertheless, there is still a room for improvement. Despite the fact that we can compute the vector of possible values for each particular element very fast ($CR[i][j]$ contains 1-bits in all the positions in which $LS[i][j]$ **can not** take the value of 1), we still need to iterate over all N values of $LS[i][j]$. Is there a way to make proper use of this information? Fortunately, yes. We can reconstruct the **for** loop in order to iterate over only such values of $LS[i][j]$ that satisfy the condition $CR[i][j]\&LS[i][j] = 0$, thus eliminating the need for **if** block in the loop body.

For this purpose we employ two more constants: $MaxN = 1 \ll N$ that contains the upper bound for $LS[i][j]$, and $AllN = MaxN - 1$ that has exactly N 1-bits in the beginning and 0-bits otherwise. We also make use of additional auxiliary integer array $L[i][j]$. The next version heavily relies on the bit twiddling tricks that make it possible to isolate the rightmost 1-bit ($y = x\&(-x)$) and to turn off the rightmost 1-bit ($y = x\&(x - 1)$). Let us present the modified inner loop structure in the pseudocode below.

```
Data: LS[N][N], CR[N][N], L[N][N], Rows[N], Columns[N], MD, AD, SquaresCnt
/* Compute vector of possible values for cell [i][j]                      */
L[i][j] = Rows[i]|Columns[j];
/* Iterate over values of cell [i][j] that do not violate any uniqueness
   constraint.                                                            */
for CR[i][j] = AllN ⊕ L[i][j]; CR[i][j]! = 0; CR[i][j] = CR[i][j]&(CR[i][j] − 1) do
    LS[i][j] = CR[i][j]&(−CR[i][j]);
    /* Mark the value as ''occupied'' and proceed                         */
    Rows[i] = Rows[i]|LS[i][j];
    Columns[j] = Columns[j]|LS[i][j];
    BODY OF INNER LOOP FOR NEXT CELL LS[i'][j'];
    Rows[i] = Rows[i] ⊕ LS[i][j];
    Columns[j] = Columns[j] ⊕ LS[i][j];
end
```

Algorithm 4. Inner loop structure in the improved bit-arithmetic-based version

Now in the **for** loop we first initialize $CR[i][j]$ with bit vector that contains 1 bits only in positions corresponding to values of $LS[i][j]$ that do not violate any uniqueness constraint. Then we iterate over them by switching off the rightmost 1 bit until $CR[i][j]$ becomes 0. For each value of $CR[i][j]$ we produce the value of $LS[i][j]$ by isolating the rightmost bit in $CR[i][j]$.

This improved algorithm version makes it possible to generate about 6×10^6 diagonal Latin squares of order 9 per second even without heuristic optimizations.

In Table 1 we present the results of experimental evaluation of our algorithm. The generation speed is measured on one core of Intel Core i7-6700 (16 Gb RAM, Windows 10, Microsoft Visual Studio 2015, x64-release). For diagonal Latin squares we fix only the first row. For ordinary Latin squares we fix both the first row and the first column. $(D)LS$ followed by number stands for (diagonal) Latin squares of specific order. The results are shown for the most relevant version of algorithm, including all modifications described above. The order

Table 1. Performance of the proposed versions of the algorithm for generation of Latin squares of small order.

Version	Problem	Squares per second
Standard	DLS9	1.8×10^6
Bit-arithmetic-based	DLS9	2.6×10^6
	DLS9	6.8×10^6
	LS8	9×10^6
Improved bit-arithmetic-based	DLS8	5.8×10^6
	LS9	8.0×10^6
	LS10	6.3×10^6
	DLS10	6.0×10^6

of cells in each case is determined according to heuristic procedure outlined in Subsect. 2.3. The performance of "Standard" and "Improved" versions for $DLS9$ was measured for the algorithm versions that use Lookahead heuristic (in other cases the corresponding optimization requires a lot of empirical evaluation and testing).

It is clear that bit arithmetic techniques make it possible to significantly increase the performance of the algorithm.

4   Enumeration of Diagonal Latin Squares of Order 9

We used the preliminary version of the presented algorithm (as it is outlined in Sect. 2, i.e. without bit arithmetic) to enumerate diagonal Latin squares of order 8 (300 286 741 708 800) [6], which is reflected in the sequence A274806 [7] in the On-Line Encyclopedia of Integer Sequences. At the time of experiment it took about 30 CPU h. Our current improved bit-arithmetic-based implementation achieves this result in 21 min.

A large part of our motivation for the present research was to improve the algorithm effectiveness to make enumerating of diagonal Latin squares of order 9 possible. The improved bit-arithmetic-based version of the algorithm, presented in the previous sections, was employed to solve this large-scale computational problem. In order to ensure the correctness of the results, we launched two separate experiments. The first one was launched in the volunteer computing project Gerasim@home [8]. The second one was performed on the computing cluster "Academician V.M. Matrosov" of Irkutsk supercomputing center SB RAS[1]. In the following subsection we describe how we decomposed the search space into a pool of computational tasks. After this both experiments are described in detail.

---

[1] http://www.hpc.icc.ru.

4.1   Decomposition of the Search Space

The considered problem of enumerating diagonal Latin squares of order 9 can be decomposed into a separate subproblems in quite natural way. First, we fixed the first row to 012345678. So, here we considered the problem of enumerating diagonal Latin squares of order 9 with fixed first row. After this we formed 1 225 884 workunits (WU) which correspond to all possible correct assignments of the first 10 cells in accordance with the order, presented on Fig. 1. It means that in each WU the values of 19 cells out of 81 were fixed (the others should be varied by our algorithm). Here by processing of a certain WU we mean the enumeration of all diagonal Latin squares of order 9 in which the values of 19 corresponding cells are equal to those specified by WU. According to our results, for some WUs the corresponding number of squares can be equal to 0.

The formed WUs can be processed by employing the embarrassing parallelism [9], so one can use both parallel and distributed computing for this purpose. As a result we have an array of 1 225 884 integer numbers (as we mentioned above, some ot them can be equal to 0). The summation of all these numbers gives us the number of reduced diagonal Latin squares of order 9. By multiplying this number to 9! = 362880 we obtain the number of diagonal Latin squares of order 9.

4.2   Experiment in a Volunteer Computing Project

The first experiment was held in the volunteer computing project Gerasim@home [8] which is based on BOINC (Berkeley Open Infrastructure for Network Computing [10]). In Fig. 2 the scheme of this experiment is shown.

Let us describe this scheme. An input data of the application is a WU file with values of 10 cells in accordance with the order from Fig. 1 (the values of the first 9 cells are constant in the algorithm implementation, so we don't send them). A number of diagonal Latin squares, which corresponds to a WU, is written to an output file. In order to decrease the influence of possible software and hardware errors we used quorum of 2. It means that each WU was generated and processed twice. According to the BOINC redundancy technique, two copies of a certain WU are sent to different volunteers. Then two results are compared and if they coincide, then the result of the corresponding WU is marked as correct. If not, then two new copies of the WU are generated.

The experiment was launched on 18 June, 2016. At that moment the performance of the project was about 2.5 teraflops. In the experiment we used x86 and x64 applications for Windows OS. The experiment ended on 17 September, 2016, so it took 3 months. According to the statistics, about 500 volunteers from 51 countries participated in it. In total, they connected to the project about 1000 hosts and, as a result, the peak performance of 5 teraflops was achieved in some time intervals (an average performance was equal to 3 teraflops).

As a result of post-processing we determined the number of reduced diagonal Latin squares of order 9: 5 056 994 653 507 584. By multiplying it to 9! we determined the number of diagonal Latin squares of order 9: 1 835 082 219 864 832 081 920. It should be noted, that in the case we didn't made the algorithm

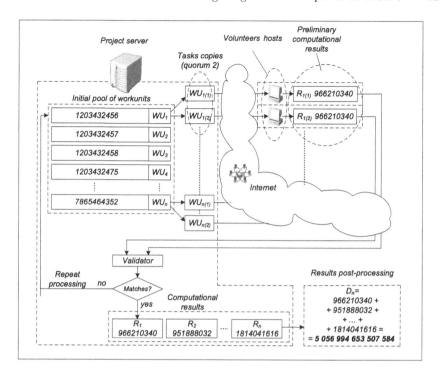

Fig. 2. Scheme of the experiment which was held in Gerasim@home

improvements, described in Sects. 2 and 3, the corresponding experiment could take about 10 years in Gerasim@home with its current performance (according to the speed of enumeration from [6]).

4.3 Experiment on a Computing Cluster

While solving an enumeration problem the correctness of a result is the most crucial feature. That is why we launched one more experiment aimed at solving the same problem. This experiment was performed on the computing cluster "Academician V.M. Matrosov" of Irkutsk supercomputing center SB RAS. Each node of this cluster is equipped with 2 16-core CPUs AMD Opteron 6276 and 64 GB of RAM. We used the same decomposition (as it was described in the previous subsection). Unlike the experiment held in Gerasim@home, here we didn't employ a redundancy technique, so each WU was processed exactly once.

We developed an MPI-program based on the improved bit-arithmetic-based version of our algorithm. In this program there is one control process, and the other ones are computing processes. First, on the control process a pool of WUs is created. Next, these WUs are processed by computing processes. Control process sends WUs to computing processes and receives results from them. Finally, control process accumulates results for all WUs and calculates their sum. The corresponding scheme of the experiment is quite similar to one used in Gerasim@home (see Fig. 2).

There is very important restriction on the cluster, which was employed for this experiment. According to administrator's policy, the time duration of an arbitrary MPI-task can't exceed 7 days. That is why we added to our MPI-program the possibility to continue a computational process from a checkpoint file. This file contains a current state of the processing of each WU from the pool. It is updated every time when new result from a computing process is obtained. At the end of the experiment, this file contains the corresponding numbers for each WU.

The experiment was started on 17 July, 2016. 19 launches of the MPI-program were performed in total. In these launches the number of cluster nodes varied from 10 to 15, and the time duration varied from 2 h to 7 days. The most of launches were performed on 15 nodes with time duration of 7 days. The experiment was finished on 17 October, 2016, so it took 3 months. The number obtained as a result of this experiment was the same as the one obtained in Gerasim@home (see the previous subsection). So, we can conclude, that the obtained result is correct with very high probability.

In contrast to a volunteer computing project, our cluster is equipped with nodes of the same type. That is why in this experiment we can compare WUs not only on the number of squares, but on the processing time too. In Table 2 we give some details on processing WUs with no squares and WUs with at least one square.

Table 2. Processing time (in seconds) of different types of WUs.

Type of WUs	Number of WUs	Min. time	Max. time	Avg. time
With no square	36254	0.84	9573.33	407.03
With at least one square	1219630	60.056	24993.2	1713.32

It should be noted, that the necessity to launch MPI-program many times to perform the whole experiment is quite uncomfortable circumstance. From this point of view, a volunteer computing project suits better for such large-scale experiments.

5   Related Work

Authors are not aware of algorithms developed specifically for enumeration of diagonal Latin squares. Usually, when it comes to Latin squares, even when the order is relatively small, say 8 or 9, the number of species representatives is so large that one does not attempt to generate them all since the amount of required memory would be colossal. Instead, there are algorithms that make it possible to generate uniformly distributed random Latin squares [2]. In paper [11] the enumeration of mutually orthogonal Latin squares of order 9 was performed. For this purpose the authors used proprietary algorithm. We are not aware of its performance and whether or not it can be adapted to work with diagonal Latin squares.

Papers [12–14] report on the number of Latin squares of orders 9, 10 and 11. The algorithms in these works are designed to take into account the fact that the space of Latin squares can be divided into relatively large main classes, and exploit the symmetrical structure of Latin squares a lot. In contrast to Latin squares, diagonal Latin squares form relatively small equivalence classes and restrict the vast majority of transformations. Thus we believe that these algorithms can not be directly applied to the considered enumeration problem. The authors of the present paper developed an algorithm for generating diagonal Latin squares of special kind [15]. Based on this algorithm a triple of diagonal Latin squares of order 10 that is the closest to being a triple of mutually orthogonal diagonal Latin squares found so far was found.

There are several examples of application of parallel and volunteer computing to the search for combinatorial designs based on Latin squares. With the help of a computing cluster there was proven that there is no finite projective plane of order 10 [16]. In the volunteer computing project SAT@home several dozen pairs of mutually orthogonal diagonal Latin squares were found [17, 18].

The most similar approach, compared to ours, is the one employed in [19]. In this paper the hypothesis about the minimal number of clues in Sudoku was proven. The authors developed the fast algorithm to enumerate and check all possible Sudoku variants. This algorithm was implemented and launched on a modern computing cluster. It took about 11 months for this cluster to check all variants. The volunteer computing project Sudoku@vtaiwan [20] was used to confirm the solution of this problem.

6   Conclusions and Future Work

In this paper we presented the fast algorithm for generating (diagonal) Latin squares of small order. It was employed to enumerate diagonal Latin squares of order 9. The obtained result is a consequence of the proper development of the enumeration algorithm and the rational employing of huge amount of computational resources. The corresponding source code is available online[2]. In the nearest future we plan to make a GPU implementation of the suggested algorithm.

References

1. Colbourn, C.J., Dinitz, J.H.: Handbook of Combinatorial Designs, 2nd edn. (Discrete Mathematics and Its Applications). Chapman & Hall/CRC, Boca Raton (2006). http://dx.doi.org/10.1201/9781420010541

---

[2] http://github.com/veinamond/LS_enum.

2. Jacobson, M.T., Matthews, P.: Generating uniformly distributed random Latin squares. J. Comb. Des. **4**(6), 405–437 (1996)
3. Karp, R.M.: Reducibility among combinatorial problems. In: Miller, R.E., Thatcher, J.W., Bohlinger, J.D. (eds.) Complexity of Computer Computations. The IBM Research Symposia Series, pp. 85–103. Springer, Boston (1972). doi:10. 1007/978-1-4684-2001-2_9
4. Knuth, D.: Dancing links. Millenn. Perspect. Comput. Sci. 187–214 (2000)
5. Golomb, S.W., Baumert, L.D.: Backtrack programming. J. ACM **12**(4), 516–524 (1965). http://dx.doi.org/10.1145/321296.321300
6. Vatutin, E., Zaikin, O., Zhuravlev, A., Manzyuk, M., Kochemazov, S., Titov, V.: Using grid systems for enumerating combinatorial objects on example of diagonal Latin squares. In: Selected Papers of the 7th International Conference Distributed Computing and Grid-technologies in Science and Education, Dubna, Russia, 4–9 July 2016. vol. 1787, pp. 486–490. CEUR-WS (2016)
7. Sloane, N.J.A.: The on-line encyclopedia of integer sequences. In: Kauers, M., Kerber, M., Miner, R., Windsteiger, W. (eds.) Calculemus/MKM -2007. LNCS, vol. 4573, p. 130. Springer, Heidelberg (2007). doi:10.1007/978-3-540-73086-6_12
8. Vatutin, E., Valyaev, S., Titov, V.: Comparison of sequential methods for getting separations of parallel logic control algorithms using volunteer computing. In: Second International Conference BOINC-based high performance computing: fundamental research and development (BOINC: FAST 2015), Petrozavodsk, Russia, 14–18 September 2015, vol. 1502, pp. 37–51. CEUR-WS (2015)
9. Foster, I.: Designing and Building Parallel Programs: Concepts and Tools for Parallel Software Engineering. Addison-Wesley Longman Publishing Co., Inc., Boston (1995)
10. Anderson, D.P., Fedak, G.: The computational and storage potential of volunteer computing. In: Sixth IEEE International Symposium on Cluster Computing and the Grid (CCGrid 2006), 16–19 May 2006, Singapore, pp. 73–80. IEEE Computer Society (2006). http://dx.doi.org/10.1109/ccgrid.2006.101
11. Egan, J., Wanless, I.M.: Enumeration of MOLS of small order. Math. Comput. **85**(298), 799–824 (2016). http://dx.doi.org/10.1090/mcom/3010
12. Bammel, S.E., Rothstein, J.: The number of 9×9 Latin squares. Discrete Math. **11**(1), 93–95 (1975). http://dx.doi.org/10.1016/0012-365X(75)90108-9
13. McKay, B.D., Rogoyski, E.: Latin squares of order 10. Electr. J. Comb. **2**(1), 1–4 (1995)
14. McKay, B.D., Wanless, I.M.: On the number of Latin squares. Ann. Comb. **9**(3), 335–344 (2005). http://dx.doi.org/10.1007/s00026-005-0261-7
15. Zaikin, O., Zhuravlev, A., Kochemazov, S., Vatutin, E.: On the construction of triples of diagonal Latin squares of order 10. Electron. Notes Discret. Math. **54**, 307–312 (2016). http://dx.doi.org/10.1016/j.endm.2016.09.053
16. Lam, C., Thiel, L., Swierz, S.: The nonexistence of finite projective planes of order 10. Canad. J. Math. **41**, 1117–1123 (1989). http://dx.doi.org/10.4153/cjm-1989-049-4
17. Zaikin, O., Vatutin, E., Zhuravlev, A., Manzyuk, M.: Applying high-performance computing to searching for triples of partially orthogonal Latin squares of order 10. In: 10th Annual International Scientific Conference on Parallel Computing Technologies Arkhangelsk, Russia, 29–31 March 2016, vol. 1576, pp. 155–166. CEUR-WS (2016)

18. Zaikin, O., Kochemazov, S., Semenov, A.: SAT-based search for systems of diagonal Latin squares in volunteer computing project sat@home. In: Biljanovic, P., Butkovic, Z., Skala, K., Grbac, T.G., Cicin-Sain, M., Sruk, V., Ribaric, S., Gros, S., Vrdoljak, B., Mauher, M., Tijan, E., Lukman, D. (eds.) 39th International Convention on Information and Communication Technology, Electronics and Microelectronics, MIPRO 2016, Opatija, Croatia, 30 May–3 June 2016, pp. 277–281. IEEE (2016). http://dx.doi.org/10.1109/MIPRO.2016.7522152

19. McGuire, G., Tugemann, B., Civario, G.: There is no 16-clue Sudoku: solving the Sudoku minimum number of clues problem via hitting set enumeration. Exp. Math. **23**(2), 190–217 (2014). http://dx.doi.org/10.1080/10586458.2013.870056

20. Lin, H.H., Wu, I.C.: Solving the minimum Sudoku problem. In: The 2010 International Conference on Technologies and Applications of Artificial Intelligence, TAAI 2010, pp. 456–461 (2010). http://dx.doi.org/10.1109/taai.2010.77

Globalizer Lite: A Software System for Solving Global Optimization Problems

Alexander V. Sysoyev$^{(\boxtimes)}$ (ID), Anna S. Zhbanova, Konstantin A. Barkalov (ID), and Victor P. Gergel (ID)

Lobachevsky State University of Nizhny Novgorod, Nizhny Novgorod, Russia
`alexander.sysoyev@itmm.unn.ru`

Abstract. In this paper, we describe the Globalizer Lite software system for solving global optimization problems. This system implements an approach to solving global optimization problems applying a block multistage scheme of dimension reduction that combines the use of Peano curve type evolvents and a multistage reduction scheme. The scheme allows for an efficient parallelization of the computations and a significant increase in the number of processors employed in the parallel solution of global optimization search problems. We also describe the synchronous and asynchronous schemes of MPI-implementation of this approach in the Globalizer Lite software system, and present a comparison of these schemes demonstrating the advantage of the asynchronous variant.

Keywords: Multidimensional multiextremal optimization · Global search algorithms · Parallel computations · Dimension reduction · Block multistage dimension reduction scheme

1 Introduction

In spite of the rapid increase in computer performance, the capacity of existing computer systems appears to be insufficient for the analysis and investigation of some important problems [37]. Some of these problems may be reduced to multidimensional problems of multiextremal nonlinear programming, for which the solution is a global extremum. The computation costs of such problems creates a need for the parallelization of algorithms and the use of parallel computer systems.

The general state of the art in the field of global optimization has been presented in a number of key monographs [10,18,22,27,38,39,41]. The development of optimization methods that use high-performance computer systems to solve time-consuming global optimization problems is a field that is currently receiving extensive attention (see, for example, [5,6,23,37,38]).

This research was supported by the Russian Science Foundation, project No. 16-11-10150, "Novel efficient methods and software tools for time consuming decision making problems using supercomputers of superior performance".

L. Sokolinsky and M. Zymbler (Eds.): PCT 2017, CCIS 753, pp. 130–143, 2017.
DOI: 10.1007/978-3-319-67035-5_10

The theoretical results obtained provide efficient solutions to many applied global optimization problems in various fields of scientific and technological applications (see, for example, [8,9,22,23,26,27]).

At the same time, the practical implementation of these algorithms for multiextremal optimization is quite limited.

Among the software for global optimization, one can mention the following systems: LGO (Lipschitz Global Optimization) [27], GlobSol [19], LINDO [21], IOSO (Indirect Optimization on the basis of Self-Organization) [7], MATLAB Global Optimization Toolkit (see, for example, [40]), TOMLAB system (see, for example, [17]), BARON (Branch-And-Reduce Optimization Navigator) [30], GAMS (General Algebraic Modeling System) (see [4]), Global Optimization Library in R (see, for example, [25]).

The list provided above is certainly not exhaustive (additional information on software systems for a wider spectrum of optimization problems can be obtained, for example, in [24,28,29]). Nevertheless, even from such a short list the following conclusions can be drawn (see also [20]):

- the collection of available global optimization software systems for practical use is insufficient;
- the availability of numerous methods through these systems allows to solve complex optimization problems in a number of cases, but this requires a rather high level of user knowledge and understanding in the field of global optimization;
- the use of parallel computing to achieve a higher efficiency in solving complex time consuming problems is limited, therefore, the computational potential of modern supercomputer systems is very poorly utilized.

In this paper, a novel Globalizer Lite software system is considered. The development of this system was conducted on the basis of the information-statistical theory of multiextremal optimization, aimed at developing efficient parallel algorithms for global search (see, for example, [36–38]).

The paper is structured as follows. In Sect. 2, we consider the general formulation of the multidimensional global optimization problem. In Sect. 3, we give the approaches to its solving based on the information-statistical theory of multiextremal optimization [38]. The Sect. 4 is devoted to the description of the Globalizer Lite software system; the implementations of the synchronous and asynchronous schemes of parallelization of the block multistage scheme will also be described here. The results of experiments comparing the efficiencies of these implementations will be presented in Sect. 5. Finally, our conclusions will be summarized in Sect. 6.

2   Formulation of the Multidimensional Global Optimization Problem

Let us consider the multidimensional multiextremal optimization problem

$$\varphi(y) \to inf, y \in D \subset R^N, \tag{1}$$

$$D = \left\{ y \in R^N : a_i \le y_i \le b_i, 1 \le i \le N \right\}, \tag{2}$$

i.e. the problem of finding the extremum values of the objective (minimized) function $\varphi(y)$ in a domain D defined by the coordinate restrictions (2) on the choice of feasible points $y = (y_1, y_2, ..., y_N)$.

If y^* is an exact solution of the problem (1)–(2), then the numerical solution of the considered problem may be reduced to constructing an estimate of the exact solution y^0 matching some notion of nearness to a point (for example, $\|y^* - y^0\| \le \varepsilon$ where $\varepsilon > 0$ is a predefined precision) based on a finite number k of computations of the optimized function values.

Regarding the class of problems considered here, the fulfillment of the following important conditions is presumed:

– The optimized function (y) can be defined by some algorithm for computing its values at the points of the domain D.
– The computation of the function value at each point is a computational-costly operation.
– The function $\varphi(y)$ satisfies the Lipschitz condition:

$$|\varphi(y_1) - \varphi(y_2)| \le L\|y_1 - y_2\|, \quad y_1, y_2 \in D, 0 < L < \infty, \tag{3}$$

which corresponds to a bounded variation of the function value for bounded variations of the argument.

Multiextremal optimization problems i.e. problems with an objective function $\varphi(y)$ possessing several local extrema in the feasible domain D, are the subject of consideration in the present paper. The dimensionality considerably affects the difficulty of solving such problems. The so called "curse of dimensionality", consisting in the exponential growth of computational costs as the dimension increases, appears in the case of multiextremal problems.

3   Approach for Solving Multidimensional Global Optimization Problems

3.1   Dimension Reduction Methods

One of the approaches to solving multidimensional global optimization problems consists in their reduction to one-dimensional problems, and then using efficient one dimensional global search algorithms for solving the reduced problems. Moreover, the reduction can be applied to the domain D in (2), which defines a one to one mapping of the hyperparallelepiped D onto the interval $[0, 1]$, and also to the function $\varphi(y)$, whose minimization can be performed on the basis of the following recursive scheme [33]:

$$\min\{\varphi(y) : y \in D\} = \min_{a_1 \le y_1 \le b_1} \min_{a_2 \le y_2 \le b_2} \ ... \ \min_{a_N \le y_N \le b_N} \varphi(y). \tag{4}$$

For the multistage scheme of dimension reduction represented by the relation (4), a generalization has been proposed in [1], namely a block multistage scheme of dimension reduction that reduces the solution of the initial multidimensional optimization problem (1)–(2) to that of a sequence of "nested" problems of lower dimension.

Thus, the initial vector y is represented as a vector of "aggregated" macro-variables,

$$y = (y_1, y_2, ..., y_N) = (u_1, u_2, ..., u_M) \tag{5}$$

where the i-th macro-variable u_i is a vector of dimension N_i consisting of components of the vector y taken sequentially, i.e.

$$u_1 = (y_1, y_2, ..., y_{N_1}),$$

$$u_2 = (y_{N_1+1}, y_{N_1+2}, ..., y_{N_1+N_2}), \tag{6}$$

$$u_i = (y_{p+1}, y_{p+2}, ..., y_{p+N_i}), \quad p = \sum_{k=1}^{i-1} N_k.$$

In addition $\sum_{k=1}^{M} N_k = N$.

Using the macro-variables, the main relation of the multistage scheme (4) can be rewritten as

$$\min_{y \in D} \varphi(y) = \min_{u_1 \in D_1} \min_{u_2 \in D_2} \ ... \ \min_{u_M \in D_M} \varphi(y), \tag{7}$$

where the subdomains $D_i, 1 \leq i \leq M$, are the projections of the initial search domain D onto the subspaces corresponding to the macro-variables $u_i, 1 \leq i \leq M$.

The fact that the nested subproblems

$$\varphi_i(u_1, ..., u_i) = \min_{u_{i+1} \in D_{i+1}} \varphi_{i+1}(u_1, ...u_i, u_{i+1}), 1 \leq i \leq M \tag{8}$$

are multidimensional is the principal difference between the block multistage scheme and the initial one. Thus, this approach can be combined with the reduction of the domain D (for example, with the evolvent based on a Peano curve) in order to use efficient methods for solving one-dimensional problems of multi-extremal programming [37].

The Peano curve $y(x)$ allows to map the interval $[0, 1]$ of the real axis onto the domain D uniquely:

$$\{y \in D \subset R^N\} = \{y(x): \ 0 \leq x \leq 1\}. \tag{9}$$

The evolvent is an approximation to the Peano curve with a precision of the order of 2^{-m}, where m is the density of the evolvent.

Mappings of this kind allow to reduce the multidimensional problem (1)–(2) to a one-dimensional one:

$$\varphi(y^*) = \varphi(y(x^*)) = \min\{\varphi(y(x)) : x \in [0, 1]\}. \tag{10}$$

3.2 Method for Solving the Reduced Global Optimization Problems

The information-statistical theory of global search formulated in [36,38] has served as a basis for the development of a large number of efficient multiextremal optimization methods (see, for example, [13–16,31–35]).

The Multidimensional Algorithm of Global Search (MAGS) established the basis for the methods applied in Globalizer Lite. The general computational scheme of MAGS can be presented as follows (see [36,38]).

Let us introduce a simpler notation for the problem being solved:

$$f(x) = \varphi(y(x)) : \ x \in [0,1].\tag{11}$$

Let us assume that $k > 1$ iterations of the methods have already been completed (an arbitrary point of the interval $[a,b]$, for example, its middle point, may be taken as the point of the first trial, x^1). Then, the next trial point is selected in the $(k+1)$-st iteration according to the following rules.

Rule 1. Renumber the points of the preceding trials, $x^1, ..., x^n$ (including the boundary points of the interval $[a,b]$), with subscripts in increasing order of coordinate values:

$$0 = x_0 < x_1 < ... < x_k < x_{k+1} = 1.\tag{12}$$

The function values $z_i = \varphi(x_i)$ have been calculated at all points $x_i, i = 1, ..., k$. At the points $x_0 = 0$ and $x_{k+1} = 1$, the function values have not been computed. For the sake of convenience, let us introduce the dummy notations z_0 and z_{k+1}.

Rule 2. Compute the quantities

$$\mu = \max_{1 \le i \le k} \frac{|z_i - z_{i-1}|}{\Delta_i}, \quad M = \begin{cases} r\mu, & \mu > 0, \\ 1, & \mu = 0, \end{cases}\tag{13}$$

where $r > 1$ is the reliability parameter of the method (specified by the user), $\Delta_i = x_i - x_{i-1}$.

Rule 3. Compute the characteristics for all intervals $(x_{i-1}, x_i), 1 < i < k+1$, according to the formulae

$$R(1) = 2\Delta_1 - 4\frac{z_1}{M},$$

$$R(i) = \Delta_i + \frac{(z_i - z_{i-1})^2}{M^2 \Delta_i} - 2\frac{z_i + z_{i-1}}{M}, 1 < i < k+1,\tag{14}$$

$$R(k+1) = 2\Delta_{k+1} - 4\frac{z_k}{M}.$$

Rule 4. Select the interval with the highest characteristic. Let us denote the index of this interval with t.

Rule 5. Execute the next trial at the point

$$
x^{k+1} = \begin{cases} \frac{x_t + x_{t-1}}{2}, t \in \{1, k+1\}, \\ \frac{x_t + x_{t-1}}{2} - \text{sign}(z_t - z_{t-1}) \frac{1}{2r} \left[\frac{|z_t - z_{t-1}|}{\mu} \right]^N, 1 < t < k+1. \end{cases} \tag{15}
$$

The algorithm is terminated if the condition $\Delta_t \leq \varepsilon$ is satisfied.
The values

$$
\varphi_k^* = \min_{1 \leq i \leq k} \varphi(x^i), \quad x_k^* = \arg \min_{1 \leq i \leq k} \varphi(x^i). \tag{16}
$$

are selected as the estimates of the globally optimized solution of the problem (1)–(2).

The computational scheme of the Parallel Multidimensional Algorithm of Global Search (PMAGS) is practically identical to that of the MAGS scheme. The differences consist in the following set of rules.

Rule 4'. Arrange the characteristics of the intervals obtained according to (14) in descending order,

$$
R(t_1) \geq R(t_2) \geq ... \geq R(t_k) \geq R(t_{k+1}), \tag{17}
$$

and select p intervals with the highest characteristics (p is the number of processors/cores used for parallel computations).

Rule 5'. Execute new trials at the points

$$
x^{k+j} = \begin{cases} \frac{x_{t_j} + x_{t_j-1}}{2}, t_j \in \{1, k+1\}, \\ \frac{x_{t_j} + x_{t_j-1}}{2} - \text{sign}(z_{t_j} - z_{t_j-1}) \frac{1}{2r} \left[\frac{|z_{t_j} - z_{t_j-1}|}{\mu} \right]^N, 1 < t_j < k+1. \end{cases} \tag{18}
$$

4   General Description of Globalizer Lite

4.1   Implementation of the Parallel Algorithm of Global Optimization

Let us consider a parallel implementation of the block multistage dimension reduction scheme described in Subsect. 3.1.

For the description of the parallelism in the multistage scheme, let us introduce a vector of parallelization degrees,

$$
\pi = (\pi_1, \pi_2, ..., \pi_M), \tag{19}
$$

where $\pi_i, 1 \leq i \leq M$, is the number of subproblems in the $(i+1)$-st nesting level being solved in parallel, arising as a result of the execution of parallel iterations at the i-th level. For the macro-variable u_i, the number π_i means the number of parallel trials in the course of the minimization of the function $\varphi_M(u_1, ..., u_M) = \varphi(y_1, ..., y_N)$ with respect to u_i for fixed values of $u_1, u_2, ..., u_{i-1}$ i.e. the number of values of the objective function $\varphi(y)$ computed in parallel.

In the general case, the quantities $\pi_i, 1 \leq i \leq M$ can depend on various parameters and can change during the optimization, but we will limit ourselves to the case when all the components of the vector π are constant.

Thus, a tree of MPI-processes is built during the solution of the problem. At every nesting level (every level of the tree), the parallel implementation of the MAGS described in Subsect. 3.2 is used. Let us remind that the parallelization is implemented not by selecting a single point for the next trial (as in the sequential version) but p points, which are placed in p intervals with the highest characteristics. Therefore, if p processors are available, p trials can be executed at these points in parallel. Moreover, the solution of the problem at the i-th level of the tree generates the subproblems for the $(i + 1)$-st level. This scheme corresponds to a method of organization of parallel computations of "master-slave" type.

When launching the software, the user specifies:

- The number of subdivision levels of the initial problem (in other words, the number of levels in the tree of processes) M;
- The number of variables (dimensions) at each level ($\sum_{k=1}^{M} N_k = N$, where N is the dimension of the problem);
- The number of MPI-processes and their distribution among the levels ($\pi = (\pi_1, \pi_2, ..., \pi_M)$).

Let us consider an example:

$$N = 10, M = 3, N_1 = 3, N_2 = 4, N_3 = 3, \pi = (2, 3, 0).$$

Thus, we have nine MPI-processes, which are arranged in a tree (Fig. 1: for each function φ_i, only the varied parameters are shown; the fixed values are not shown). According to N_1, N_2, N_3, we have the following macro-variables: $u_1 = (y_1, y_2, y_3), u_2 = (y_4, y_5, y_6, y_7), u_3 = (y_8, y_9, y_{10})$. Each node solves a problem from the relation (10). The root (level #0) solves the problem with respect to the first N_1 variables of the initial N-dimensional problem. The iteration generates a problem of the next level at any point. The nodes of level #1 solve the problems with respect to N_2 variables for fixed values of the first N_1 variables, etc.

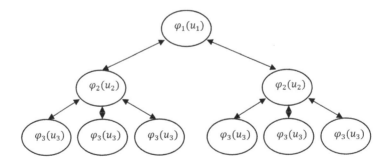

Fig. 1. Scheme of organization of parallel computations.

4.2   Synchronous Scheme of Computations

Let p be the number of child processes of a given node of the tree of MPI processes.

At each iteration, we select p points of the next trials. The data is transferred to the child processes. We wait until all child processes are completed and the solutions of the subproblems are obtained. Now we compute the current estimate of the extremum and select the points for the next trials.

The main disadvantage of the synchronous scheme is that the child processes of some nodes, having completed their trials earlier than others, would stay idle waiting for the completion of the whole iteration. The asynchronous scheme considered below is free from this disadvantage.

4.3   Asynchronous Scheme of Computations

At the first iteration, we proceed in the same way as in the synchronous scheme: we generate the trial points according to the number of child processes, then send the scheduled trial point to each child process, and wait until all child processes complete their subproblems.

At the second iteration, we send a single point to each child process. This is be-cause for the correct functioning of the asynchronous scheme, a large enough number of intervals generated during the solution should be available. In some cases, the first iteration is not enough to ensure this, therefore, the second iteration is executed separately from all the next ones according to a pure synchronous scheme. At that, the transition to the asynchronous mode takes place, and we wait until only one child process is completed, without waiting for the completion of the rest, and obtain the data from this process.

At the next iteration, we select only one trial point and send it to the child process from which a solution was obtained at the preceding iteration. Then, we wait until each child process completes its own subproblem, and we obtain the data from this one. These operations are repeated until the required precision of the solution is reached.

4.4   Globalizer Lite System Architecture

Globalizer Lite, the system considered in this paper, expands the family of global optimization software systems successively developed by the authors during the past several years. One of the first developments was the SYMOP multiextremal optimization system [11], which has been successfully applied for solving many optimization problems. A special place is occupied by the ExaMin system (see, for example, [1]), which was developed and used extensively to investigate the application of novel parallel algorithms to the solution of global optimization problems using high performance multiprocessor computing systems.

The program architecture of the Globalizer Lite system is shown in Fig. 2.

The structural components of the systems are:

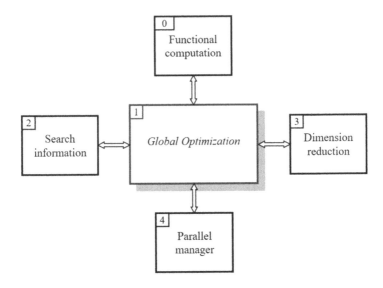

Fig. 2. Program architecture of the Globalizer Lite system.

- Block 0 consists of the procedures for computing the functional values (criteria and constraints) for the optimization problem being solved; this is an external block with respect to the Globalizer Lite system, and is provided by the user through a predefined interface.
- Block 1 forms the optimization subsystem and solves the global optimization problems.
- Block 2 is a subsystem for accumulating and processing search information; this is one of the basic subsystems. On one hand, the amount of search information for complex optimization problems may be quite large, but on the other hand, the efficiency of global optimization methods depends to a great extent on how completely all the available search data is used.
- Block 3 contains the dimension reduction procedures based on Peano evolvents. The optimization block solves the reduced (one-dimensional) optimization problems. Block 3 provides interaction between the optimization block and the initial multidimensional optimization problem.
- Block 4 is responsible for managing the parallel processes when performing the global search.

5   Results of Numerical Experiments

In this section, we present the results of computational experiments. The first series of experiments was aimed at comparing the synchronous and asynchronous schemes of the block multistage reduction by comparison of various distributions of dimensions over the levels of the tree of MPI-processes, and finding the most efficient distribution.

The computational experiments considered below were first carried out using the ExaMin system (see [1–3, 12]), and then were reproduced using the Globalizer Lite system.

The Rastrigin function (20), the Rosenbrock function (21) and the Neumeier function (22) were used in the experiments. This functions are classic test functions in optimization theory. The Rastrigin function is a fairly difficult problem due to a large search space and a large number of local minima. The global minimum of the Rosenbrock function lies inside a long and narrow valley. It is considered that the search for the global minimum of this functions is a non-trivial task.

$$\varphi(y) = A_n + \sum_{i=1}^{n} [y_i^2 - A\cos(2\pi y_i)], \quad A = 10, \quad y_i \in [-5.12, 5.12]. \tag{20}$$

$$\varphi(y) = \sum_{i=1}^{n-1} [(1 - y_i)^2 + 100(y_{i+1} - y_i^2)^2], \quad y \in R^N. \tag{21}$$

$$\varphi(y) = \sum_{k=1}^{n} (b_k - \sum_{i=1}^{n} y_i^{k+1})^2, \quad b[4] = \{8, 18, 44, 114\}. \tag{22}$$

From the results of the experiments, one can see that the asynchronous version works faster in most cases. In average, a 1.5-2-fold speedup was achieved (Table 1).

Table 1. Results of the test problems solution: Rastrigin function (4-dimensional)

Dimension	Sync. scheme, time (sec)	Async. scheme, time (sec)	Sync. scheme, number of trials of root process	Async. scheme, number of trials of root process	Speedup
(1, 3)	5.874	5.819	33	29	1.009
(2, 2)	0.156	0.041	522	144	3.771
(3, 1)	0.069	0.007	3429	307	10.105

Also, when comparing the schemes one can note that the root-level process (the root of the tree) performs fewer iterations in the asynchronous scheme than it does in the synchronous one (Table 2).

The total computation time of the algorithm strongly depends on the problem being solved. Comparing the data obtained on the three test functions, one can conclude that, for equal numbers of processes, it is more profitable to leave the largest dimension at the zero level when distributing the dimensions among the levels of the tree. Even though in this case the root performs more operations

Table 2. Results of the test problems solution: Rosenbrock function (4-dimensional)

Dimension	Sync. scheme, time (sec)	Async. scheme, time (sec)	Sync. scheme, number of trials of root process	Async. scheme, number of trials of root process	Speedup
(1, 3)	68.473	69.708	69	70	0.982
(2, 2)	12.832	7.236	2691	1931	1.773
(3, 1)	1.391	0.535	88428	48930	2.599

Table 3. Results of the test problems solution: Neumeier function (4-dimensional)

Dimension	Sync. scheme, time (sec)	Async. scheme, time (sec)	Sync. scheme, number of trials of root process	Async. scheme, number of trials of root process	Speedup
(1, 3)	9.054	8.445	39	35	1,072
(2, 2)	0.241	0.163	396	258	1,480
(3, 1)	0.0609	0.032	3705	2274	1,878

than in other situations, the total time required to solve the problem is considerably reduced. Thus, the distribution of the dimensions among the levels is an important factor determining the computation time of the algorithm (Table 3).

Also, it is worth noting that increasing the number of levels in the tree may not result in the desired speedup of the algorithm. This is caused by an increasing amount of data being transmitted between processes. However, if the time required to compute the objective function is rather long, then this variant of tree construction is fully justified.

The next series of experiments was aimed both at comparing various structures of MPI-process trees and at determining the effect of the tree structure on the rate of problem solving.

The results of the experiments are presented in Table 4. The number of child processes at each level is shown in the first column. So, the sequence (2, 3, 0) means that the root of the MTP-tree has 2 child processes, and every node of the first level in the tree has 3 child processes (Fig. 1).

According to the data obtained in the experiments, a clear dependence is observed for the synchronous scheme between the time required to solve the problem and the number of trials of the root process.

From the results of the experiments, one can point out that (4, 2, 0) is the most profitable structure for the synchronous scheme. It presents the minimum number of trials of the root process in the MPI-tree and the best solution time. This can be ex-plained by the fact that for the functions considered here, which are computed relatively quickly, two child processes at the second level, which would compute the function directly, are quite enough. In this regard, the structures (1, 11, 0) and (2, 5, 0) are less efficient.

Table 4. Results of the test problems solution: Rastrigin function (5-dimensional)

Number of child processes at each level	Sync. scheme, time (sec)	Async. scheme, time (sec)	Sync. scheme, number of trials of root process	Async. scheme, number of trials of root process
(1, 11, 0)	3.270	1.369	180	123
(2, 5, 0)	3.024	1.479	266	131
(3, 3, 0)	7.0824	1.698	552	150
(4, 2, 0)	1.789	2.552	156	214
(6, 1, 0)	2.836	1.517	228	136
(3, 1, 2, 0)	22.037	9.449	552	173
(2, 1, 4, 0)	75.341	7.430	266	93

The structures having four levels are not efficient for the Rastrigin function, since the time costs of data transfer appear to exceed essentially the times of computing the function itself.

6 Conclusion

In this paper, we have examined the Globalizer Lite global optimization software system for implementing a general scheme of parallel solution of globally optimized decision making problem. We have considered the possibility of speeding up the search of the global optimum when solving multidimensional multiextremal optimization problems using a parallel block multistage scheme of dimension reduction.

We have also considered the general principles of organization of synchronous and asynchronous parallel computations. These schemes were implemented using MPI. Finally, we studied the efficiency of the proposed block multistage schemes when applied to problems with various times of computation of the optimized function.

References

1. Barkalov, K., Gergel, V.: Parallel global optimization on GPU. J. Glob. Optim. **66**, 3–20 (2015). doi:10.1007/s10898-016-0411-y
2. Barkalov, K., Gergel, V., Lebedev, I.: Use of xeon phi coprocessor for solving global optimization problems. In: Malyshkin, V. (ed.) PaCT 2015. LNCS, vol. 9251, pp. 307–318. Springer, Cham (2015). doi:10.1007/978-3-319-21909-7_31
3. Barkalov, K., Gergel, V., Lebedev, I.: Solving global optimization problems on GPU cluster. In: Simos, T.E. (ed.) ICNAAM 2015, AIP Conference Proceedings, vol. 1738, Article No. 400006 (2016). doi:10.1063/1.4952194

4. Bussieck, M.R., Meeraus, A.: General algebraic modeling system (GAMS). In: Kallrath, J. (ed.) Modeling Languages in Mathematical Optimization. APOP, vol. 88, pp. 137–157. Springer, Boston (2004). doi:10.1007/978-1-4613-0215-5_8

5. Censor, Y., Zenios, S.A.: Parallel Optimization: Theory, Algorithms, and Applications. Oxford University Press, Oxford (1998). doi:10.1016/S0898-1221(98)90713-1

6. Ciegis, R., Henty, D., Kgstrm, B., Zilinskas, J. (eds.): Parallel Scientific Computing and Optimization: Advances and Applications. Springer, Heidelberg (2009). doi:10.1007/978-0-387-09707-7

7. Egorov, I.N., Kretinin, G.V., Leshchenko, I.A., Kuptzov, S.V.: IOSO optimization toolkit - novel software to create better design. In: 9th AIAA/ISSMO Symposium on Multidisciplinary Analysis and Optimization, Atlanta, Georgia (2002). doi:10.2514/6.2002-5514. www.iosotech.com/text/2002_4329.pdf

8. Fasano, G., Pintr, J.D. (eds.): Modeling and Optimization in Space Engineering. Springer, Heidelberg (2013). doi:10.1007/978-1-4614-4469-5

9. Floudas, C.A., Pardalos, M.P.: State of the Art in Global Optimization: Computational Methods and Applications. Kluwer Academic Publishers, Dordrecht (1996). doi:10.1007/978-1-4613-3437-8

10. Floudas, C.A., Pardalos, M.P.: Recent Advances in Global Optimization. Princeton University Press, Princeton (2016)

11. Gergel, V.P.: A software system for multiextremal optimization. Eur. J. Oper. Res. **65**(3), 305–313 (1993). doi:10.1016/0377-2217(93)90109-z

12. Gergel, V.: A unified approach to use of coprocessors of various types for solving global optimization problems. In: 2nd International Conference on Mathematics and Computers in Sciences and in Industry, pp. 13–18 (2015). doi:10.1109/MCSI.2015.18

13. Gergel, V., Grishagin, V., Gergel, A.: Adaptive nested optimization scheme for multidimensional global search. J. Glob. Optim. **66**(1), 35–51 (2016). doi:10.1007/s10898-015-0355-7

14. Gergel, V., Lebedev, I.: Heterogeneous parallel computations for solving global optimization problems. Procedia Comput. Sci. **66**, 53–62 (2015). doi:10.1016/j.procs.2015.11.008

15. Gergel, V., Sidorov, S.: A two-level parallel global search algorithm for solution of computationally intensive multiextremal optimization problems. In: Malyshkin, V. (ed.) PaCT 2015. LNCS, vol. 9251, pp. 505–515. Springer, Cham (2015). doi:10.1007/978-3-319-21909-7_49

16. Grishagin, V.A., Strongin, R.G.: Optimization of multiextremal functions subject to monotonically unimodal constraints. Eng. Cybern. **5**, 117–122 (1984)

17. Holmstrm, K., Edvall, M.M.: The TOMLAB optimization environment. In: Kallrath, J. (ed.) Modeling Languages in Mathematical Optimization. APOP, vol. 88, pp. 369–376. Springer, Boston (2004). doi:10.1007/978-1-4613-0215-5_19

18. Horst, R., Tuy, H.: Global Optimization: Deterministic Approaches. Springer, Heidelberg (1990)

19. Kearfott, R.B.: Globsol user guide. Optim. Methods Softw. **24**, 687–708 (2009). doi:10.1080/10556780802614051

20. Liberti, L.: Writing global optimization software. In: Liberti, L., Maculan, N. (eds.) Global Optimization: From Theory to Implementation. NOIA, vol. 84, pp. 211–262. Springer, Boston (2006). doi:10.1007/0-387-30528-9_8

21. Lin, Y., Schrage, L.: The global solver in the LINDO API. Optim. Methods Softw. **24**, 657–668 (2009). doi:10.1080/10556780902753221

22. Locatelli, M., Schoen, F.: Global Optimization: Theory, Algorithms, and Applications. SIAM, Philadelphia (2013). doi:10.1137/1.9781611972672

23. Luque, G., Alba, E.: Parallel Genetic Algorithms. Theory and Real World Applications. Springer, Heidelberg (2011). doi:10.1007/978-3-642-22084-5
24. Mongeau, M., Karsenty, H., Rouz, V., Hiriart-Urruty, J.B.: Comparison of public-domain software for black box global optimization. Optim. Methods Softw. **13**(3), 203–226 (2000). doi:10.1080/10556780008805783
25. Mullen, K.M.: Continuous global optimization in R. J. Stat. Softw. **60**(6), 1–45 (2014). doi:10.18637/jss.v060.i06
26. Pardalos, M.P., Zhigljavsky, A.A., Zilinskas, J.: Advances in Stochastic and Deterministic Global Optimization. Springer, Cham (2016). doi:10.1007/978-3-319-29975-4
27. Pintr, J.D.: Global Optimization in Action (Continuous and Lipschitz Optimization: Algorithms, Implementations and Applications). Kluwer Academic Publishers, Dordrecht (1996). doi:10.1007/978-1-4757-2502-5
28. Pintr, J.D.: Software development for global optimization. In: Pardalos, P.M., Coleman, T.F. (eds.) Lectures on Global Optimization. Fields Institute Communications, vol. 55, pp. 183–204 (2009). doi:10.1090/fic/055/08
29. Rios, L.M., Sahinidis, N.V.: Derivative-free optimization: a review of algorithms and comparison of software implementations. J. Glob. Optim. **56**(3), 1247–1293 (2012). doi:10.1007/s10898-012-9951-y
30. Sahinidis, N.V.: BARON: a general purpose global optimization software package. J. Global Optim. **8**(2), 201–205 (1996). doi:10.1007/bf00138693
31. Sergeyev, Y.D.: An information global optimization algorithm with local tuning. SIAM J. Optim. **5**(4), 858–870 (1995). doi:10.1137/0805041
32. Sergeyev, Y.D.: Multidimensional global optimization using the first derivatives. Comput. Maths. Math. Phys. **39**(5), 743–752 (1999)
33. Sergeyev, Y.D., Grishagin, V.A.: A parallel method for finding the global minimum of univariate functions. J. Optim. Theory Appl. **80**(3), 513–536 (1994). doi:10.1007/bf02207778
34. Sergeyev, Y.D., Grishagin, V.A.: Parallel asynchronous global search and the nested optimization scheme. J. Comput. Anal. Appl. **3**(2), 123–145 (2001)
35. Sergeyev, Y.D., Strongin, R.G., Lera, D.: Introduction to Global Optimization Exploiting Space-Filling Curves. Springer, New York (2013)
36. Strongin, R.G.: Numerical Methods in Multiextremal Problems (Information-Statistical Algorithms). Nauka, Moscow (1978). (In Russian)
37. Strongin, R.G., Gergel, V.P., Grishagin, V.A., Barkalov, K.A.: Parallel Computations for Global Optimization Problems. Moscow State University, Moscow (2013). (In Russian)
38. Strongin, R.G., Sergeyev, Y.D.: Global Optimization with Non-convex Constraints: Sequential and Parallel Algorithms. Kluwer Academic Publishers, Dordrecht (2000). doi:10.1007/978-1-4615-4677-1
39. Törn, A., Žilinskas, A. (eds.): Global Optimization. LNCS, vol. 350. Springer, Heidelberg (1989). doi:10.1007/3-540-50871-6
40. Venkataraman, P.: Applied Optimization with MATLAB Programming. Wiley, Hoboken (2009)
41. Zhigljavsky, A.A.: Theory of Global Random Search. Kluwer Academic Publishers, Dordrecht (1991). doi:10.1007/978-94-011-3436-1

Optimized Algorithms for Solving Structural Inverse Gravimetry and Magnetometry Problems on GPUs

Elena N. Akimova[1,2(✉)] [iD], Vladimir E. Misilov[1,2], and Andrey I. Tretyakov[1,2]

[1] Krasovskii Institute of Mathematics and Mechanics, Ural Branch of RAS,
16, S. Kovalevskaya Street, Yekaterinburg, Russia
[2] Ural Federal University, 19, Mira Street, Yekaterinburg, Russia
aen15@yandex.ru, out.mrscreg@gmail.com, fr1z2rt@gmail.com

Abstract. In this article, we construct new variants of iteratively regularized linearized gradient-type methods for solving structural inverse gravimetry and magnetometry problems, namely the regularized conjugate gradient method, the modified regularized conjugate gradient method, and the hybrid regularized conjugate gradient method.

The main idea of the modification is to calculate the Jacobian matrix of the integral operator at a fixed point, without updating it during the entire iteration process.

We also developed memory-optimized and time-efficient parallel algorithms and programs on the basis of the constructed modified methods. The memory optimization uses the block-Toeplitz structure of the Jacobian matrix. The algorithms were implemented on GPUs using the NVIDIA CUDA technology. We performed an efficiency and speedup analysis, and solved a model problem with synthetic disturbed data.

Keywords: Nonlinear gradient-type methods · Parallel algorithms · Gravimetry and magnetometry problems · Toeplitz matrix · GPU

1 Introduction

Solving structural inverse gravimetry and magnetometry problems has a great importance in the study of the Earth's crust structure. The solution consists in finding the interface between layers of different densities or magnetizations using known density or magnetization contrasts and a gravitational or magnetic field [6,10].

The problems are ill-posed. Therefore, it is necessary to use iterative regularization methods [5].

This work was partly supported by the Russian Foundation for Basic Research (project no. 15-01-00629a), and the Ural Branch of the Russian Academy of Sciences (project no. 15-7-1-3).

L. Sokolinsky and M. Zymbler (Eds.): PCT 2017, CCIS 753, pp. 144–155, 2017.
DOI: 10.1007/978-3-319-67035-5_11

Effective methods to determine structural boundaries were constructed in [1–4,7,11], namely the regularized linearized steepest descent method, the linearized conjugate gradient method and the componentwise gradient method.

The modified regularized linearized steepest descent method is proposed and studied in [12]. The main idea of the modification is to calculate the Jacobian matrix of the integral operator at a fixed point without updating it at each iteration.

In this paper, we construct a regularized variant of the linearized conjugate gradient method based on the Tikhonov regularization, a modified variant of this method and a hybrid conjugate gradient method.

We construct parallel algorithms on the basis of the modified regularized linearized steepest descent method and the methods constructed in this paper. Additionally, we implement a memory optimization based on the block-Toeplitz structure of the Jacobian matrix.

Real observations are performed on large areas. To increase the accuracy and degree of detail, it is essential to use larger grids producing high dimensional data sets. The use of modern computing technologies and parallel computations allows to reduce computation time.

In this paper, we implement parallel algorithms for GPUs using the NVIDIA CUDA technology. Parallel programs are used to solve the model problem with a large grid on the NVIDIA Tesla GPUs of the Uran supercomputer, installed at the Institute of Mathematics and Mechanics of the Ural Branch of the Russian Academy of Sciences. We also investigate the efficiency and speedup of the parallel algorithms.

2   Problems Statement

Let us introduce a Cartesian coordinate system where the $x0y$ plane coincides with the Earth's surface, and the z axis is directed downwards, as shown in Fig. 1. Assume that the lower half-space consists of two layers with constant densities σ_1 and σ_2 divided by the sought surface, described by a bounded function $\zeta = \zeta(x, y)$ and $\lim\limits_{|x|+|y|\to\infty} (h - \zeta(x, y)) = 0$ for some h. The ζ function must satisfy the following equation

$$
f\Delta\sigma \int\limits_{-\infty}^{\infty}\int\limits_{-\infty}^{\infty}\left\{ \frac{1}{((x - x')^2 + (y - y')^2 + \zeta^2(x', y'))^{1/2}} \right.
$$
$$
\left. - \frac{1}{((x - x')^2 + (y - y')^2 + h^2)^{1/2}} \right\} dx'dy' = \Delta g(x, y, 0), \tag{1}
$$

where f is the gravitational constant, $\Delta\sigma = \sigma_2 - \sigma_1$ is the density contrast, $\Delta g(x, y, 0)$ is an anomalous gravitational field measured at the Earth's surface.

A preliminary processing of gravitational data with the aim of extracting the anomalous field from the measured gravitational data is performed using a technique suggested by Martyshko and Prutkin [9]. This technique was numerically implemented in the Uran supercomputer [8].

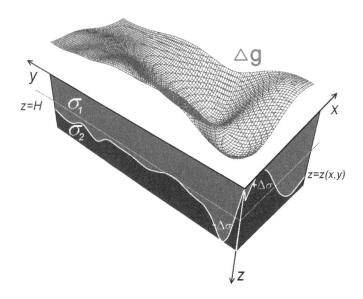

Fig. 1. Two-layer medium for the gravimetry problem.

For the magnetometry problem, the lower half-space consists of two layers with constant vertical magnetizations J_1 and J_2 divided by the sought surface $\zeta = \zeta(x, y)$, as shown in Fig. 2. The function ζ must satisfy the equation

$$
\Delta J \int\limits_{-\infty}^{\infty} \int\limits_{-\infty}^{\infty} \left\{ \frac{\zeta(x', y')}{((x - x')^2 + (y - y')^2 + \zeta^2(x', y'))^{3/2}} \right.
$$
$$
\left. - \frac{h}{((x - x')^2 + (y - y')^2 + h^2)^{3/2}} \right\} dx' dy' = \Delta Z(x, y, 0), \tag{2}
$$

where $\Delta J = J_2 - J_1$ is the magnetization contrast, $\Delta Z(x, y, 0)$ is the vertical component of the anomalous magnetic field measured at the Earth's surface.

Equations (1) and (2) are nonlinear two-dimensional integral equations of the first kind.

After the discretization of the area $\Pi = \{(x, y) : a \leqslant x \leqslant b, c \leqslant y \leqslant d\}$ into $n = M \times N$ grid and the approximation of the integral operators using quadrature rules, we have the right-hand part vector F and the approximation of the solution vector z of dimension n. Equations (1) and (2) take the form

$$
f \Delta\sigma \Delta x \Delta y \sum_{j=1..n} \left[\frac{1}{\sqrt{(x_i - x_j)^2 + (y_i - y_j)^2 + z_j^2}} \right.
$$
$$
\left. - \frac{1}{\sqrt{(x_i - x_j)^2 + (y_i - y_j)^2 + h^2}} \right] = F_i, \quad i = 1..n; \tag{1a}
$$

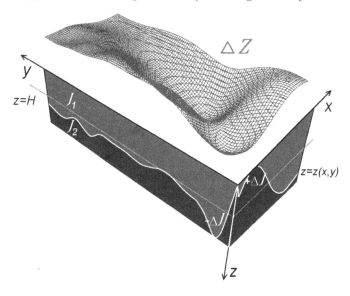

Fig. 2. Two-layer medium for the magnetometry problem.

$$\frac{\Delta J}{4\pi}\Delta x \Delta y \sum_{j=1..n} \left[\frac{-h}{\left((x_i - x_j)^2 + (y_i - y_j)^2 + h^2\right)^{3/2}} \right.$$
$$\left. + \frac{z_j}{\left((x_i - x_j)^2 + (y_i - y_j)^2 + z_j^2\right)^{3/2}} \right] = F_i, \quad i = 1..n. \tag{2a}$$

We can rewrite these equations as

$$A(z) = F. \tag{3}$$

3 Numerical Methods for Solving the Problems

3.1 Modified Regularized Linearized Steepest Descent Method

The main idea of the modified regularized linearized steepest descent method developed by Vasin [12] is to calculate the Jacobian matrix of the nonlinear operator at a initial point without updating it throughout the entire iteration process. As a result, the method becomes more economical in terms of the number of operations executed at each iteration step. Moreover, numerical experiments show that, in many cases, this modification is superior to the unmodified method in terms of the computer time required for achieving the same accuracy of the desired solution.

The modified regularized linearized steepest descent method (MRLSDM) has the following form:

$$z^{k+1} = z^k - \psi \frac{\left\| S_\alpha^0(z^k) \right\|^2}{\left\| A'(z^0)S_\alpha^0(z^k) \right\|^2 + \alpha \left\| S_\alpha^0(z^k) \right\|} S_\alpha^0(z^k), \tag{4}$$

where $S_\alpha^0 = A'(z^0)^T(A(z) - F) + \alpha(z - z^0)$, z^0 is the initial approximation of the solution, z^k is the approximation of the solution at the k-th iteration, $k \in \mathbb{N}$, α is a regularization parameter, and ψ is a damping factor.

The condition $\|A(z^k) - F\|/\|F\| < \varepsilon$ for some sufficiently small ε is taken as a termination criterion.

3.2   Regularized Linearized Conjugate Gradient Method

Let us construct the regularized variant of the linearized conjugate gradient method [3] by using the Tikhonov regularization, replacing Eq. (3) with

$$S_\alpha(z) = A'(z)^T(A(z) - F) + \alpha(z - z^0) = 0, \tag{5}$$

where z^0 is the initial approximation of the solution, and α is the regularization parameter.

Let us construct a process in the form

$$z^{k+1} = z^k + \psi \, \chi^k \, p^k,$$

where

$$p^k = S_\alpha(z^k) + \beta^k p^{k-1},$$
$$p^0 = S_\alpha(z^0),$$
$$\beta^k = \max \left\{ \frac{\langle S_\alpha(z^k), (S_\alpha(z^k) - S_\alpha(z^{k-1})) \rangle}{\|S_\alpha(z^{k-1})\|^2}, 0 \right\}.$$

To find an approximate expression for the step size χ^k, we use a linear approximation of the operator A:

$$\chi_k \approx \arg\min_\chi \left\{ \|A(z^k) + A'(z^k)(z^k + \chi p^k - z^k) - F\|^2 \right.$$
$$\left. + \alpha \|z^k + \chi p^k - z^0\|^2 \right\}$$
$$= \arg\min_\chi \left\{ \|(A(z^k) - F)\|^2 + 2\chi \langle A'(z^k)^T(A(z^k) - F), p^k \rangle \right.$$
$$+ \chi^2 \|A'(z^k)p^k\|^2 + \alpha \|z^k - z^0\|^2$$
$$\left. + 2\alpha\chi \langle (z^k - z^0), p^k \rangle + \alpha\chi^2 \|p^k\|^2 \right\}$$
$$= \arg\min_\chi \left\{ \|(A(z^k) - F)\|^2 + \alpha \|z^k - z^0\|^2 \right.$$
$$\left. + \chi^2 \|A'(z^k)p^k\|^2 + \alpha\chi^2 \|p^k\|^2 + 2\chi \langle S_\alpha, p^k \rangle \right\}.$$

Using the necessary condition for minimum,

$$\frac{\partial}{\partial \chi}\Big(\|(A(z^k) - F)\|^2 + \alpha\,\|z^k - z^0\|^2 + \chi^2\,\|A'(z^k)p^k\|^2$$
$$+ \alpha\chi^2\,\|p^k\|^2 + 2\chi\langle S_\alpha, p^k\rangle\Big) = 0;$$
$$2\chi\,\|A'(z^k)p_k\|^2 + 2\alpha\chi\,\|p^k\|^2 + 2\langle S_\alpha, p^k\rangle = 0,$$

we obtain

$$\chi^k \approx \widetilde{\chi^k} = -\frac{\langle p^k, S_\alpha(z^k)\rangle}{\|A'(z^k)p^k\|^2 + \alpha\,\|p^k\|^2}.$$

Thus, the regularized linearized conjugate gradient method (RLCGM) takes the following form:

$$z^{k+1} = z^k - \psi\frac{\langle p^k, S_\alpha(z^k)\rangle}{\|A'(z^k)p^k\|^2 + \alpha\,\|p^k\|^2}p^k,$$
$$p^k = S_\alpha(z^k) + \beta^k p^{k-1},$$
$$p_0 = S_\alpha(z^0),$$
$$\beta_k = \max\left\{\frac{\langle S_\alpha(z^k), (S_\alpha(z^k) - S_\alpha(z^{k-1}))\rangle}{\|S_\alpha(z^{k-1})\|^2}, 0\right\},$$
$$S_\alpha(z) = A'(z)^T(A(z) - F) + \alpha(z - z^0),$$

$$(6)$$

where z^0 is the initial approximation of the solution, z^k is the approximation of the solution at the k-th iteration, $k \in \mathbb{N}$, α is a regularization parameter, and ψ is a damping factor.

3.3   Modified Regularized Linearized Conjugate Gradient Method

Let us modify the RLCGM method (6) using the same idea of calculating the Jacobian matrix only at a fixed point z^0. As a result, we obtain the modified regularized linearized conjugate gradient method (MRLCGM):

$$z^{k+1} = z^k - \psi\frac{\langle p^k, S_\alpha^0(z^k)\rangle}{\|A'(z^0)p^k\|^2 + \alpha\,\|p^k\|^2}p^k,$$
$$p^k = S_\alpha^0(z^k) + \beta^k p^{k-1},$$
$$p^0 = S_\alpha^0(z^0),$$
$$\beta^k = \max\left\{\frac{\langle S_\alpha^0(z^k), (S_\alpha^0(z^k) - S_\alpha^0(z^{k-1}))\rangle}{\|S_\alpha^0(z)^{k-1}\|^2}, 0\right\},$$
$$S_\alpha^0(z) = A'(z^0)^T(A(z) - F) + \alpha(z - z^0).$$

$$(7)$$

3.4 Hybrid Regularized Linearized Conjugate Gradient Method

Let us construct the hybrid method by recalculating the Jacobian matrix at each j-th iteration. This method should have slightly higher speed of convergence than the modified method.

$$z^{k+1} = z^k - \psi \frac{\langle p^k, S_\alpha^k(z^k) \rangle}{\left\| \widetilde{A_k} p^k \right\|^2 + \alpha \left\| p^k \right\|^2} p^k,$$

$$p^k = S_\alpha^k(z^k) + \beta^k p^{k-1},$$

$$p^0 = S_\alpha^k(z^0),$$

$$\beta^k = \max \left\{ \frac{\langle S_\alpha^k(z^k), (S_\alpha^k(z^k) - S_\alpha^k(z^{k-1})) \rangle}{\| S_\alpha^k(z)^{k-1} \|^2}, 0 \right\}, \tag{8}$$

$$S_\alpha^k(z) = \widetilde{A_k}^T (A(z) - F) + \alpha(z - z^0),$$

$$\widetilde{A} = \underbrace{A'(z^0), A'(z^0), ...,}_{j} \underbrace{A'(z^j), A'(z^j), ...,}_{j} \underbrace{A'(z^{2j}), A'(z^{2j}), ...,}_{j} ...$$

4 Matrix Structure and Storage Method

Storing the matrix A' can be very memory consuming for large grids; hence, it is worthwhile investigating the matrix structure to optimize the storage method.

Let us assume that A' is a block matrix. Then, its elements can be defined as

$$a_{k,p,l,q} = a_{(k-1)M+p,(l-1)M+q}$$
$$= f \Delta\sigma \Delta x \Delta y \left(\frac{-z_{(k-1)M+p}}{((x_k - x_l)^2 + (y_p - y_q)^2 + z_{(k-1)M+p}^2)^{3/2}} \right),$$

or

$$a_{k,p,l,q} = a_{(k-1)M+p,(l-1)M+q}$$
$$= f \Delta\sigma \Delta x \Delta y \left(\frac{(x_k - x_l)^2 + (y_p - y_q)^2 - 2z_{(k-1)M+p}}{((x_k - x_l)^2 + (y_p - y_q)^2 + z_{(k-1)M+p}^2)^{5/2}} \right),$$

where $k, l = 1..M$ are the block indices and $p, q = 1..N$ are the indices of the elements inside each block.

Apparently, when $z^0 \equiv h$, the matrix elements depend only on the terms $(x_k - x_l)^2 + (y_p - y_q)^2$.

Note that

$$(y_{p+1} - y_{q+1})^2 = (y_p + \Delta y - y_q - \Delta y)^2 = (y_p - y_q)^2,$$
$$(x_{k+1} - x_{l+1})^2 = (x_k + \Delta x - x_l - \Delta y)^2 = (x_k - x_l)^2.$$

The former equation means that $p = q \Rightarrow a_{k,p,l,q} = a_{k,p+1,l,q+1}$; *i.e.* in each block, each descending diagonal from left to right is constant. The latter equation means that $k = l \Rightarrow a_{k,p,l,q} = a_{k+1,p,l+1,q}$; *i.e.* each diagonal of blocks is constant as well. In other words, the matrix A' is symmetric Toeplitz-block-Toeplitz. Its scheme structure is shown in Fig. 3.

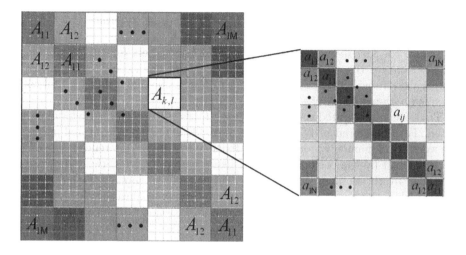

Fig. 3. Matrix structure.

An obvious approach is to store only the first row of this matrix. Then each subsequent row can be obtained by the following procedure:

(1) shifting the rows of elements inside each block rightwise by one element and complementing each row from the left with the symmetrically positioned element;

(2) shifting the entire row of blocks rightwise by one block and complementing it with the symmetrically positioned block.

This method requires only $O(MN)$ memory to store an $MN \times MN$ matrix.

5   Parallel Implementation and Automatic Parameter Adjustment For GPUs

The parallel algorithms were implemented on NVIDIA M2090 GPUs using the CUDA technology.

Note that storing the Jacobian matrix for a $2^9 \times 2^9$ grid takes 2 MB when using the optimized method, while the full storage takes more than 512 GB.

In the unmodified methods, we use the on-the-fly calculation of the elements of the Jacobian matrix; *i.e.*, the value of the matrix element is computed when calling this element without storing it in memory.

The costliest operation is computing the values of the integral operator and its Jacobian matrix. It consists of four nested loops. When using multiple GPUs, two outer loops are distributed to the GPUs, and two inner loops are executed on each GPU. The CPU transfers data between the host memory and GPUs, and calls the kernel functions.

The adjustment of the kernel execution parameters for the grid size is an important problem. The CUDA Occupancy API helps to estimate the required parameters using heuristics methods. This API has an essential drawback: it works only with one-dimensional grids.

To use the Occupancy API for generating parameters for two-dimensional grids, which are the ones required in our problem, one should apply either a complicated transformation of parameters or a fake one-dimensional kernel with the same parameters.

Thus, we implemented the original method for automatic adjustment of parameters. For the reference 128×128 grid and M2090 GPUs, the optimal parameters were found manually. For grid sizes divisible by 128, the reference parameters are multiplied by the coefficient. When using multiple GPUs, the x dimension is divided by the number of GPUs; *i.e.* the number of threads in the block is reduced while the number of blocks in the grid remains constant.

This imposes some constraints on the input data and GPUs configuration:

– grid size must be divisible by 128 (128, 256, 512, 1024 ...);
– GPUs number must be a power of 2 (1, 2, 4, 8, ...).

6   Numerical Experiments

We consider the model problem of finding the interface between two layers.

Figure 4 shows the model gravitational field $\Delta g(x, y, 0)$. This field was obtained by solving the direct gravimetry problem using the surface z^* shown in Fig. 5 with the asymptotic plane $H = 6$ km and the density jump $\Delta \sigma = 0,1$ g/cm^3.

A noise with a 10% peak value was added to the gravitational field.

The problem was solved on the Uran supercomputer nodes with eight NVIDIA Tesla M2090 GPUs by five methods:

– regularized linearized steepest descent method (RLSDM) (see [11]);
– regularized linearized conjugate gradient method (RLCGM) (6);
– modified regularized linearized steepest descent method (MRLSDM) (4);
– modified regularized linearized conjugate gradient method (MRLCGM) (7);
– hybrid regularized linearized conjugate gradient method (HRLCGM) (8).

The reconstructed interface z is shown in Fig. 6.

The termination criterion for all five methods was $\varepsilon = 0.0087$. The parameters $\psi = 1$ and $\alpha = 0.1$ were used for all five methods.

The interfaces reconstructed by five methods are similar. The relative error of the solution is $\delta = \|z - z^*\|/\|z^*\| < 0.01$.

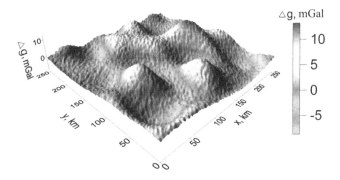

Fig. 4. Model gravitational field.

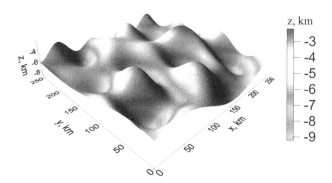

Fig. 5. Original surface z^*.

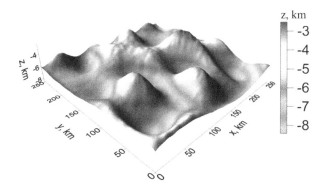

Fig. 6. Reconstructed surface z.

Table 1. Comparison of methods

Method	Number of iterations	Execution time, min
RLSDM (see [11])	60	25
RLCGM (6)	50	23
MRLSDM (4)	70	18
MRLCGM (7)	70	18
HRLCGM (8)	70	18.5

Table 1 shows the average execution times for 10 runs of the five methods for a 512×512 grid.

Speedup and efficiency coefficients are used to analyse the scaling of the parallel algorithms. The speedup is $S_m = T_1/T_m$, where T_1 is the execution time of the program running on one GPU, and T_m is the execution time for m GPUs. The efficiency is $E_m = S_m/m$. The ideal values are $S_m = m$ and $E_m = 1$, but the real values are lower because of the overhead.

Table 2 shows the average execution times, speedup and efficiency for the MRLCGM method on a 512×512 grid for various number of GPUs.

The experiments showed that the constructed modified algorithms are very effective. New algorithms are more economical in terms of operations and time at each iteration step. The modification reduced the time required to perform one iteration by approximately 45%. For this model problem, the hybrid method and the modified methods showed the same performance in terms of iterations number. The MRLCG method shows the best results in terms of execution time. The parallel algorithms demonstrate a good scaling; the efficiency is more than 90% for eight GPUs.

Table 2. Speedup and efficiency of the parallel MRLCGM algorithm

Number m of GPUs	Execution time, min	Speedup S_m	Efficiency E_m
1	18	1	1
2	9.5	1.88	0.94
4	5	3.75	0.93
8	2.5	7.24	0.9

7   Conclusions

We constructed a regularized variant of the linearized conjugate gradient method based on the Tikhonov regularization, a modified variant of this method and a hybrid conjugate gradient method.

We developed parallel algorithms on the basis of the modified regularized linearized steepest descent method and the methods constructed in this paper.

The parallel algorithms were implemented on GPUs using the CUDA technology. A model problem with large grids was solved. The parallel algorithms demonstrated good scaling.

References

1. Akimova, E.N., Martyshko, P.S., Misilov, V.E.: Algorithms for solving the structural gravity problem in a multilayer medium. Dokl. Earth Sci. **453**(2), 1278–1281 (2013). doi:10.1134/S1028334X13120180
2. Akimova, E.N., Martyshko, P.S., Misilov, V.E.: Parallel algorithms for solving structural inverse magnetometry problem on multicore and graphics processors. In: Proceedings of 14th International multidisciplinary scientific GeoConference SGEM 2014, vol. 1(2), pp. 713–720 (2014)
3. Akimova, E.N., Martyshko, P.S., Misilov, V.E.: A fast parallel gradient algorithm for solving structural inverse gravity problem. In: AIP Conference Proceedings, vol. 1648, 850063 (2015). doi:10.1063/1.4913118
4. Akimova, E.N., Misilov, V.E.: A fast componentwise gradient method for solving structural inverse gravity problem. In: Proceedings of 15th International Multidisciplinary Scientific GeoConference SGEM 2015, vol. 3(1), pp. 775–782 (2015)
5. Bakushinskiy, A., Goncharsky, A.: Ill-Posed Problems: Theory and Applications. Springer Science & Business Media, Netherlands (1994). doi:10.1007/978-94-011-1026-6
6. Malkin, N.R.: On solution of inverse magnetic problem for one contact surface (the case of layered masses). DAN SSSR, Ser. A, vol. 9, pp. 232–235 (1931)
7. Martyshko, P.S., Akimova, E.N., Misilov, V.E.: Solving the structural inverse gravity problem by the modified gradient methods. Izv. Phys. Solid Earth **52**(5), 704–708 (2016). doi:10.1134/S1069351316050098
8. Martyshko, P.S., Fedorova, N.V., Akimova, E.N., Gemaidinov, D.V.: Studying the structural features of the lithospheric magnetic and gravity fields with the use of parallel algorithms. Izv. Phys. Solid Earth **50**(4), 508–513 (2014). doi:10.1134/S1069351314040090
9. Martyshko, P.S., Prutkin, I.L.: Technology of depth distribution of gravity field sources. Geophys. J. **25**(3), 159–168 (2003)
10. Numerov, B.V.: Interpretation of gravitational observations in the case of one contact surface. Doklady Akad. Nauk SSSR, pp. 569–574 (1930)
11. Vasin, V.V.: Irregular nonlinear operator equations: Tikhonov's regularization and iterative approximation. Inverse Ill-Posed Prob. **21**(1), 109–123 (2013). doi:10.1515/jip-2012-0084
12. Vasin, V.V.: Modified steepest descent method for nonlinear irregular operator equations. Dokl. Math. **91**(3), 300–303 (2015). doi:10.1134/S1064562415030187

The High-Performance Parallel Algorithms for the Numerical Solution of Boundary Value Problems

Vadim Volokhov[1], Sergey Martynenko[1,2](✉), Pavel Toktaliev[1,2],
Leonid Yanovskiy[1,2], Dmitriy Varlamov[1], and Alexander Volokhov[1]

[1] Institute of Problems of Chemical Physics of RAS, Academician Semenov avenue 1,
Chernogolovka, Moscow Region 142432, Russian Federation
{vvm,Martynenko,Toktaliev,Yanovskiy,dima,vav}@icp.ac.ru
[2] Central Institute of Aviation Motors n.a. P.I.Baranov, Aviamotornaya, 2,
Moscow 111116, Russian Federation
http://www.icp.ac.ru/en/

Abstract. The boundary and the initial boundary value problems form the basis of numerous mathematical models. Ultimately, the discrete (linearized) boundary and the initial boundary value problems are reduced to the systems of linear algebraic equations with sparse and ill-conditioned coefficient matrix. In modern applications (such as computational fluid dynamics) the number of equations in the system can reach about 10^{12} and higher. Just the numerical solution of such systems requires significant computational effort, so an actual problem of modern computational mathematics is working-out, theoretical analysis and testing of high-performance parallel algorithms. The article discusses algebraic, geometric and combined ways to formation of the parallel algorithms. In this work we presented advantages and disadvantages of each ways, the estimate of parallelism's acceleration and efficiency, the comparison of volume of computational work compared with the optimal sequential algorithm, and the results of computational experiments. The peculiarities of parallel algorithms' implementation by using of software and hardware structures for parallel programming were discussed in this work.

Keywords: Initial boundary value problems · Parallel algorithms

1 Introduction

A promising and challenging trend in numerical simulation and scientific computing is the use of parallelism in numerical algorithms. The background to this trend

---

The activity is a part of the research work "Supercomputer simulation of physical and chemical processes in the high-speed direct-flow propulsion jet engine of the hypersonic aircraft on solid fuels" supported by Russian Science Foundation (project no. 15-11-30012).

L. Sokolinsky and M. Zymbler (Eds.): PCT 2017, CCIS 753, pp. 156–165, 2017.
DOI: 10.1007/978-3-319-67035-5_12

is the fact that most high performance computers are now parallel systems [1]. Following [2], there are two common measures of parallelism that we now introduce:

Definition 1. The speedup $\bar{S}$ and the efficiency $\bar{\mathcal{E}}$ of a parallel algorithm is

$$\bar{S} = p\bar{\mathcal{E}} = \frac{T(1)}{T(p)}, \tag{1}$$

where $T(1)$ is an execution time for a single processor and $T(p)$ is an execution time using p processors.

The execution time $T(p)$ can be represented as

$$T(p) = \frac{1}{p}T(1) + T^*(p),$$

where $T^*(p)$ is time for the communication and data transfer between the processors. Assuming that the execution for a single processor takes too much time compared to time for the communication and data transfer

$$T(1) \gg pT^*(p) \quad \text{or} \quad T^*(p) \rightarrow 0,$$

we have

$$\bar{S} = p\bar{\mathcal{E}} = \frac{T(1)}{\frac{1}{p}T(1) + T^*(p)} \rightarrow p, \quad \text{for} \quad T^*(p) \rightarrow 0.$$

It results in almost full parallelism, i.e.

$$\bar{\mathcal{E}} \rightarrow 1 \quad \text{for} \quad T^*(p) \rightarrow 0$$

independently on p. Often numerically inefficient algorithms give (much) better parallel efficiencies than more sophisticated and numerically efficient ones since those are much easier to parallelize [1]. In order to avoid this paradox (high parallel efficiency for numerically inefficient methods and low parallel efficiency for numerically efficient ones), the following measures of parallelism can be used:

Definition 2. The speedup $\tilde{S}$ and the efficiency $\tilde{\mathcal{E}}$ of a parallel algorithm is

$$\tilde{S} = p\tilde{\mathcal{E}} = \frac{\tilde{T}(1)}{T(p)}, \tag{2}$$

where $\tilde{T}(1)$ is an execution time for a single processor processor of fastest sequential algorithm and $T(p)$ is an execution time using p processors.

Assume that some classical multigrid method (for example, V-cycle [1]) is used as the fastest sequential solver. Remember that the multigrid methods are typically optimal in the sense that the number of arithmetic operations needed to solve a (discrete) problem is proportional to the number N of unknowns in the problem considered. We can assume that $\tilde{T}(1) = C_{\mathrm{v}} N \log \varepsilon$, where $\log \varepsilon$ accounts the discretization accuracy. On the other hand, some parallel method can has close-to-optimal algorithmic complicity, i.e. $T(1) = CN \log N \log \varepsilon$. In this case

$$T(p) = \frac{1}{p} T(1) + T^*(p) = \frac{1}{p} CN \log N \log \varepsilon + T^*(p),$$

and the speedup $\tilde{S}$ and the efficiency $\tilde{\mathcal{E}}$ (2) become

$$\tilde{S} = p\tilde{\mathcal{E}} = \frac{C_{\mathrm{v}} N \log \varepsilon}{\frac{1}{p} CN \log N \log \varepsilon + T^*(p)}.$$

If the solver with close-to-optimal algorithmic complicity is highly parallelizable $(T^*(p) \rightarrow 0)$, we have

$$\tilde{S} = p\tilde{\mathcal{E}} \rightarrow p \frac{C_{\mathrm{v}}}{C} \frac{1}{\log N}.$$

It means that $\tilde{\mathcal{E}} = O(1/\log N)$ independently on p. It is clear that $\tilde{\mathcal{E}} \rightarrow 0$ as $N \rightarrow +\infty$.

As a rule, optimal solver cannot be parallelized efficiently in practice, but efficiency $\tilde{\mathcal{E}}$ of a parallel algorithm with close-to-optimal complicity is not high enough. This paper describes two approaches to development of efficient parallel multigrid algorithms. The first relates to a decomposition of the given problem into a number of subproblems on subdomains with an overlap. This (algebraic) approach is effective if the computational grid is sufficiently fine. The second relates to a decomposition of the given problem into a number of subproblems on the domain without an overlap. This (geometric) approach is effective for coarse grids.

In this paper we discuss different approaches to parallel computing. Developed parallel algorithms are applied to simulation of the hydrocarbon fuel flows in cooling systems of the ramjet engines. The mathematical model is developed for prediction of coke deposit formation on solid walls and for study an opportunity to control the deposit phenomenon. Presented mathematical model is based on the Navier-Stokes and Maxwell equations in the electric potential form.

2   Decomposition of the Discrete Problems

Grid partitioning is a natural approach for parallel solution of the discrete boundary value problems on parallel computers. In this approach, the original grid is split into p subgrids, such that p available processors can jointly solve the underlying discrete problem. Figure 1 represents the 2D grid partitioning on four overlapped parts. Each subgrid (and the corresponding "subproblem", i.e. the equations and the unknowns located in the subgrid) is assigned to a different process such that each process is responsible for the computations in its part of the domain. The grid partitioning idea is widely independent of the particular boundary value problem to be solved and of the particular parallel architecture to be used. It is applicable to general d-dimensional domains, structured and unstructured grids, linear and nonlinear equations and systems of partial differential equations [1].

2.1   Algebraic Decomposition

The grid partitioning shown on Fig. 1 results in four independent subproblems. The subproblems can be solved in parallel. After that the data exchange in the overlapped areas as shown on Fig. 2. Unfortunately, convergence rate of such parallel algorithms will be very slow. The convergence rate of iterative algorithms for solving boundary value problems depends on the speed of information transfer about the boundary conditions into the region. The grid partitioning poses barriers on way of the information transfer and deteriorates the convergence rate.

The damped Jacobi iterations can be used as a smoother in parallel algorithms. There are two obvious reasons why an algorithm and/or a parallel system may perform unsatisfactorily: load imbalance and communication overhead [1]. Load imbalance means that some processors have to do much more work than most of the others. In this case, most of the processors have to wait for others to finish their computation before a data exchange can be carried out. Communication overhead means that the communication and data transfer between the processors takes too much time compared to the effective computing time.

Communication overhead on the very coarse grids (as compared to the time for arithmetic computations) increases and may finally dominate over the arithmetic work. This can result in a significant loss of efficiency for the overall parallel multigrid algorithm.

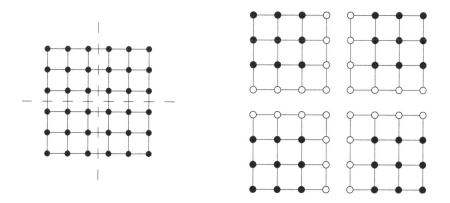

Fig. 1. Example of 2D grid partitioning.

Note that the grid partitioning is often considered as a geometric decomposition of the computational grid. Discrete boundary value problems can be split into p subproblems in algebraic manner.

2.2 Geometric Decomposition

To overcome problem on the communication overhead and load imbalance (i.e. processor idleness) on very coarse grids, the geometric decomposition has been proposed in [3]. The decomposition is based on the grid partitioning without an overlap. It result in 3^{dl^*} independent subproblems, where $d = 2, 3$ is dimension of the problem and l^* is parallelization depth or serial number of the grid level for switching from one approach of parallelization to another.

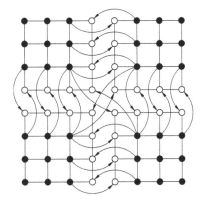

Fig. 2. Scheme of data exchange in the overlapped areas.

Application of algebraic decomposition on finer grids $(0 \leqslant l < l^*)$ and geometric decomposition on coarser grids $(l \geqslant l^*)$ makes it possible to construct highly efficient parallel multigrid [3].

In this part of work, we implement numerical multigrid technique to problem of coke precursors formation in heated channel. Series of simplifications were used for the quantitative description of the main characteristics of the process of coke precursor deposition on the solid surfaces of the channel taking into account heat exchange and liquid-phase oxidation in the channel using a model substance imitating the real hydrocarbon fuel [4]. Pentane was chosen as a model fuel. The geometry of the considered region of the fuel is shown in Fig. 4b. The region considered consists of the channel 200 mm in length, 10 mm in width, and 1 mm in depth and of the electrodes mounted in the medium part on the inner bottom and upper walls of the channel. It was assumed that near-surface reactions that occur on the smooth walls do not change the geometry of the walls and their thermal state. The constant in time potential difference was fed to the electrodes.

2.3   Governing Equations

Flow equations. The system of Reynolds equations for a multispecies system closed by the Spallart-Almaras turbulence model was used as a flow model. The system of the Navier-Stokes equations with allowance for the electromagnetic effect and the presence of neutral and charged particles were applied in the following form (the equations of continuity, momentum, and energy are presented):

$$\frac{\partial \rho}{\partial t} + \nabla(\rho U) = 0,$$

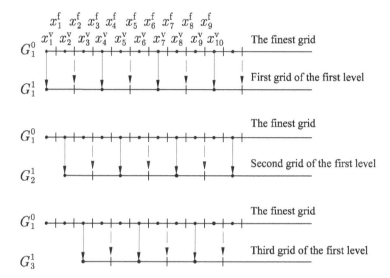

Fig. 3. Example of 1D grid partitioning without an overlap.

$$\frac{\partial(\rho U)}{\partial t} + \nabla(\rho U \times U) = -\nabla p + \nabla \tau + F, \qquad F = qE$$

$$C_p\left(\frac{\partial(\rho T)}{\partial t} + \nabla(\rho U T)\right) = \nabla(\lambda \nabla T) + \sum_i J_i C_{pi} \nabla T + \dot{Q} + \sum_{r=elastic} \varepsilon_r \dot{\omega}_r + \sum_{r=ions} \varepsilon_r \dot{\omega}_r,$$

where ρ is the density, U is the velocity vector, p is the pressure, τ is the tensor of viscous stresses, q is the volume charge, T is the temperature, and F is the Lorentz force (magnetic induction $B = 0$), E is the electric field intensity, λ is the thermal conductivity coefficient, C_p is the mixture heat capacity under constant pressure coefficient, C_{pi} is heat capacity under constant pressure coefficient of ith specie, $\varepsilon_r \dot{\omega}_r$ is heat effect and reaction rate of nonelectronic and electron elastic collisions reaction, respectively. The energy equation in the right part contains the terms of the source related to elastic collisions of the electronic species, ohmic heating of the ionic species, and the heat effect of the volume and surface reactions (see Fig. 3).

The source term related to the interaction of the ionic and neutral species ($\dot{Q}$) can be expressed by analogy to the ohmic heating

$$\dot{Q} = \sum_{q_i > 0} q_i n_i \mu_i \boldsymbol{E} \cdot \boldsymbol{E}.$$

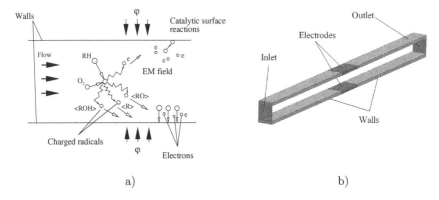

Fig. 4. Scheme of processes (a) and geometry (b).

To determine the concentrations of the species (ionic and neutral), we solved the transfer equations in the following form (with allowance for ion drift):

$$\frac{\partial(\rho Y_j)}{\partial t} + \nabla(\rho \boldsymbol{U} Y_j) = -\nabla \hat{\boldsymbol{J}}_i,$$

$$\hat{\boldsymbol{J}}_i = -\left(\rho D_{j,M} + \frac{\mu_t}{\mathsf{Sc}_t}\right) \nabla Y_j + \rho \boldsymbol{U}_d Y_j + \boldsymbol{J}_i^c,$$

where Y_j is the mass concentration of the jth species, $\hat{\boldsymbol{J}}_i$ is the mass flux, $D_{j,M}$ is the mass diffusion coefficient of the jth species in a mixture, μ_t is dynamic turbulent viscosity coefficient, $\mathsf{Sc}_t = 0.7$ is the Schmidt turbulent number, $\boldsymbol{U}_d$ is the ion drift velocity, and $\boldsymbol{J}_i^c$ is the correction mass flow necessary for the fulfillment of continuity equation for individual species. The transfer coefficients for individual species of the mixture are taken as polynomial temperature dependences. Maxwell equations were reduced to electrostatic form:

$$-\nabla \cdot \varepsilon \nabla \varphi = e \left(\sum_i q_i n_i - n_e\right),$$

$$\boldsymbol{E} = -\nabla \varphi,$$

where $\boldsymbol{E}$ is the electric field strength and ε is the electric permeability of the medium.

Chemical kinetic equations. The temperatures $T < 550\,\mathrm{K}$ were taken as the main temperature range for which the liquid-phase oxidation of hydrocarbon (LOH) fuel with dissolved oxygen predominates. The characteristic pressures in the system $P \sim 5\,\mathrm{MPa}$ corresponding to the conditions fulfilled in tracts of fuel systems. The simplified model based on degenerate chain branching characteristic of partial oxidation is used as a kinetic model of coke formation. The initial stage of chain formation is the interaction of the starting hydrocarbon with dissolved oxygen. In the absence of initiating additives, the rate of radical formation in the oxidation reactions is low. The degenerate branching character of the LOH reactions is usually related to the accumulation of hydroperoxides that easily decompose at the O–O bond, resulting in an additional chain nucleation. The chain mechanism can be presented as follows:

$$RH + O_2 \xrightarrow{k_0} \dot{R} + HO_2, \qquad 2R\dot{H} + O_2 \xrightarrow{k_\infty} 2\dot{R} + H_2O_2,$$

$$\dot{R} + O_2 \xrightarrow{k_1} R\dot{O}_2, \qquad R\dot{O}_2 + RH \xrightarrow{k_2} ROOH + \dot{R},$$

$$ROOH \xrightarrow{k_3} R\dot{O} + \dot{O}H, \qquad ROOH + RH \xrightarrow{k_4} R\dot{O} + \dot{R} + H_2O,$$

$$2ROOH \xrightarrow{k_5} R\dot{O}_2 + R\dot{O} + H_2O, \qquad \dot{R} + ROOH \xrightarrow{k_6} ROH + R\dot{O},$$

$$R\dot{O} + RH \xrightarrow{k_7} ROH + \dot{R}, \qquad H_2O + RH \xrightarrow{k_8} H_2O_2 + \dot{R},$$

$$\dot{O}H + RH \xrightarrow{k_9} H_2O + \dot{R}, \qquad \dot{R} + H_2O_2 \xrightarrow{k_{10}} ROH + OH,$$

$$\dot{R} + RO_2 \xrightarrow{k_{11}} ROOR, \qquad R\dot{O}_2 + R\dot{O}_2 \xrightarrow{k_{12}} ROH + RCOR_2 + O_2,$$

$$\dot{R} \xrightarrow{k_w}, \qquad R\dot{O} \xrightarrow{k_w},$$

$$R\dot{O}_2 \xrightarrow{k_w}, \qquad \dot{O}H \xrightarrow{k_w},$$

$$H_2O \xrightarrow{k_w}, \qquad H_2O_2 \xrightarrow{k_w}.$$

The radicals in the equations are designated as point above symbol, whereas stoichiometric coefficients of the electronic species are not indicated. The reaction constants k_1–k_{12} and k_w are taken from the work [5]. The mechanism presented covers both directions of chain development and includes 14 species and 17 reactions. However, the mechanism ignores the acceleration of solid deposit formation due to the acceleration of bimolecular reactions on the wall in the presence of primary deposits. High-molecular-weight products (HMP) similar to the products of the volume recombination of radicals (alcohols ROH, ketones R–R, aldehydes ROOR) are formed due to the bimolecular recombination on the wall. It was accepted for simplicity that the three first reactions of radical chain termination on the wall are bimolecular and directly result in the formation of coke precursors corresponding, in turn, by weight to solid deposits.

Numerical procedure and results. As first stage of implementing multigrid technique hybrid calculations was adopted. Transport equations for species, energy and electronic temperature were solved without multigrid technique, but pressure correction elliptic equation in SIMPLE procedure and electric potential equation was solved separately on hierarchical numerical mesh. Details of numerical procedure for multigrid part of solution can be found in [6].

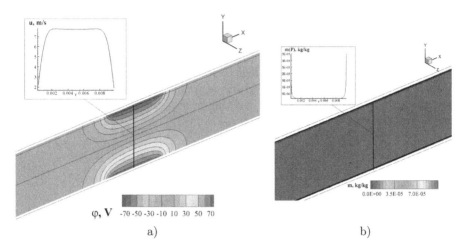

Fig. 5. Electric potential field and profile of axial velocity (a), mass fraction of high-molecular products (b) with profile across the channel, $Re = 4000, \Delta\varphi = 150\,V$, adiabatic walls, $T_{in} = 600\,K$, $t = 10$ s.

For numerical experiments shared memory system 2*Intel Xeon E5-2630V4 10 cores 2.2 GHz, RAM DDR4 ECC 128 GB with QPI (up to 38.4 Gb/s) was used. Charged volumes with different mobilities are formed in the system when considering the block of chemical transformations in addition to the equations of flow and electrostatic field. Effect of the Lorentz force on the charged volumes results in convection and diffusion and in the mechanism of displacement of liquid volumes to the region of highest electrostatic field strengths. The calculations for $t \sim 3600$ (t is time), $T_{in} = 600K$ (T_{in} is the temperature of the mixture at the inlet of the channel), and $Re = 4000$ showed that the formation of a discrete charged phase as ions at the presented parameters of the flow exerts almost no effect on the global strength distribution. Therefore, further, to shorten the necessary resources, we accepted the simplification of the steady regime: the regime was considered steady if the total mass concentration of the reaction products at the outlet of the channel at consecutive time moments did not exceed 1%. The field of the weight fraction of HMP, more exactly, the sum of weight fractions of alcohols ROH and aldehydes ROOR formed due to the reactions in the channel volume, is presented in Fig. 5b.

The profile of the chosen weight concentration over the channel width along the medium line is also shown. As can be seen from Fig. 5 the main amount of HMP is formed at the presented time moment in the boundary layer near the walls, and the characteristic value of the fraction for this moment is 10^6 kg kg^{-1}.

Table 1. Scalabilities of cases

Cores	Standard approach, s	Multigrid approach, s
1	30.1	26.4
3	15.2	13.0
5	10.1	9.5
9	5.6	4.2

Cases scalabilities for multigrid and non-multigrid cases are represented in Table 1 for 3 mln. cells mesh. As can be seen multigrid approach gives better results, but, performance are depend strongly on multiplicity of the cores number.

3   Conclusions

The presented method of numerical solution of boundary value problems has close to optimal algorithmic complicity and it can effectively parallelize computations. Currently, the main difficulty is to expand the multigrid method on computational girds of arbitrary mesh topology.

References

1. Trottenberg, U., Oosterlee, C.W., Schüller, A.: Multigrid. Academic Press, London (2001)
2. Ortega, J.: Introduction to Parallel and Vector Solution of Linear Systems. Plenum Press, New York (1988)
3. Martynenko, S.I.: Multigrid Technique: Theory and Applications. FIZMATLIT, Moscow (2015). (in Russian)
4. Toktaliev, P.D., Martynenko, S.I., Yanovskiy, L.S., Volokhov, V.M., Volokhov, A.V.: Features of model hydrocarbon fuel oxidation for channel flow in the presence of electrostatic field. RAS Rep. Chem. Ser. **65**(8), 2011–2017 (2016). (in Russian)
5. Emmanuel, N.M., Zaikov, G.E., Maizus, Z.K.: Medium Role in Chain Reactions of Organic Components Oxidation. Nauka, Moscow (1973). (in Russian)
6. Martynenko, S.I., Volokhov, V.M., Toktaliev, P.D.: Pseudomultigrid Gauss-Seidel method for large scale and high performance computing. In: European Congress on Computational Methods in Applied Sciences and Engineering, vol. 1, pp. 1237–1259 (2016)

Supercomputer Simulation

Complex of Models, High-Resolution Schemes and Programs for the Predictive Modeling of Suffocation in Shallow Waters

Aleksandr Sukhinov[1], Albert Isayev[2], Alla Nikitina[2(✉)],
Aleksandr Chistyakov[1], Vladimir Sumbaev[3], and Alena Semenyakina[4]

[1] Don State Technical University, Rostov-on-Don, Russia
sukhinov@gmail.com, cheese_05@mail.ru
[2] Polytechnic Institute – Branch of Don State Technical University in Taganrog,
Rostov-on-Don, Russia
alis_11111@mail.ru, nikitina.vm@gmail.com
[3] South Federal University, Rostov-on-Don, Russia
valdec4813@mail.ru
[4] Kalyaev Scientific Research Institute of Multiprocessor Computer Systems
at Southern Federal University, Taganrog, Russia
j.a.s.s.y@mail.ru

Abstract. The paper covers the development and research of mathematical models of the algal bloom, causing suffocations in shallow waters on the basis of modern information technologies and computational methods, by which the accuracy of predictive modeling of the ecology situation of coastal systems is increased. Developed model takes into account the follows: the water transport; microturbulent diffusion; gravitational sedimentation of pollutants and plankton; nonlinear interaction of plankton populations; biogenic, temperature and oxygen regimes; influence of salinity. The computational accuracy is significantly increased and computational time is decreased at using schemes of high order of accuracy for discretization of the model. The practical significance is the software implementation of the proposed model, the limits and prospects of it practical use are defined. Experimental software was developed based on multiprocessor computer system and intended for mathematical modeling of possible progress scenarios of shallow waters ecosystems on the example of the Azov Sea in the case of suffocation. We used decomposition methods of grid domains in parallel implementation for computationally laborious convection-diffusion problems, taking into account the architecture and parameters of multiprocessor computer system. The advantage of the developed software is also the use of hydrodynamical model including the motion equations in the three coordinate directions.

This paper was partially supported by the task No. 2014/174 of the basic part of state task of Russian Ministry of education, the program of fundamental researches of the Presidium of RAS No. 43 Fundamental problems of mathematical modeling, and partial financial support of RFFR for projects No. 15-01-08619, No. 15-07-08626, No. 15-07-08408, No. 16-37-00129.

L. Sokolinsky and M. Zymbler (Eds.): PCT 2017, CCIS 753, pp. 169–185, 2017.
DOI: 10.1007/978-3-319-67035-5_13

Keywords: Mathematical model · Water bloom · Suffocation · Phytoplankton · Multiprocessor computer system · Computational experiments

1   Introduction

The Azov Sea is the great shallow water in stretch. Such waters are suffered the great anthropogenic influence. However, most of them is the unique ecological systems of fish productivity. The biogenic matters are entered in the shallow waters with the river flows which causing the growth of the algae – water bloom. The suffocation periodically occurs in shallow waters in summer. Because there is the significant decrease of dissolved oxygen in them, consumed in the decomposition of organic matter, due to the high temperature. The fish is suffering the oxygen starvation and the mass dying of suffocation.

The results of satellite monitoring of the Earth are used in this paper to control the quality modeling of processes of hydrodynamics and biological kinetics [1,2]. The satellite monitoring data of the Azov Sea, obtained by SRC Planeta, are given in Fig. 1 [3]. The analysis of satellite data reveals the water areas of suffocations.

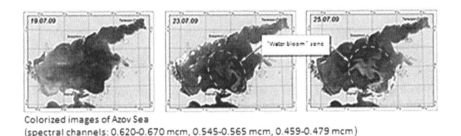

Colorized images of Azov Sea
(spectral channels: 0.620-0.670 mcm, 0.545-0.565 mcm, 0.459-0.479 mcm)

Fig. 1. The wide areas of the water bloom in the Azov Sea.

The dynamics of the development of the water bloom is given in Fig. 2. It caused the suffocation in the south-eastern part of the Azov Sea on July 16, 2013.

Designations in Fig. 2 are follows: 1 – the removal of river or estuary water out of the sleeves and throats of the delta of the Kuban river; 2 – water bloom area; 3 – waters with higher degree of turbidity after the desintegration of the water bloom area by the winds and waves.

The 3D spatially heterogeneous mathematical model was desintegration performed for the reconstruction of the water bloom. This process caused the suffocation in the South-Eastern part of the Azov Sea in July 2013. The information about the wind velocity and direction in the Temryuk Bay in July 2016, provided the meteorological station in Kerch city (WMO_ID 33983) and shown in Fig. 3, was used for this model as input data.

Water temperature in the computational domain for the simulated time interval is shown in Fig. 4.

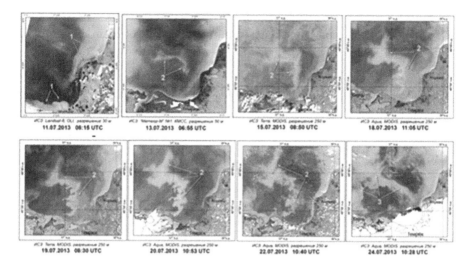

Fig. 2. Dynamic of the water bloom process in the south-eastern part of the Azov Sea.

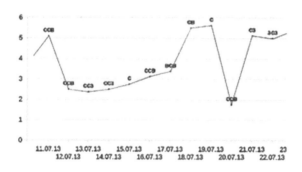

Fig. 3. Wind velocity and direction on July 16, 2013.

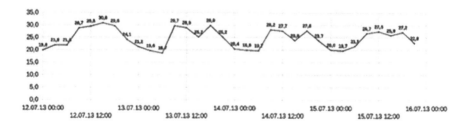

Fig. 4. Water temperature on July 16, 2013.

2 Hydrodynamic Mathematical Model

The Navier-Stokes motion equations are initial equations of hydrodynamics of shallow water:

$$u'_t + uu'_x + vu'_y + wu'_z = -\frac{1}{\rho}p'_x + \left(\mu u'_x\right)'_x + \left(\mu u'_y\right)'_y + \left(\nu u'_z\right)'_z + 2\Omega(v\sin\theta - w\cos\theta),$$

$$v'_t + uv'_x + vv'_y + wv'_z = -\frac{1}{\rho}p'_y + \left(\mu v'_x\right)'_x + \left(\mu v'_y\right)'_y + \left(\nu v'_z\right)'_z - 2\Omega u\sin\theta, \qquad (1)$$

$$w'_t + uw'_x + vw'_y + ww'_z = -\frac{1}{\rho}p'_z + \left(\mu w'_x\right)'_x + \left(\mu w'_y\right)'_y + \left(\nu w'_z\right)'_z + 2\Omega u\cos\theta + g\left(\rho_0/\rho - 1\right);$$

– continuity equation was written for the case of variable density:

$$\rho'_t + (\rho u)'_x + (\rho v)'_y + (\rho w)'_z = 0, \qquad (2)$$

where $u = \{u, v, w\}$ are velocity vector components; p is an excess pressure above the undisturbed fluid hydrostatic pressure; ρ is a density; Ω is an Earth's angular velocity; θ is an angle between the angular velocity vector and the vertical vector; μ,ν are horizontal and vertical components of turbulent exchange coefficient.

We consider the system of Eqs. (1) and (2) with the following boundary conditions:

– at the entrance (the mouth of Don and Kuban rivers):

$$u(x, y, z, t) = u(t), v(x, y, z, t) = v(t), p'_n(x, y, z, t) = 0, u_n(x, y, z, t) = 0,$$

– the lateral boundary (beach and bottom):

$$\rho_v\mu(u')_n(x, y, z, t) = -\tau_x(t), \quad \rho_v\mu(v')_n(x, y, z, t) = -\tau_y(t),$$

$$u_n(x, y, z, t) = 0, \quad p'_n(x, y, z, t) = 0,$$

– the upper boundary:

$$\rho\mu(u')_n(x, y, z, t) = -\tau_x(t), \quad \rho\mu(v')_n(x, y, z, t) = -\tau_y(t),$$

$$w(x, y, t) = -\omega - p'_t/\rho g, \ p'_n(x, y, t) = 0, \qquad (3)$$

– at the output (Kerch Strait):

$$p'_n(x, y, z, t) = 0, u'_n(x, y, z, t) = 0,$$

where ω is a liquid evaporation intensity; τ_x, τ_y are tangential stress components (Van-Dorn law); ρ_v is a suspension density.

Tangential stress components for free surface are in the form:

$$\tau_x = \rho_a C_p\left(|w|\right)w_x\,|w|\,,\tau_y = \rho_a C_p\left(|w|\right)w_y\,|w|\,,$$

where w is a wind velocity vector relative to the water; ρ_a is an atmosphere density,

$$C_p\left(x\right) = \begin{cases} 0.0088; & x < 6.6\ m/s \\ 0.0026; & x \geq 6.6\ m/s \end{cases} - \text{non-dimensional coefficient.}$$

Tangential stress components for bottom are in the form:

$$\tau_x = \rho C_p\left(|u|\right)u\,|u|,\ \tau_y = \rho C_p\left(|u|\right)v\,|u|\ .$$

We can define the coefficient of the vertical turbulent exchange with inhomogeneous depth on the basic of the measured velocity pulsation:

$$\nu = C_s^2 \Delta^2 \frac{1}{2}\sqrt{\left(\frac{\partial u}{\partial z}\right)^2 + \left(\frac{\partial v}{\partial z}\right)^2}, \tag{4}$$

where Δ is a grid scale; C_s is a non-dimensional empirical constant, defined on the basis of attenuation process calculation of homogeneous isotropic turbulence.

Grid method was used for solving the problem (1)–(3) [4]. The approximation in a time variable based on splitting schemes into physical processes [5–9] using the pressure correction method.

3 Mathematical Model of Water Bloom Processes of Shallow Waters

The spatially heterogeneous model of water bloom (WB) is described by equations:

$$S_{i,t} + u\frac{\partial S_i}{\partial x} + v\frac{\partial S_i}{\partial y} + (w - w_{gi})\frac{\partial S_i}{\partial x} = \mu_i \Delta S_i + \frac{\partial}{\partial z}\left(\nu_i \frac{\partial S_i}{\partial z}\right) + \psi_i\,. \tag{5}$$

(5) are equations of changes the concentration of impurities, index i indicates the substance type, S_i is the concentration of i-th impurity, $i = \overline{1,6}$; 1 is the total organic nitrogen (N); 2 are phosphates (PO_4); 3 is a phytoplankton; 4 is a zooplankton; 5 is a dissolved oxygen (O_2); 6 is a hydrogen sulfide (H_2S); u, v, w are components of water flow velocity vector; ψ_i is a chemical-biological source (drain) or a summand that describes the aggregation (clumping-declumping) if the corresponding component is a suspension.

The WB model takes into account the water transport; microturbulent diffusion; gravitational sedimentation of pollutants and plankton; nonlinear interaction of planktonic populations; nutrient, temperature and oxygen regimes; influence of salinity.

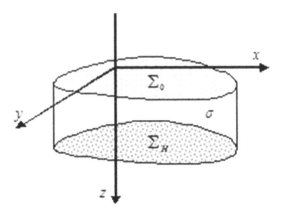

Fig. 5. Diagram of the computational domain $\bar{G}$.

Computational domain $\bar{G}$ (Fig. 5) is a closed area, limited by the undisturbed water surface Σ_0, bottom $\Sigma_H = \Sigma_H(x, y)$, and the cylindrical surface, the undisturbed surface σ for $0 < t \leq T_0$. $\Sigma = \Sigma_0 \cup \Sigma_H \cup \sigma$ – the sectionally smooth boundary of the domain G [10–14].

We consider the system (5) with the following boundary conditions:

$$S_i = 0 \; on \; \sigma, \; if \; U_n < 0; \; \frac{\partial S_i}{\partial n} = 0 \; on \; \sigma, \; if \; U_n \geq 0;$$
$$S'_{i,z} = \phi(S_i) \; on \; \Sigma_0; \; S'_{i,z} = -\varepsilon_i S_i \; on \; \Sigma_H, \tag{6}$$

where ε_i is the absorption coefficient of the i-th component of the bottom material.

We has to add the following initial conditions to (5):

$$S_i|_{t=0} = S_{i0}(x, y, z), \; i = \overline{1, 6}. \tag{7}$$

Water flow velocity fields, calculated according to the model (1)–(3), are used as input data for the model (5)–(7). The discretization of models (1)–(4), (5)–(7) was performed on the basis of the high-resolution schemes which are described in [15].

4   Parallel Implementation

The grid equations, obtained in the finite-difference approximations of tasks (1)–(3), (5)–(7), can be present in the matrix form [16]:

$$Ax = f, \tag{8}$$

where A is a linear, positive definite operator (A > 0). We use the implicit iterative process for solving problem (8):

$$B\frac{x^{m+1} - x^m}{\tau_{m+1}} + Ax^m = f. \tag{9}$$

In Eq. (9) m is the number of iteration, $\tau > 0$ is an iterative parameter, and B is an invertible operator (a stabilizer). The inverting of the operator B in Eq. (9) should be significantly easier than the directly inverting of the original operator A in Eq. (8). We construct B with using the additive representation of operator A_0, i.e., the symmetric part of the operator A:

$$A_0 = R_1 + R_2, \quad R_1 = R_2^*, \tag{10}$$

where $A = A_0 + A_1, \quad A_0 = A_0^*, \quad A_1 = -A_1^*$.
    The operator-stabilizer can be written as follows:

$$B = (D + \omega R_1)D^{-1}(D + \omega R_2), \quad D = D^* > 0, \quad \omega > 0, \tag{11}$$

where D is some, generally diagonal, operator.
    Relations (10) and (11) define the modified alternating triangular method (MATM) for solving the problems if the operators R_1, R_2 are defined and methods of determining the parameters τ_{m+1}, ω and the operator D are specified.
    The algorithm of the adaptive modified alternating triangular method of minimal corrections for calculating the grid equations with nonself-adjoint operators is in the form:

$$r^m = Ax^m - f, B(\omega_m)w^m = r^m, \tilde{\omega}_m = \sqrt{\frac{(Dw^m, w^m)}{(D^{-1}R_2w^m, R_2w^m)}}, \tag{12}$$

$$s_m^2 = 1 - \frac{(A_0w^m, w^m)^2}{(B^{-1}A_0w^m, A_0w^m)(Bw^m, w^m)} , \quad k_m = \frac{(B^{-1}A_1w^m, A_1w^m)}{(B^{-1}A_0w^m, A_0w^m)},$$

$$\theta_m = \frac{1 - \sqrt{\frac{s_m^2 k_m}{(1+k_m)}}}{1 + k_m(1 - s_m^2)}, \tau_{m+1} = \theta_m \frac{(A_0w^m, w^m)}{(B^{-1}A_0w^m, A_0w^m)},$$

$$x^{m+1} = x^m - \tau_{m+1}w^m, \omega_{m+1} = \tilde{\omega}_m,$$

where r^m is the residual vector, w^m is the correction vector, the diagonal part of the operator A is used as the operator D [17, 18].
    The estimation of convergence rate of this method is in the form:

$$\rho \leq \frac{\nu^* - 1}{\nu^* + 1}, \nu^* = \nu\left(\sqrt{1+k} + \sqrt{k}\right)^2, k = \frac{(B^{-1}A_1\omega^m, A_1\omega^m)}{(B^{-1}A_0\omega^m, A_0\omega^m)},$$

where ν is the condition number of the operator C_0, $C_0 = B^{-1/2}A_0B^{-1/2}$.

4.1   Parallel Implementation of the Modified Alternating Triangular Method

We describe the parallel algorithms, which are used for solving the problems (1)–(3), (5)–(7), with different types of domain decomposition.

Algorithm 1. Each processor is received its computational domain after the partition of the initial computational domain into two coordinate directions, as shown in Fig. 6. The adjacent domains overlap by two layers of nodes in the perpendicular direction to the plane of the partition [19].

The residual vector and it uniform norm are calculated after that as each processor will receive the information for its part of the domain. Then, each processor determines the maximum element in module of the residual vector and transmits its value to all remaining calculators. Now receiving the maximum element on each processor is enough to calculate the uniform norm of the residual vector [20].

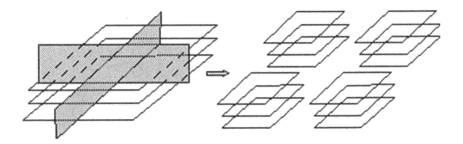

Fig. 6. Domain decomposition.

The parallel algorithm for calculating the correction vector is in the form:

$$(D + \omega_m R_1)D^{-1}(D + \omega_m R_2)w^m = r^m,$$

where R_1 is the lower-triangular matrix, and R_2 is the upper-triangular matrix. We should solve consistently the next two equations for calculating the correction vector:

$$(D + \omega_m R_1)y^m = r^m, (D + \omega_m R_2)w^m = Dy^m.$$

At first, the vector y^m is calculated, and the calculation is started in the lower left corner. Then, the correction vector w^m is calculated from the upper right corner. The calculation scheme of the vector y^m is given in Fig. 7 (the transferring elements after the calculation of two layers by the first processor is presented).

In the first step of calculating the first processor work on with the top layer. Then the transfer of overlapping elements is occurred to the adjacent processors. In the next step the first processor work on with the second layer, and its neighbors – the first. The transfer of elements after calculating two layers by the first processor is given in Fig. 7. In the scheme for the calculation of the vector y^m only the first processor does not require additional information and can independently work on with its part of the domain. Other processors are waiting the results from the previous processor, while it transfers the calculated values of the grid functions for the grid nodes, located in the preceding positions of this line. The process continues until all the layers will be calculated. Similarly, we can solve the systems of linear algebraic equations (SLAE) with the upper-triangular matrix for calculating the correction vector. Further, the scalar products are defined (12), and the transition is proceeded to the next iteration layer.

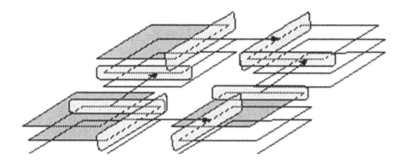

Fig. 7. Scheme of calculation the vector y^m.

We performed the theoretical estimate of the time. It's required to perform the MATM step for SLAE with seven-diagonal matrix with using decomposition in two spatial directions on a cluster of distributed calculations. All computational domain is distributed among processors (n is the total number of processors, $n = n_x \cdot n_y$, $n_x \geq n_y$), i.e. each of them received the domain by the size N/n, $N = N_x N_y N_z$, where N_x, N_y, N_z is the number of nodes in the spatial directions; t_0 is an execution time of one arithmetic operation; t_x is a response times (latency); t_n is the time, required to transfer the floating point numbers

Therefore, we obtain the theoretical estimates [1] of acceleration $S_{(1)}$ and efficiency $E_{(1)}$ of the parallel Algorithm 1:

$$S_{(1)} = \frac{n}{1 + (\sqrt{n} - 1)\left(\frac{36}{50 N_z} + \frac{4n}{50 t_0}\left(t_n\left(\frac{1}{N_x} + \frac{1}{N_y}\right) + \frac{t_x \sqrt{n}}{N_x N_y}\right)\right)},$$

$$E_{(1)} = \frac{S_{(1)}}{n} = \frac{1}{1 + (\sqrt{n} - 1)\left(\frac{36}{50 N_z} + \frac{4n}{50 t_0}\left(t_n\left(\frac{1}{N_x} + \frac{1}{N_y}\right) + \frac{t_x \sqrt{n}}{N_x N_y}\right)\right)}.$$

We considered the case of the problem solution with the rectangular domain. The domain has a complex shape in the case of real water. At the same time the real acceleration is less than its theoretical estimation. The dependence of the acceleration, obtained in the theoretical estimates, can be used as the upper estimate of the acceleration for parallel implementation of the MATM algorithm by the domain decomposition in two spatial directions. ' We describe the domain decomposition in two spatial directions with using the *k-means* algorithm.

Algorithm 2. The *k-means* method was used for geometric partition of the computational domain for the uniform loading of MCS calculators (processors). This method is based on the minimization of the functional of the total variance of the element scatter (nodes of the computational grid) relative to the gravity center of subdomain: $Q = Q^{(3)}$. Let X_i – the set of computational grid nodes, included in the i-th subdomain, $i \in \{1, ..., m\}$, m – the given number of subdomains. $Q^{(3)} = \sum_i \frac{1}{|X_i|} \sum_{x \in X_i} d^2(x, c_i) \rightarrow \min$, where $c_i = \frac{1}{|X_i|} \sum_{x \in X_i} x$ – the center of the subdomain X_i, and $d(x, c_i)$ – the distance between the calculated node and the center of the grid subdomain in the Euclidean metric. The *k-means* method converges only when all subdomain will be approximately equal.

The algorithm of *k-means* method.

(1) The initial centers of subdomains are selected with using maximum algorithm.

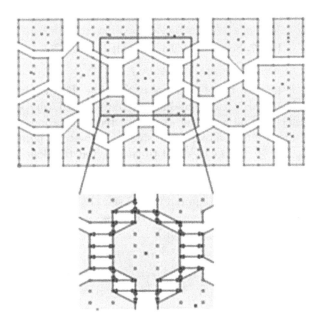

Fig. 8. Domain decomposition.

(2) All calculated nodes are divided into m Voronoi's cells by the method of the nearest neighbor, i.e. the current calculated grid node $x \in X_c$, where X_c is a subdomain, which is chosen according to the condition $\|x - s_c\| = \min_{1 \le i \le m} \|x - s_i\|$, where the s_c is the center of the subdomain X_c.

(3) New centers are calculated by the formula: $s_c^{(k+1)} = \frac{1}{|X_i^{(k)}|} \sum_{x \in X_i^{(k)}} x$.

(4) The condition of the stop is checked $s_c^{(k+1)} = s_c^{(k)}$, $k = 1, ..., m$. If the condition of the stop is not performed, then the transition to the algorithm step 2.

The result of the k-means method for model domains is given in Fig. 8 (arrows are indicated exchanges between subdomains). All points in the boundary of each subdomains are required to data exchange in the computational process. The Jarvis's algorithm was used for this aim (the task of constructing the convex hull). The list of the neighboring subdomains for each subdomain was created, and an algorithm was developed for data transfer between subdomains.

Theoretical estimates of the acceleration and efficiency of the Algorithm 2 were obtained similarly to the corresponding estimates of the Algorithm 1:

$$S_{(2)} = \frac{n \cdot \chi}{1 + (\sqrt{n} - 1)\left(\frac{36}{50N_z} + \frac{4n}{50t_0}\left(t_n\left(\frac{1}{N_x} + \frac{1}{N_y}\right) + \frac{t_x\sqrt{n}}{N_xN_y}\right)\right)},$$

$$E_{(2)} = \frac{S_{(2)}}{n} = \frac{\chi}{1 + (\sqrt{n} - 1)\left(\frac{36}{50N_z} + \frac{4n}{50t_0}\left(t_n\left(\frac{1}{N_x} + \frac{1}{N_y}\right) + \frac{t_x\sqrt{n}}{N_xN_y}\right)\right)},$$

where χ is the ratio of the number of computational nodes to the total number of nodes (computational and fictitious).

Parallel algorithms of the adaptive alternating triangular method was implemented on multiprocessor computer system (MCS) SFU. The peak performance of the MCS is 18.8 TFlops. The system includes 8 computational racks. The computational field of MCS is designed on the basis of the HP BladeSystem c-class

Table 1. Comparison of acceleration and efficiency of algorithms

n	$t_{(1)}$	$S_{(1)}^t$	$S_{(1)}$	$t_{(2)}$	$E_{(2)}^t$	$E_{(2)}$
1	7.491	1.0	1.0	6.073	1.0	1.0
2	4.152	1.654	1.804	3.121	1.181	1.946
4	2.549	3.256	2.938	1.811	2.326	3.354
8	1.450	6.318	5.165	0.997	4.513	6.093
16	0.882	11.928	8.489	0.619	8.520	9.805
32	0.458	21.482	16.352	0.317	15.344	19.147
64	0.266	35.955	28.184	0.184	25.682	33.018
128	0.172	54.618	43.668	0.117	39.013	51.933

infrastructure with integrated communication modules, power and cooling systems. The computational nodes are 512 same type 16-core HP ProLiant BL685c Blade-servers, each of which has the four 4-core AMD Opteron 8356 2.3 GHz processors and the operative memory in the volume of 32 GB. The total number of cores in the complex is equaled to the 2048, the total amount of RAM – 4 TB.

The comparison of the developed parallel Algorithms 1 and 2 for the solution (1)–(3), (5)–(7) was performed. The results are given in the Table 1.

In Table 1: n is the number of processors; $t_{(k)}, S_{(k)}, E_{(k)}$ are the processing time, the acceleration and efficiency of the k-th algorithm; $S_{(k)}^t, E_{(k)}^t$ are the theoretical estimates of the efficiency and acceleration of the k-th algorithm, $k = \{1, 2\}$.

According to the Table 1 we can conclude that the developed algorithms based on the decomposition method in two spatial directions and *k-means* method can be effectively used for solving hydrodynamics problems in the case the sufficiently large number of computational nodes.

The graphs of accelerations of Algorithms 1 and 2 for solving the WB problem (5)–(7), obtained theoretically and practically, are given in Fig. 9.

The estimation is used for comparison the performance values of the Algorithms 1 and 2, obtained practically:

$$\delta = \sqrt{\sum_{k=1}^{n} \left(E_{(2)k} - E_{(1)k} \right)^2} \Big/ \sqrt{\sum_{k=1}^{n} E_{(2)k}^2}. \tag{13}$$

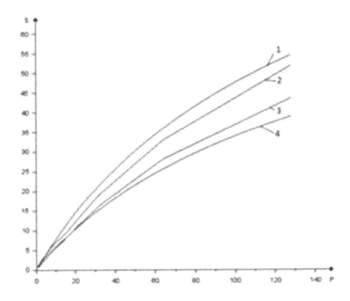

Fig. 9. Graphs of accelerations for the developed parallel algorithms.

In Fig. 9: 1 – the theoretical estimation of acceleration Algorithm 1; 2 – the acceleration of the Algorithm 2, obtained practically; 3 – the acceleration of the Algorithm 1, obtained practically; 4 – the theoretical estimations acceleration of Algorithm 2.

The value δ in the Eq. (13) was calculated by:

$$\delta = \sqrt{\sum_{k=1}^{n} \left(E_{(2)k} - E_{(1)k}\right)^2} \Bigg/ \sqrt{\sum_{k=1}^{n} E_{(2)k}^2} = 0.154.$$

On the basis the data, presented in the Table 1 and the Eq. (13), the comparison of the developed algorithms is shown that the use of the Algorithm 2 increased the efficiency for the problem (1)–(3) on 15%.

5   Software Complex Description

The Azov3d software complex was designed for solving problems (1)–(3), (5)–(7) on MCS and the calculating water flow fields, the concentrations of pollutants, phytoplankton and zooplankton in the complex areas (the Azov Sea and the Taganrog Bay).

The software complex is designed for MCS of the Southern Federal University for mathematical modeling of possible scenarios of ecological situations of coastal systems on the example of the Azov-Black Sea basin. This complex contain the computational modules with the help of which we can: to take into account factors that affect to the pollutant spread in coastal systems (weather conditions, the influence of coastline and bottom topography); to research the dependence of pollutant concentrations, the degree and size of the affected water area from the intensity of water transport, hydrophysical parameters, climatic and meteorological factors. Its features include the high performance, reliability and the high accuracy of modelling results.

New computational modules (units) can be integrated to the Azov3d software complex). The complex includes the module, which calculated the SLAE solutions in the discretization by the following methods: Jacobi; minimal corrections; steepest descent; Seidel; high relaxation; adaptive modified alternating triangular method (MATM) of the variational type.

The sequentially condensed rectangular grids by dimensions $251 \times 351 \times 15$, $502 \times 702 \times 30$, $1004 \times 1404 \times 60$ were used for the mathematical modeling of hydrobiological and hydrodynamic processes in the three-dimensional complex domain – the Azov Sea.

The calibration and verification of the developed WB model were performed on the basis of the environmental data of the Azov Sea. They were obtained during the scientific-research expeditions that are conducted by the scientists of the SFU since 2000. The expedition data processing was digitized, classified for use in various model problems of hydrobiology of the sea.

The software complex includes: the control unit, the oceanographic and meteorological database, interface systems, input-output and visualization systems. High-level language C++ and MPI technology were used in the development of this software.

6   Results of Numerical Experiments

A series of numerical experiments of the modeling the water bloom processes was performed in the Azov Sea for the period from April 1 to October 31, 2013. The results of the numerical experiment for reconstruction of the suffocation caused by the phytoplankton bloom in July 2013 is given in Fig. 10.

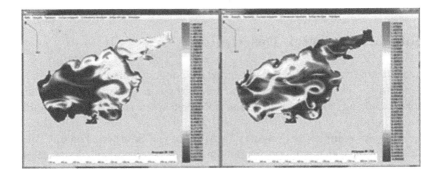

Fig. 10. Phytoplankton concentration change in Azov Sea.

The developed software complex implements the designed scenarios for the changing ecological situation in the Azov Sea using the numerical realization of plankton evolution problems of biological kinetics. The comparison the similar works in the mathematical modeling of hydro-biological processes was performed in this work.

In result of the data analyze, shown in the Fig. 11, we obtained their qualitative conformity.

The verification criterion of the developed models (1)–(3), (5)–(7) was an estimate of the error modeling taking into account the available field data measurements, calculated according to the formula:

$$\delta = \sqrt{\sum_{k=1}^{n} (S_{k\ nat} - S_k)^2} / \sqrt{\sum_{k=1}^{n} S_{k\ nat}^2},$$

where $S_{k\ nat}$ – the value of the harmful algae concentration, obtained through field measurements; S_k – the value of the harmful algae concentration, calculated by the model (1)–(3). The concentrations of pollutants and plankton calculated

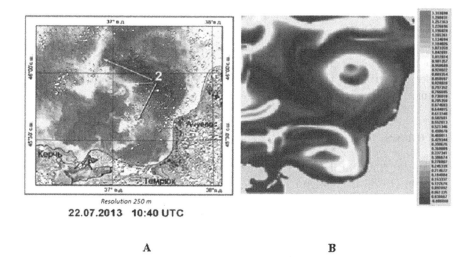

Fig. 11. Comparison of the software complex results with the satellite data. A: satellite photo of the Azov Sea by RC Planeta; B: the software complex result (the variation of the phytoplankton concentration).

for different wind situations were taken into consideration, if the relative error did not exceed 30%.

The analysis of the same software complexes for shallow waters has shown that the accuracy of the predictive changes in pollutant concentrations of plankton in shallow waters has been increased at 10–15% depending on the chosen problem of the biological kinetics.

7   Conclusion

The model of hydrodynamics and water bloom were proposed in this paper. They used for the reconstruction of suffocation, occurred on July 16, 2013 in the south-eastern part of the Azov Sea. The numerical implementation of the developed models was performed on the multiprocessor computer system with distributed memory. The theoretical values of the acceleration and efficiency of parallel algorithms was calculated. The developed experimental software is designed for mathematical modeling of possible scenarios of development of ecosystems of shallow waters on the example of Azov-Black Sea basin. The decomposition methods of grid domains were used in parallel implementation for computationally laborious convection-diffusion problems, taking into account the architecture and parameters of multiprocessor computer system. The maximum acceleration value was achieved with using 128 computational nodes and equaled to 43 times. Two algorithms were developed in the parallel algorithm implementation for solving the problem on the MCS and the data distribution between the processors. Using *k-means* method, the algorithm efficiency of the problem was increased at 15% compared with the algorithm, based on a standard partition the computational domain.

Due to the application on the MCS, the calculation time was decreased and the accuracy was preserved that required for modeling of hydrobiological processes occurring in the shallow waters. It is important in aquatic ecology problems.

References

1. Samarsky, A.A., Nikolaev, E.S.: Methods of solving grid equations, p. 588. Science, Moscow (1978). (in Russian)
2. Sukhinov, A.I., Chistyakov, A.E.: Adaptive modified alternating triangular iterative method for solving grid equations with non-selfadjoint operator. Math. Model. **24**(1), 3–20 (2012). (in Russian)
3. State Research Center Planeta. http://planet.iitp.ru/english/index_eng.htm
4. Samarskiy, A.A.: Theory of Difference Schemes. M. Nauka (1989). (in Russian)
5. Konovalov, A.N.: The method of steepest descent with adaptive alternately-triangular preamplification. Diff. Eqn. **40**(7), 953 (2004). (in Russian)
6. Konovalov, A.N.: The theory of alternating-triangular iterative method. Siberian Math. J. **43**(3), 552 (2002). (in Russian)
7. Sukhinov, A.I., Chistyakov, A.E., Shishenya, A.V.: Error estimate of the solution of the diffusion equation on the basis of the schemes with weights. Math. Model. **25**(11), 53–64 (2013). (in Russian)
8. Chetverushkin, B., Gasilov, V., Iakobovski, M., Polyakov, S., Kartasheva, E., Boldarev, A., Abalakin, I., Minkin, A.: Unstructured mesh processing in parallel CFD project GIMM. In: Parallel Computational Fluid Dynamics, pp. 501–508. Elsevier, Amsterdam (2005)
9. Petrov, I.B., Favorsky, A.V., Sannikov, A.V., Kvasov, I.E.: Grid-characteristic method using high order interpolation on tetrahedral hierarchical meshes with a multiple time step. Math. Model. **25**(2), 42–52 (2013). (in Russian)
10. Sukhinov, A.I., Chistyakov, A.E., Semenyakina, A.A., Nikitina, A.V.: Parallel realization of the tasks of the transport of substances and recovery of the bottom surface on the basis of high-resolution schemes. Comput. Methods Program. New Comput. Technol. **16**(2), 256–267 (2015). (in Russian)
11. Chistyakov, A.E., Hachunts, D.S., Nikitina, A.V., Protsenko, E.A., Kuznetsova, I.: Parallel Library of iterative methods of the SLAE solvers for problem of convection-diffusion-based decomposition in one spatial direction. Modern Prob. Sci. Educ. (1–1), 1786 (2015). (in Russian)
12. Sukhinov, A.I., Nikitina, A.V., Semenyakina, A.A., Protsenko, E.A.: Complex programs and algorithms to calculate sediment transport and multi-component suspensions on a multiprocessor computer system. Eng. J. Don **38**(4(38)), 52 (2015). (in Russian)
13. Nikitina, A.V., Abramenko, Y.A., Chistyakov, A.E.: Mathematical modeling of oil spill in shallow water bodies. Inf. Comput. Sci. Eng. Educ. **3**(23), 49–55 (2015). (in Russian)
14. Chistyakov, A.E., Nikitina, A.V., Ougolnitsky, G.A., Puchkin, V.M., Semenov, I.S., Sukhinov, A.I., Usov, A.B.: A differential game model of preventing fish kills in shallow water bodies. Game Theory Appl. **17**, 37–48 (2015)

15. Sukhinov, A.I., Nikitina, A.V., Semenyakina, A.A., Chistyakov, A.E.: A set of models, explicit regularized high-resolution schemes and programs for predictive modeling of consequences of emergency oil spill. In: Proceedings of the International Scientific Conference on Parallel Computational Technologies (PCT 2016), pp. 308–319 (2016). (in Russian)
16. Nikitina, A.V., Semenyakina, A.A., Chistyakov, A.E.: Parallel implementation of the tasks of diffusion-convection-based high-resolution schemes. Vestnik Comput. Inf. Technol. **7**(145), 3–8 (2016). (in Russian)
17. Sukhinov, A.I., Chistyakov, A.E., Semenyakina, A.A., Nikitina, A.V.: Numerical modeling of an ecological condition of the Azov Sea with application of high-resolution schemes on the multiprocessor computing system. Comput. Res. Model. **8**(1), 151–168 (2016). (in Russian)
18. Sukhinov, A.I., Nikitina, A.V., Semenyakina, A.A., Chistyakov, A.E.: Complex of models, explicit regularized high-resolution schemes and applications for predictive modeling of after-math of emergency oil spill. In: CEUR Workshop Proceedings, vol. 1576, pp. 308–319, 10th Annual International Scientific Conference on Parallel Computing Technologies, PCT 2016; Arkhangelsk; Russian Federation; 29 March 2016 through 31 March 2016; Code 121197 (2016). (in Russian)
19. Sukhinov, A.I., Nikitina, A.V., Chistyakov, A.E., Semenov, I.S., Semenyakina, A.A., Khachunts, D.S.: Mathematical modeling of eutrophication processes in shallow waters on multiprocessor computer system. In: CEUR Workshop Proceedings, vol. 1576, pp. 320–333, 10th Annual International Scientific Conference on Parallel Computing Technologies, PCT 2016; Arkhangelsk; Russian Federation; 29 March 2016 through 31 March 2016; Code 121197 (2016). (in Russian)
20. Nikitina, A.V., Sukhinov, A.I., Ougolnitsky, G.A., Usov, A.B., Chistyakov, A.E., Puchkin, M.V., Semenov, I.S.: Optimal management of sustainable development at the biological rehabilitation of the Azov Sea. Math. Model. **28**(7), 96–106 (2016). (in Russian)

High-Performance Simulation of Electrical Logging Data in Petroleum Reservoirs Using Graphics Processors

Vyacheslav Glinskikh[1,2]([⊠]), Alexander Dudaev[1,2]([⊠]), and Oleg Nechaev[1]([⊠])

[1] Trofimuk Institute of Petroleum Geology and Geophysics,
Siberian Branch of the Russian Academy of Sciences, Novosibirsk, Russia
{glinskikhvn,dudaevar,nechayevov}@ipgg.sbras.ru
[2] Novosibirsk State University, Novosibirsk, Russia
http://www.ipgg.sbras.ru

Abstract. The work is concerned with the development of numerical algorithms for solving direct problems of borehole geoelectrics by applying high-performance computing on GPUs. The numerical solution of the direct 2D problem is based on the finite-element method and the Cholesky decomposition for solving a system of linear equations. The software implementations of the algorithm are made by means of the NVIDIA CUDA technology and computing libraries making it possible to decompose the equation system and find its solution on CPU and GPU. The analysis of computing time as a function of the matrix order has shown that in the case at hand the computations are the most effective when decomposing on GPU and finding a solution on CPU. We have estimated the operating speed of CPU and GPU computations, as well as high-performance CPU–GPU ones. Using the developed algorithm, we have simulated electrical logging data in realistic models.

Keywords: Graphics processing units · Parallel algorithm · Finite-element method · Direct 2D problem · Electrical logging data

1 Introduction

Contemporary geophysical methods are widely used to obtain comprehensive information about oil and gas reservoirs, starting with the search for prospective objects and estimation of their reserves, and ending with field management. The reconstruction of the electrical resistivity of rocks plays an important role when studying geological environments. The evaluation of oil and gas content in reservoirs is conducted through electrical resistivity values based on borehole measurements accomplished with electrical logging tools.

Interpreting data measured in some geological environments requires a specialized mathematical description and the corresponding computational algorithms. Among these environments one can mention stratified sedimentary rocks represented by interbedded thin layers with various material composition

© Springer International Publishing AG 2017
L. Sokolinsky and M. Zymbler (Eds.): PCT 2017, CCIS 753, pp. 186–200, 2017.
DOI: 10.1007/978-3-319-67035-5_14

and electrophysical properties, such as highly conductive clay rocks and resistive oil sands. Such media are electrically anisotropic, since the resistivity values in the bedding plane and in the vertical direction differ greatly. The lack of electrical anisotropy during the interpretation of electrical logging data leads to significant errors in the determination of oil content. Therefore, a solution to the problem of determining the electrical anisotropy of rocks according to electrical logging data is extremely relevant and has great practical significance.

The simulation of electrical logging data in models of anisotropic media calls for the development of special algorithms and software. The finite-difference and finite-element methods are the main techniques for simulating electrical fields in spatially inhomogeneous media [1–3]. As far as is known, the numerical solution of electrodynamic problems with the use of grid methods results in solving systems of linear algebraic equations (SLE) with high-dimensional sparse matrices; direct and iterative solution methods are effectively used for this task. However, the use of these solutions for fast data interpretation in real time happens to be inefficient due to their low performance and high resource intensity.

The use of multiprocessor computer systems and computing clusters is one of the widely applied methods for reducing calculation time. However, their utilization for prompt solutions of geophysical interpretation objectives directly at the well is extremely challenging. Another method is associated with the use of NVIDIA graphics processors and Intel Xeon Phi coprocessors for speeding up calculations, and is widely applied in solving present-day problems in various scientific fields, including well logging [4–7].

A further increase in processing speed and effectiveness of numerical solutions of electrodynamic problems based on grid methods is connected with the creation of highly efficient parallel algorithms. To a great extent, these algorithms create a need for the development or application of fast computational libraries based on different methods for solving SLE. To find a solution to an SLE, we use in this study the direct method based on the Cholesky decomposition and subsequent solution of two subsidiary systems with triangular matrices.

Traditionally, when it comes to the development of such high-performance parallel algorithms, much attention is given to the efficiency of matrix decomposition. For this purpose, both computational algorithms and corresponding data structures are developed [8]. This is because solving the subsidiary SLE takes much less computational resources compared to the original matrix decomposition. The developed parallel algorithms for matrix decomposition are also implemented on graphics processors (GPU). However, unlike central processors (CPU), GPUs make it possible to implement only a rather narrow class of algorithms effectively, which leads to the necessity of creating methods using both CPUs and GPUs [8–10].

When it is necessary to solve a large number of SLE with the same matrix but different right-hand sides, computational costs of solving the system significantly exceed the costs of matrix decomposition. It will be illustrated further in the article that this may lead to a low efficiency of the solution if we use only GPUs.

This gives rise to a need for developing a method to solve such SLE that employs cooperative computing on CPUs and GPUs.

Ultimately, these approaches make it possible to apply resource-intensive computational tasks for the practical purpose of processing electrical logs, providing an improved accuracy in the determination of reservoir parameters, primarily those of anisotropic. This work is dedicated to the development and implementation of an algorithm for solving the direct 2D electrical logging problem in anisotropic models on the basis of the finite-element method and high-performance computing on GPUs.

2   Solution of the Direct 2D Electrical Logging Problem

As already mentioned, the determination of the resistivity of rocks around a well for further estimation of their hydrocarbon saturation plays a significant part when investigating geological sections. One of the best-known ways of addressing this problem is to record electrical potentials induced by a direct current source in the borehole. Electrical logging holds a special place among geophysical well logging methods [11]. An electrical logging borehole tool comprises several probes including a coaxially placed current electrode A and two closely-spaced measuring electrodes M and N. The probes measure the apparent resistivity, which is calculated according to the formula for a homogeneous medium:

$$\rho_k = k\frac{\varphi_M - \varphi_N}{I_A}, \tag{1}$$

where ρ_k is the apparent resistivity, k is a geometrical constant of a probe depending on the distances between the electrodes, φ_M is the potential measured at the electrode M, φ_N is the potential measured at the electrode N, and I_A is the strength of current flowing through the current electrode A.

The measurement results are well logs characterized by the apparent resistivity values as a function of depth (distance) along the borehole. Electrical logging makes use of three-electrode probes, which have the following designation: A2.0M0.5N. Here the distance between the electrodes A and M equals 2.0 m, whereas that between M and N is equal to 0.5 m (Fig. 1). The present article provides an analysis of the logs corresponding to probes with lengths from 0.4 to 8.0 m.

Fig. 1. Scheme of the electrical logging tool

The basic ideas and principles of physical measurements by this method, established as early as in the previous century, are fundamental, with their further development largely depending on the existing mathematical framework.

This implies considering various effects, including electrical anisotropy due to thin layering of sedimentary rocks, which requires the development of appropriate algorithms and software.

We examine a solution of the direct electrical logging problem within the framework of a 2D anisotropic model (Fig. 2). The model approximates a case when a vertical well penetrates a geological section represented by a thin-layer formation with plane-parallel horizontal boundaries. In the context of an anisotropic model, the formation is characterized by two values of resistivity: in a horizontal plane and in the vertical direction. Such an anisotropic model is transversely isotropic. In this case, the formation resistivity is described by the diagonal tensor

$$\hat{\rho} = \begin{bmatrix} \rho_h & 0 & 0 \\ 0 & \rho_h & 0 \\ 0 & 0 & \rho_v \end{bmatrix}.$$

Here ρ_h is the value of the resistivity in a horizontal plane, whereas ρ_v is its value in the vertical direction. Resistivity is the reciprocal of conductivity:

$$\rho_h = \frac{1}{\sigma_h}, \ \rho_v = \frac{1}{\sigma_v}.$$

The electrical anisotropy factor is defined as follows:

$$\Lambda^2 = \frac{\rho_v}{\rho_h} = \frac{\sigma_h}{\sigma_v}.$$

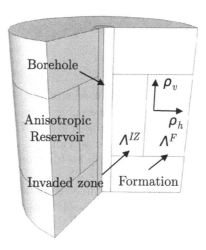

Fig. 2. 2D anisotropic earth model

In the near-borehole environment, there may be an invaded zone, which is formed during the displacement of the formation fluid by drilling mud filtrate.

It is separated from the well and reservoir by a coaxial cylindrical boundary. In this case, each region (invaded zone and formation) is characterized by its own set of values of horizontal and vertical resistivity (indices IZ and F; see Fig. 2).

We should note that the method consisting in the registration of responses from eddy currents in the medium, which are generated by a monochromatic inductive source in the borehole, is extensively used for studying electrophysical properties of rocks. In induction logging, eddy currents excited in the borehole environment are coaxial with the borehole and do not intersect the formation boundaries. Therefore, the measured electromagnetic responses, in contrast to resistivity logging signals, do not contain information on the vertical resistivity and, correspondingly, on the electrical anisotropy. Induction logging data are used to determine the geoelectric parameters of a 2D isotropic model [12–16].

Conventionally, a solution to the direct electrical logging problem or a simulation of electrical logs is associated with the determination of the apparent resistivity from a given distribution function of both the horizontal and vertical resistivity of the medium described by a fixed vector of the model parameters. As follows from the definition of the apparent resistivity (1), to find its value it is necessary to know the values of the electrical potentials at the measuring electrodes M and N.

The distribution of the electrical potential in the simulated area is described by the following boundary value problem:

$$\mathrm{div}\, \sigma\, \mathrm{grad}\, \varphi = 0, \tag{2}$$

$$\varphi|_{\Gamma^0} = 0, \tag{3}$$

$$\sigma \frac{\partial \varphi}{\partial \overrightarrow{n}}\Big|_{\Gamma^1} = 0, \tag{4}$$

$$\sigma \frac{\partial \varphi}{\partial \overrightarrow{n}}\Big|_{\Gamma^2} = j, \tag{5}$$

where φ is the potential of the electrical field strength $(\overrightarrow{E} = -\,\mathrm{grad}\,\varphi)$, σ is the electrical conductivity $(\sigma = \rho^{-1})$, Γ^0 is the outer boundary of the area in which the electrical potential is considered to be close to zero, Γ^1 is the dielectric surface of the tool, Γ^2 is the current electrode surface, and j is the current density. It is assumed that the tool housing has an infinite length and is oriented along the borehole axis.

Since we are considering an axially symmetric model with cylindrical boundaries, it is convenient to convert Cartesian coordinates to cylindrical coordinates. Then the differential operators take the form

$$\mathrm{grad}\, u = \left(\frac{\partial u}{\partial r}, \frac{1}{r}\frac{\partial u}{\partial \phi}, \frac{\partial u}{\partial z} \right),$$

$$\mathrm{div}\, \overrightarrow{V} = \frac{1}{r}\frac{\partial r V_r}{\partial r} + \frac{1}{r}\frac{\partial V_\phi}{\partial \phi} + \frac{\partial V_z}{\partial z}.$$

We will assume further that the simulated area has axial symmetry, i.e. the vertical borehole axis and that of the probe coincide. In this case, the boundary value problem (2)–(5) takes the following form:

$$\frac{1}{r}\frac{\partial}{\partial r}\left(r\sigma_h\frac{\partial\varphi}{\partial r}\right) + \frac{\partial}{\partial z}\left(\sigma_v\frac{\partial\varphi}{\partial z}\right) = 0, \tag{6}$$

$$\varphi|_{\Gamma^0} = 0, \tag{7}$$

$$\sigma_h\frac{\partial\varphi}{\partial r}\bigg|_{\Gamma^1} = 0, \tag{8}$$

$$\sigma_h\frac{\partial\varphi}{\partial r}\bigg|_{\Gamma^2} = j. \tag{9}$$

We introduce the functional spaces

$$H^1(\Omega) = \left\{u \in L^2(\Omega)\,|\mathrm{grad}\,u \in L^2(\Omega)\right\},$$

$$H_0^1(\Omega) = \left\{u \in H^1(\Omega)\,|\,u|_{\Gamma^0} = 0\right\},$$

where Ω is a computational space, namely a rectangle, Γ^0 is a part of the computational space boundary on which the first boundary conditions are given.

Let us define the following scalar product for the elements of these spaces:

$$(u,v) = \int_\Omega uv\,d\Omega.$$

To solve the boundary value problem (6)–(9), we will use the finite-element method [17].

Let us use the defined scalar product to multiply the expression (6) by some function $v \in H_0^1(\Omega)$, applying the integration by parts formula, and bearing in mind that the function v is equal to zero in a part of the computational domain boundary. As a result, we obtain the following variational formulation of the boundary value problem (6)–(9):

Find a function $\varphi \in H_0^1(\Omega)$ such that for any $v \in H_0^1(\Omega)$ the following relation is fulfilled:

$$\int_\Omega \left[\sigma_h\frac{\partial\varphi}{\partial r}\frac{\partial v}{\partial r} + \sigma_v\frac{\partial\varphi}{\partial z}\frac{\partial v}{\partial z}\right] r\,dr\,dz = \int_{\Gamma^2} jv\,d\Gamma^2. \tag{10}$$

We will seek an approximate solution φ^h of the variational formulation (10) as an expansion in the set of basis functions ψ_i forming a finite-dimensional subspace of the space $H_0^1(\Omega)$. As a basis, we will use bilinear basis functions [17]. As a result, the approximate solution φ^h is completely determined by the vector x consisting of the weight coefficients, which can be found by solving the following SLE:

$$Ax = b.$$

The elements of the matrix as well as those of the vector on the right side can be calculated in the following manner:

$$[A]_{i,j} = \int\limits_{\Omega} \left[\sigma_h \frac{\partial \psi_i}{\partial r} \frac{\partial \psi_j}{\partial r} + \sigma_v \frac{\partial \psi_i}{\partial z} \frac{\partial \psi_j}{\partial z} \right] r \, dr \, dz,$$

$$[b]_i = \int\limits_{\Gamma^2} j \psi_i \, d\Gamma^2.$$

To solve the resulting system, we will apply the Cholesky decomposition:

$$A = LL^t, \tag{11}$$

where L and L^t are upper and lower triangular matrices.

It should be noted that when the tool is moved along the well, only the locations of the boundaries Γ^1 and Γ^2 in the whole boundary value problem (6)–(9) will change, which in turn leads to a change in the right-hand side b of the equation, whereas the A-matrix elements are invariable.

Thus, to simulate measurements at several points in the borehole, the decomposition of the matrix into a product of two triangular matrices can be performed just once. Afterwards, it may be used for finding several solutions of the SLE.

In such a way, the direct problem solution allows for the calculation of the measured apparent resistivity at a given depth along the borehole from a given vector of the anisotropic model parameters. The result of solving a corresponding number of the direct problems is a set of synthetic apparent resistivity data of various probes represented as a set of well logs. At the end of the article, we will analyze the results of high-performance simulation of electrical logs in a realistic oil reservoir model.

3   Performance on CPUs and GPUs in an Electrical Logging Problem

On the basis of specialized computing on personal computer GPUs, we developed parallel algorithms for high-performance electrical logging data simulation in oil and gas wells. Using these algorithms, we estimated the computation speed on a CPU (Intel Core i7-4770) and GPU (NVIDIA GeForce Titan), including cooperative computing on the CPU and GPU. The development of the parallel algorithms was performed using the NVIDIA CUDA technology [18] and the computational libraries cuSOLVER and CHOLMOD, which have an advanced application programming interface (API).

Further, we perform an analysis of computation time for the solution of the direct electrical logging problem at a given point along the borehole for one probe. In addition, employing (11), we decompose the system matrix into two triangular matrices L and L^t, which is needed for finding one solution of the SLE. It is determined that the main computational costs in an initial sequential algorithm are associated with the matrix decomposition, and reach 83% of the total execution time of the program during the solution of a single SLE. For this reason, it is advisable to perform these calculations in parallel on the GPU.

We developed a parallel algorithm for high-performance simulation of electrical logging data based on the cuSOLVER computational library integrated in the CUDA Toolkit 7 and its higher versions (for a detailed description, see [19]). When using the cuSOLVER library, both the matrix decomposition and the search of one or several solutions of the SLE are performed directly on the GPU. Furthermore, the version of the cuSOLVER library that allows for calling them separately is implemented in the CUDA Toolkit 7.5. It is shown that when solving the direct electrical logging problem with the algorithm developed here, the computing time needed for matrix decomposition on the GPU is reduced by a factor of 4.

Using the NVIDIA Visual Profiler, we analyzed the relations of time expenditures when performing GPU computing. For instance, the data transfer time from Host to Device and back is about 1% of all the operations on the GPU. The analysis of the triangular matrices L and L^t and sorting the elements takes up to 5% of the total time. The most resource-intensive task is matrix decomposition, which takes about 70% of the total computing time. Finding a solution to the SLE using the two triangular matrices is about 15% in this computational experiment, which focuses on the solution of one direct electrical logging problem.

Afterwards, we analyzed the computational time needed for solving direct electrical logging problems along a borehole profile of length 10 m, with a discretization interval of 0.1 m and 5 probes. In this task, one matrix decomposition is performed, and 505 solutions of the system are found subsequently.

We estimated the computing time corresponding to the matrix decomposition and finding of solutions to the SLE, as well as the total time on the CPU and GPU depending on the order of the matrix (Table 1). The size of the coefficient matrix of the SLE is determined by the computational grid density, the choice of which depends on the size of heterogeneous areas of the environment and their relative resistivity contrast.

Table 1. Computation time on CPU and GPU for matrix decomposition and SLE solution, and total time depending on matrix size

Matrix size th. e.	Computing time, s					
	CPU			GPU		
	Decomposition	Solution of SLE	Total	Decomposition	Solution of SLE	Total
10	0.01	0.28	0.46	0.02	13.33	13.83
39	0.06	1.44	2.13	0.10	51.55	52.58
154	0.87	11.11	14.51	0.51	209.29	212.60
613	12.92	82.73	106.03	3.23	1061.79	1075.15

The results presented in Table 1 show that, as the order of the matrix increases, time expenditures on matrix decomposition on the CPU are larger than those on the GPU. For instance, it is more reasonable to use the CPU when the number of elements in a matrix is below 10^5. Otherwise, the GPU should be used; in the last case the acceleration is by a factor of 4. This is due to the fact that the considered matrices are very sparse, and the decomposition is performed rapidly for matrices of small size. Let us note that a high performance on GPU is achieved when using denser matrices obtained by numerical solution of other logging problems.

The graphics device memory size imposes a significant constraint on GPU computations. The widely accepted CSR format is used to store the matrix. In this format, the total memory size for the matrix is determined by the size of three vectors. Two of them contain integer-valued elements whose size is defined by the number of matrix rows and the number of its non-zero elements, respectively. The third vector contains the SLE coefficients in single or double precision.

Matrix decomposition also requires memory. For example, the memory size required for the decomposition of a matrix with about 10^4 elements is approximately 16 times larger than the amount of memory necessary to store it. In the case of 10^6 elements, the difference is a factor of 3. Moreover, double precision calculations on the GPU take about 37 MB in the first case and 3 GB in the second. Therefore, to implement computations on the GPU, it is necessary to provide a sufficient amount of memory to store a matrix in the CSR format and perform its decomposition.

In contrast to matrix decomposition, finding solutions to an SLE on a GPU takes several times more time than it takes on a CPU (see Table 1). For instance, for a matrix with 10^4 elements, CPU computing is 48 times faster than GPU computing, whereas for a matrix with 10^6 elements, the corresponding difference is a factor 13. As noted above, the data transfer time is very small, that is why significant time expenditures on finding solutions to an SLE on a GPU are due to very sparse matrix and GPU computing features. The latter include delays associated with reference to the global/shared memory and synchronization of computing threads. In general, it is seen that the gain from CPU utilization decreases as the matrix size increases, but this happens quite slowly. Based on the found pattern, GPU computing will be faster only for matrices with more than 10^9 to 10^{10} elements. All the evidence points to the fact that finding solutions to SLE is the most effective when this is parallelized on CPUs.

It ought to be noted that when solving one direct electrical logging problem (calculations at a given point along the borehole profile for one probe), including a single matrix decomposition and a single cycle of finding a solution to an SLE, the use of a GPU is several times more efficient. Our calculations have shown that when finding several (up to 10) solutions to an SLE, making use of a GPU will be reasonable. This corresponds to the solution of two direct electrical logging problems at two points along the borehole profile for 6 probes, which is of no concern for practical use. However, when dealing with a large number of solutions of direct electrical logging problems (calculations on a profile along a borehole tens to hundreds of meters in length, for all the probes), the time expenditures on finding solutions to an SLE become considerable, and CPU calculations are performed much faster than those on a GPU.

The performed analysis shows that calculations will be effective if matrix decomposition is preformed on GPUs, and finding solutions to SLE on CPUs, thereby carrying out a cooperative CPU–GPU computing. For the development of a parallel algorithm, we apply the CHOLMOD library [20], which is part of the computational library package SuiteSparse [21], comprising solvers for sparse matrices employing NVIDIA graphics processors. Our selection of the CHOLMOD library is due to the fact that it allows matrix decomposition and finding a solution to an SLE to be done separately, with matrix decomposition being carried out by cooperative CPU–GPU computing. Thus, CPU and GPU calculations are performed simultaneously. As a consequence of this, the CPU does not go into a long period of waiting for the completion of the matrix decomposition on the GPU, as is the case when using the cuSOLVER library. It also allows one not to store all the data on the GPU, but only those that are necessary at the moment, which creates the possibility to perform calculations using matrices of even higher order than those considered in this paper.

Next, we show the results of computational performance estimates when using the developed parallel algorithms compared to the initial sequential algorithm for the CPU (Fig. 3). For the implementation of parallel computing on the GPU, we apply the cuSOLVER library, while for cooperative CPU–GPU computing, we used the CHOLMOD library. The estimates were obtained during a high-performance simulation of electrical logging data in a realistic petroleum reservoir model when solving the problem with the largest size of the SLE matrix (613 thousand elements).

It can be seen from Fig. 3 that the matrix decomposition by cooperative CPU–GPU computing is performed 4.1 times faster than on CPU, and 1.1 times faster than on GPU. Moreover, the time required to find all the solutions of an SLE by cooperative CPU–GPU computing is shorter by a factor of 1.6 compared to that of the initial sequential algorithm on CPU, and by a factor of 20.2 compared to that of the parallelized algorithm on GPU. It was found that the total CPU–GPU computation time is shorter than that on CPU by a factor of 1.6, and shorter than that on GPU by a factor of 16.3.

Finally, we estimated the computing time on CPU and various NVIDIA GPUs for the solution of a direct 2D electrical logging problem (Table 2). The calculations were performed on the GTX 680, GTX 960 and NVIDIA TITAN, and also on two GTX 680 GPUs. The results showed that performing the decomposition of the SLE matrix completely on the GPU (as implemented in the cuSOLVER) is not the optimal decision, since it does not provide any speed gain. As already stated, when performing a matrix decomposition, it is essential to have space for both matrix storage and the calculations themselves. Therefore, for large matrices requiring a huge amount of memory (for a matrix containing more than 10^6 elements, $10\,GB$ of free space on the videocard are required),

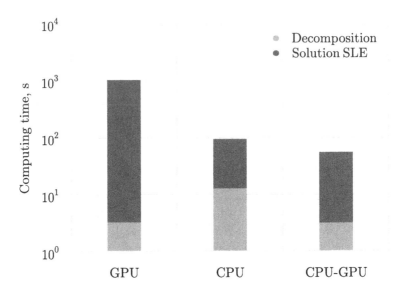

Fig. 3. Computation time on CPU and GPU, and cooperative CPU–GPU computing time, when solving a direct 2D electrical logging problem

the use of the CHOLMOD library will be more effective, since it stores the data only at the time of calculation. A comparative analysis of the calculation time for such a matrix on various GPUs shows that cards less powerful than the NVIDIA GTX TITAN are also capable of delivering high performance. On the whole, Table 2 shows that the decomposition of an SLE matrix containing more than 10^6 elements is executed 20 to 25 times faster on CPU–GPU than it is on CPU.

Table 2. Computation time on CPU and different NVIDIA GPUs for the solution of a direct 2D electrical logging problem

Computing device	NVIDIA GTX			CPU
	960	680	TITAN	
Time, s	10.28	10.33	8.32	207.29

Thus, the performance obtained for cooperative CPU–GPU computing demonstrates the high efficiency of the developed algorithm and its software implementations, pointing to their possible use for rapid data processing and interpretation based on high-performance simulation.

4   High-Performance Simulation of Electrical Logs

Using the developed computational algorithm and implemented software on the basis of cooperative CPU–GPU computing, we conducted a high-performance simulation of electrical logs in a realistic model of a petroleum reservoir typical of the Ob River region.

We consider the case of a low-resistivity argillaceous reservoir, ubiquitous in Western Siberia. The study of such reservoirs is an important task in petroleum geophysics. The presence of highly conductive clay bands in a highly resistive oil-containing reservoir results in a decrease in the formation resistivity.

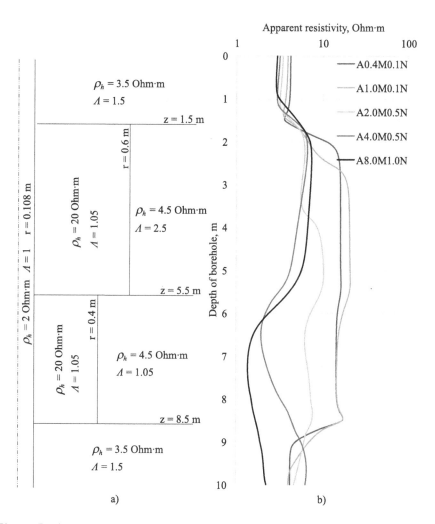

Fig. 4. Realistic anisotropic model of a petroleum reservoir (a), and the synthetic logs of the apparent resistivity of the electrical logging probes (b), calculated using cooperative CPU–GPU computing

Consequently, when processing electrical logging data, an oil-saturated layer will be identified as water-saturated owing to its low resistivity.

We provide an anisotropic model that describes thin lamination of sandy-argillaceous interlayers saturated with oil at the top and with water at the bottom, and enclosed in argillaceous deposits. The geoelectrical parameters of the anisotropic model (namely those of the borehole, fluid-saturated layers, host rocks) are shown in Fig. 4a.

As previously mentioned, during the process of drilling permeable formations, drilling mud filtrates into them, leading to the displacement of the formation fluid (oil-water mixture) and development of a modified near-borehole zone. The change in resistivity in the invaded zone is related to redistribution of formation water mineralization and that of mud filtrate. Behind this zone, an undisturbed formation is located. The invasion zones and layers have their own horizontal resistivity ρ_h and electrical anisotropy coefficient Λ.

We analyzed the apparent resistivity logs (for the probes A0.4M0.1N, A1.0M0.1N, A2.0M0.5N, A4.0M0.5N and A8.0M1.0N) presented in Fig. 4b. The logs are significantly differentiated and reflect the vertical heterogeneity of the geological section, which allows to discern various layers and interpret data correctly. In general, the presented logs correspond to a typical model of spatially heterogeneous fluid-saturated rocks. It should be noted that the results obtained are completely identical if one applies the sequential algorithm.

Thus, the fast algorithm developed by us will allow in the future to carry out rapid numerical simulations of electrical logs and examine the features of their behavior in complex models of geologic media, including anisotropic ones. This fact is extremely important for both the creation of new logging technologies and hardware systems and the express interpretation of practical log data in real time.

5   Conclusion

We developed and implemented into software an algorithm for high-performance simulation of electrical logs from oil and gas wells using graphics processors. The algorithm is based on the solution of the direct 2D electrical logging problem in anisotropic models of environment by means of the grid-based finite-element method. The solution of a SLE with a sparse matrix of high order is performed by Cholesky decomposition.

The software implementations of the algorithm used the NVIDIA CUDA technology and the cuSOLVER and CHOLMOD computing libraries, which allow to decompose a matrix and finding a solution to an SLE on CPUs and GPUs both separately and jointly. We analyzed the computation time required for decomposing a matrix and finding a solution to an SLE on CPU and GPU as well as the total time, depending on the order of the matrix and the NVIDIA GPUs that were used. It was found that, regarding the considered task, the most efficient approach consists in decomposing the matrices on GPUs and finding the solutions of SLE on CPUs.

We made estimates of computing speed on CPU and GPU, including high-performance CPU–GPU computing. It was established that the use of cooperative CPU–GPU computing allows to improve performance, which is not the case with similar computations on either CPU or GPU. For instance, matrix decomposition on CPU–GPU is 4 times faster than on CPU. In addition, finding solutions to SLE on CPU–GPU is 20 times faster than on GPU. The total CPU–GPU computation time is shorter than that on CPU by a factor of 1.5, and shorter than that on GPU by a factor of 16.

The high-performance simulation of electrical logging data was conducted in a realistic model of an anisotropic petroleum reservoir. The high-performance computing carried out in our study, processing speed estimates and simulation are indications of the high efficiency of the developed algorithm regarding a wide range of practical problems.

References

1. Eremin, V.N., Nechaev, O.V., Haberhauer, S., Shokina, N., Shurina, E.P.: Parallel realization of mathematical modelling of electromagnetic logging processes using VIKIZ probe complex. Comput. Technol. **12**(6), 18–33 (2007). (In Russian)
2. Epov, M.I., Shurina, E.P., Nechaev, O.V.: 3D forward modeling of vector field for induction logging problems. Geol. Geophys. **48**(9), 770–774 (2007)
3. Surodina, I.V., Epov, M.I.: High-frequency induction data affected by biopolymer-based drilling fluids. Geol. Geophys. **53**(8), 817–822 (2012)
4. Glinskikh, V.N., Epov, M.I., Labutin, I.B.: Electromagnetic logging data simulation on GPUs. Comput. Technol. **13**(6), 50–60 (2008). (In Russian)
5. Glinskikh, V.N., Bulantceva, Y.: Mathematical simulation of electromagnetic logs using high-performance co-processor Intel Xeon Phi Vestnik of NGU. Math. Mech. Inform. **14**(4), 11–22 (2014). (In Russian)
6. Glinskikh, V.N., Gorbatenko, V.A.: Electromagnetic logging data inversion on GPU. Comput. Technol. **20**(1), 25–37 (2015). (In Russian)
7. Labutin, I.B., Surodina, I.V.: Algorithm for sparse approximate inverse preconditioners in the conjugate gradient method. Reliable Comput. **19**, 120–126 (2013)
8. Davis, T.A., Rajamanickam, S., Sid-Lakhdar, W.M.: A survey of direct methods for sparse linear systems. Acta Numerica **25**, 383–566 (2016)
9. Mittal, S., Vetter, J.S.: A survey of CPU-GPU heterogeneous computing techniques. ACM Comput. Surv. **47**(4), 69:1–69:35 (2015)
10. Sao, P., Vuduc, R., Li, X.S.: A distributed CPU-GPU sparse direct solver. In: Silva, F., Dutra, I., Santos Costa, V. (eds.) Euro-Par 2014. LNCS, vol. 8632, pp. 487–498. Springer, Cham (2014). doi:10.1007/978-3-319-09873-9_41
11. Dakhnov, V.N.: Electrical and Magnetic Borehole Survey Techniques. Nedra, Moscow (1981). 344 p. (In Russian)
12. Epov, M.I., Glinskikh, V.N.: Fast two-dimensional simulation of high-frequency electromagnetic field in induction logging. Russ. Geol. Geophys. **44**(9), 904–915 (2003)
13. Epov, M.I., Glinskikh, V.N.: Linearization of relative parameters of a high-frequency magnetic field in two-dimensional conducting media. Russ. Geol. Geophys. **45**(2), 247–257 (2004)

14. Glinskikh, V.N., Epov, M.I.: Locally nonlinear approximation of high-frequency electromagnetic field for logging applications. Russ. Geol. Geophys. **47**(8), 930–936 (2006)
15. Glinskikh, V.N., Nikitenko, M.N., Epov, M.I.: Numerical modeling and inversion of electromagnetic logs in the wells drilled with biopolymer and oil-mud. Russ. Geol. Geophys. **54**(11), 1409–1416 (2013)
16. Glinskikh, V.N., Nikitenko, M.N., Epov, M.I.: Processing high-frequency electromagnetic logs from conducting formations: linearized 2D forward and inverse solutions with regard to eddy currents. Russ. Geol. Geophys. **54**(12), 1515–1521 (2013)
17. Shaidurov, V.V.: Multigrid Methods for Finite Elements. Nauka, Moscow (1989). 288 p. (In Russian)
18. CUDA C Programming Guide. Design Guide. NVIDIA CUDA. http://docs.nvidia.com/cuda/pdf/CUDA_C_Programming_Guide.pdf
19. cuSOLVER. http://docs.nvidia.com/cuda/cusolver
20. Chen, Y., Davis, T.A., Hager, W.W., Rajamanickam, S.: Algorithm 887: CHOLMOD, supernodal sparse Cholesky factorization and update/downdate. ACM Trans. Math. Softw. **35**(3), 22:1–22:14 (2008)
21. Davis, T.A.: SuiteSparse: a suite of sparse matrix software. http://www.suitesparse.com

The Hardware Configuration Analysis for HPC Processing and Interpretation of the Geological and Geophysical Data

Ekaterina Tyutlyaeva$^{(\boxtimes)}$, Sergey Konyukhov, Igor Odintsov,
and Alexander Moskovsky

ZAO RSC Technologies, Kutuzovskiy Av., 36, Building 23, 121170 Moscow, Russia
{xgl,s.konyuhov,igor_odintsov,moskov}@rsc-tech.ru
http://www.rscgroup.ru

Abstract. The problems which arise during the gas-oil exploration process and require high-performance computing resources can be divided in two groups.

The first group is seismic data processing, the second group is 3D reservoir simulating for the exploration process optimization.

For the each group of problems typical applications used in real technological process were chosen, and their behavior was examined on different computational architectures.

Performed analysis shows that the applications of the first group have good scalability potential on the studied computational platforms, meanwhile for the applications of the second group the limit of the performance increasing is reached relatively fast.

Keywords: Performance analysis · Architecture comparison · Profiling

1 Introduction

The parallel execution performance of applications[1] depends on both algorithmic structure of used numerical methods and features of computational hardware.

Therefore, both factors should be taken into account simultaneously. This is an internally controversial challenge, because the target application group should be studied as general as possible, but at the same time, any procedure of optimization is problematic without explicit consideration of all levels of computational hardware, such as [1]:

- vector or "single instruction multiple data" (SIMD) instructions;
- in-core multiple execution devices and the out-of-order execution logics;
- multicore CPUs and independent instruction streams;

---

[1] In this article the parallel execution performance is understood, primarily, as the execution time, while the issues related to the accuracy and correctness are the subject of separate consideration.

© Springer International Publishing AG 2017
L. Sokolinsky and M. Zymbler (Eds.): PCT 2017, CCIS 753, pp. 201–214, 2017.
DOI: 10.1007/978-3-319-67035-5_15

– the multiple CPUs within computational node;
– the multiple nodes connected via different computation networks.

In other words, the co-design of dedicated computational hardware is a typical "chicken-egg" problem. The possible bootstrapping procedure may be formulated as follows.

As a first step, among multiple real applications used in specific subject area the most suitable ones are carefully selected, which can serve as a basis for further detailed investigation.

Then the dynamic behavior of chosen applications is studied to identify their common features using the performance analysis tools. Any typical computational architecture can be used at this stage. It is possible to reduce the impact of the computational architecture on the results by reproducing the analysis on the different environments. However, the large variety of the testing configurations could be impractical, because of substantial amount of time needed to carry out studies and difficulties of interpreting the results.

The requirements found for specification of the computational hardware allow making preliminary selection from a number of possible computation configurations and/or architectures.

Further, the selected applications and computational hardware are tested together to clarify their real performance and scalability potential. For this purpose it is possible to use either real computational nodes or simulated virtual environments.

Each of these two approaches has its advantages and disadvantages. Simplicity of the test procedure, possibillity to use large amount of input data, and reliable understanding of the application behavior belong to the advantages of the real testing. However, the number of available real testbenches is still rather limited.

Simulation of the application execution on different architectures allows to increase the number of available hardware variants, albeit the simulation is labor demanding procedure because of huge computational costs and log-files sizes. Besides it requires limited scale of input data. Finally empirical and heuristic models for hardware emulation reduce the analysis quality.

At the final stage of development the testing results are used to make further adjustments.

In the present work the scheme described above was used to design of supercomputer specializing on geological data processing.

Currently, the highest priority in the subject area is given to exploration and development of the offshore and deepwater fields. Such computations are characterized by the large amount of input data (from hundreds of terabytes up to tens of petabytes) and large model complexity that requires large computational resources. Moreover, data processing rate should be relatively high. For this reason the development of specialized computational hardware is a rather complicated task.

In this work analysis of dynamic behavior of selected geophysical applications was performed on the range of computational architectures, including both

common architectures, such as x86, and prospective one, vliw. The main emphasis was put on the computations, the I/O requirements, which play important
role during data processing, need to be investigated further.

2   Tested Applications

For this paper we have chosen following real applications: the seismic
data processing module (wemig/cazmig) and 3D reservoir simulating module
(tNavigator). The applications are computation intensive and memory bound.
Relatively small input data were used to minimize the I/O impact on the testing
results.

2.1   Module wemig/cazmig

The applications of the module wemig/cazmig implement typical seismic data
processing methods, based on Fourier transformation for solving the wave equation. At the same time, the applications are characterized by acceptable level
of computation complexity, which allows to use different techniques for testing
different computational platforms.

The application wemig implements a 2D-seismic migration method using
reverse-time wavefield continuation in frequency/space domains and depth
imaging [2].

Pseudocode of the application wemig using MPI-only technique of code parallelization is represented on the Fig. 1.

```
1. Read input  traces  (sorted CMP gathers{common midpoint Y, remove H, time T})
2. Fourier transformation of the input data on the time axis T
3. Transpose Fourier spectra{frequency F, remove W, midpoint Y}
4. Loop on frequencies F: // MPI
    1. Loop on depths Z:
        1. Loop on reference velocities V_ref:
            1. Fourier transformation on midpoints Y and removes H
            2. Phase shift for plane waves
            3. Inverse Fourier transformation on midpoints Y and removes H
            4. Thin-lens correction
        2. Evaluate the wavefield on the current depth by interpolation of
        wavefields obtained for different reference velocities
        3. Produce depth image on the current depth
        // Accumulation the results from all MPI-processes
        // using MPI_reduce
5.  Transpose depth image (depth Z, midpoint Y)
6.  Save depth image to disk
```

Fig. 1. Pseudocode of the application wemig

The cazmig migration algorithm proposed by Cazdag [3] and based on 3D
data migration. In this method all computations are performed in the frequency
domain, where the source and the receiver positions are aligned with the phase
shift by the Fourier coefficients rotation operation.

Hybrid MPI+OMP programming model is used for the application `cazmig` module.

The algorithm of the application `cazmig` is much more resource demanding as compared to the algorithm of the application `wemig`.

Binary codes were generated without any architecture-dependent optimization of the source codes except for the architecture-specific compilation options at the re-compilation phase.

We have used the cross-compilation procedure for KNC and KNL, and native compilation for Haswell, Broadwell and Elbrus.

Source codes were compiled with architecture-specific optimization keys, such as `"-xCORE-AVX2 -qopt-prefetch"` for Haswell/Broadwell, `"-xMIC-AVX512 -qopt-prefetch"` for Intel Xeon Phi 7120D, `"-mmic"` for for Intel Xeon Phi 7250, and `"-mcpu=elbrus-4c -mptr64 -ffast -ffast-math"` for Elbrus architectures.

The "c++11" standard features were used for Intel architectures, while these features are unavailable for the current release of Elbrus compiler.

The library `boost` was compiled with the highest available optimization keys for all architectures.

The MPICH2 library based on lcc compiler 1.21.07:Aug-10-2016 version was used on the Elbrus testbench. The Intel MPI Library versions 2016 and 2017 was used on the Intel platforms.

The Fourier transformations was done using the FFTW [12] library on the Elbrus testbench and Math Kernel Library [13] on the Intel test benches.

2.2 Module `tNavigator`

The reservoir simulator `tNavigator` is designed for running dynamic reservoir simulations on engineers laptops, servers, and HPC clusters [4]. This module solves equations of fluid flow dynamics in porous media. Set of the state variables includes molar densities and pressure that permits formulation of the general compositional model, where the black oil model is a special case. The general compositional model takes into account the following factors:

- Darcy' law for fluid flow dynamics;
- PVT tables with multiple PVT relative permeability regions;
- absolute permeability as a function of pressure;
- different options for aquifers;
- networks and gas gathering systems;
- tracer models.

Due to license restrictions only binary code of the module `tNavigator` adapted for x86 architecture was used. For that reason, on the e2k (vliw) architecture the module launching is possible only in binary compilation mode, that adds overhead costs and decreases the computation performance.

Furthermore, for the Intel Knight Landing processor (KNL) it is possible to execute the `tNavigator` simulator without re-compilation, in contrast with the previous generation of Intel Xeon Phi processors (Intel Knight Corner, KNC).

Since the architecture-dependent instruction sets, e.g. such as AVX-512F, AVX-512CD, AVX-512ER, and AVX-512FP, cannot be been used without the re-compilation with architecture specific optimization keys, so the performance results obtained on the KNL architecture are clearly not optimal.

Finally, we have used two different input data sets distinct in size and complexity.

3   Tested Hardware

Performance analysis of chosen applications was carried out on the following computational architectures:

– E5 series Intel Xeon processors;
– Intel Xeon Phi processors;
– Elbrus-4C processor from the MCST company.

Detailed specification information is listed in the Table 1.

Table 1. CPUs technical specifications

Characteristics	Specifications				
Model	E5-2697 v3	E5-2698 v4	Xeon Phi 7120D	Xeon Phi 7250	Elbrus-4C
Architecture	x86_64	x86_64	x86_64 (MIC)	x86_64 (MIC)	e2k (VLIW)
Clock speed, GHz	2.6	2.2	1.238	1.400	0.8
Number of cores	14	20	61	68	4
Number of threads	28	40	244	272	4
Peak performance (double precision), GFLOPS	582.4	665.6	2416.6	3046.4	25.6
Max memory Bandwidth, GB/sec	68.0	76.8	35.2	76.8	38.4

The Intel Xeon [8,9] Xeon Phi [10,11] series CPUs provide a high energy efficiency, multicore and simultaneous multi-threading support (so called hyper-threading), vector instruction sets (SSE and AVX), and the DDR4 and GDDR5 support.

The Elbrus microprocessor family belongs to the VLIW architecture class. The main VLIW feature is explicit instruction-level parallelism. On the compilation stage different program instructions that allow simultaneous execution are composed into Very-Long Instruction Words treated by CPU pipeline as single instructions[2].

---

[2] Very-Long Instruction Words are executed one per CPU cycle.

Table 2. Testbenches specifications

Codename	CPU	# Cores	Memory	GB per Core
Haswell_64 GB	Intel Xeon E5-2697 v3	2x 14	8x DRAM Micron 8 GB DDR4/2133 MHz	2.28
Haswell_128 GB	Intel Xeon E5-2697 v3	2x 14	8x DRAM Samsung 16 GB DDR4/2133 MHz	4.57
Broadwell_64 GB	Intel Xeon E5-2698 v4	2x 20	8x DRAM Micron 8 GB DDR4/2133 MHz	1.6
Broadwell_128 GB	Intel Xeon E5-2698 v4	2x 20	8x DRAM Samsung 16 GB DDR4/2133 MHz	3.2
KNC	Intel Xeon Phi 7120D	61	SDRAM Intel 16 GB GDDR5/2750 MHz	0.26
KNL	Intel Xeon Phi 7250	68	MCDRAM Intel 16 GB + 6x DRAM Micron 32 GB DDR4/2133 MHz	2.8
Elbrus	Elbrus-4C	4x 4	12x DRAM Micron 4 GB DDR3/1600 MHz	3

Number of CPUs and RAM along with their characteristics define the theoretical performance of a computational node. In the Table 2 codenames and specifications of studied testbenches are listed.

4   Applications Analysis

On all studied architectures the tracing was performed by the Paraver utility [5]. The set of profiling tools including Extrae [6], IPM [15] was used to describe the dynamical behavior of wemig/cazmig module.

Additionally, the runtime analysis for the application cazmig was done using MPI performance Snapshot [7].

4.1   CPU Workload

The tracing of the application wemig was done on the Haswell_64 GB testbench. The CPU workload is equally distributed during the runtime. (see Fig. 2).

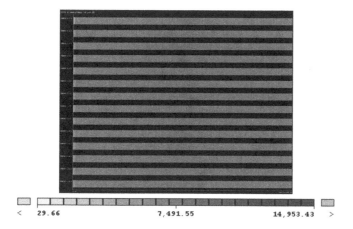

Fig. 2. MIPS trace of the application `wemig` (Haswell testbench; $N_{MPI} = 14$)

In the Table 3 different aspects of execution for the application `cazmig` are reflected. In this table 'MPI time' denotes the mean time per process spent in the serial and OMP parts of application code; 'OpenMP' denotes the mean time per process spent in the OpenMP parallel regions; 'I/O wait' denotes the total time, which the application was stalled due to I/O demand; 'I/O operations' denotes the time the application spends waiting for an I/O operation to complete.

Table 3. The application `cazmig` execution features

Characteristic	Haswell_128 GB	Broadwell_128 GB	KNL s
Calculation time %	95.92	94.10	90.52
MPI time %	4.08	5.90	9.48
OpenMP time%	86.58	83.13	83.12
I/O wait, sec	0.53	0.00	903.02
I/O operations, %	0.00	0.00	0.73

The computational load distribution for MPI processes are reflected in Fig. 3. Presented data indicate that on the Haswell and the Broadwell architectures the application `wemig` behaves similar[3], while KNL and Elbrus results differ significantly.

The MPI processes are loaded most uniformly on the Intel KNL architecture. The relative idle time ranges from 40% to 50 % percents of total process runtime.

---

[3] The GFLOPS metric wasn't measured because of architecture-specific difficulties of GFlops measuring on vector CPUs.

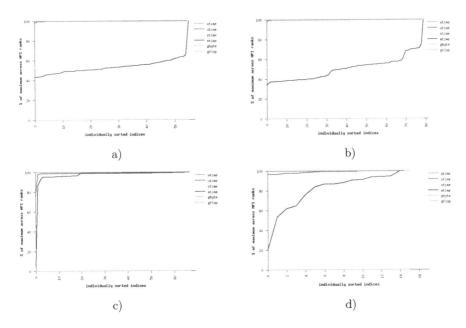

a)   b)

c)   d)

Fig. 3. The Rrelative computational load distribution of MPI processes for the application wemig on the (a) Haswell_128, (b) Broadwell_128, (c) KNL, (d) Elbrus Architectures; where wtime — wallclock time, utime — user time, stime system time, mtime the MPI time, gbyte the memory usage, gflop the computation performance per process.

The results of applications profiling reflect good scalability potential of these applications. Algthough, the workload is heavy and requires high computational demands to CPU according to MIPS metrics.

4.2   Memory Usage

Additional assessments of the module wemig/cazmig runtime were undertaken with regard to memory usage using Intel MPI Performance Snapshot [7]. The analysis results show, in particular, the higher requirements to the RAM 4.

The application wemig was analyzed with 56 MPI ranks execution, while application cazmig was executed using 14 MPI ranks and each MPI process consists of 14 OpenMP threads (Table 4).

Table 4. Memory usage

Application	Memory Bound, %		MB per process	
	Min	Max	Mean	Peak
wemig	31	31	36.48	35.58
cazmig	46	46	879.54	735.39

The memory bound values are relatively high what seems to indicate that the processor pipeline was stalled due to demands to load or store instructions.

4.3   MPI Communications

The MPI_reduce operation over the data located in different address spaces of different MPI processes, occupies most of the time during parallel execution 4.

The MPI communications pattern of the application cazmig is similar to wemig, but the heterogeneous parallel programming model is used.

The MPI data transfer percentages are maximal for the KNL testbench, while for Haswell and Broadwell testbenches MPI data transfers are accounted for less then 11% of total execution time (see the Table 5). The minimal relative data transfer rates were observed on the Elbrus testbench.

Analogous results for the application cazmig are presented in the Table 5.

Table 5. The Communication/Computation ratio (%) for the Applications wemig and cazmig

Application	Haswell_128 GB	Broadwell_128 GB	KNL	Elbrus
wemig %	4.5/95.5	10.9/89.1	33/67	3/97
cazmig%	4.1/95.9	5.9/94.1	9.5/90.5	4.3/95.7

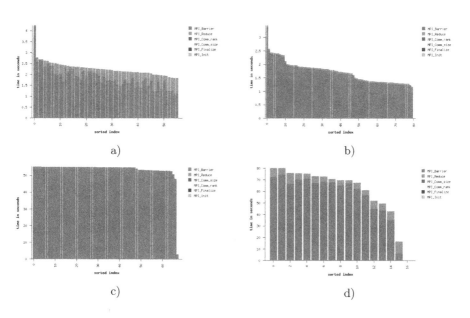

a)   b)

c)   d)

Fig. 4. MPI processes communication balance for the application wemig on the (a) Haswell_128 GB, b) Broadwell_128 GB, (c) KNL, (d) Elbrus architectures

The MPI calls pattern and the task distribution display high scalability potential for all studied testbenches. So it can be assumed that total CPU usage would be uniform with increasing the number of nodes and the data volume.

The MPI stack time distribution for the studied testbenches in shown on the Fig. 4.

5   Benchmarking

5.1   Module wemig/cazmig

Absolute values of the execution time for the application wemig are listed in the Table 6. This table includes execution time for two cases. The first case corresponds to the minimal parallel execution possible case with 2 MPI-processes; the second case is the maximum workload with the maximum number of MPI ranks listed below, which depends on the number of physical cores and the multi-threading technologies [14].

According to the preliminary investigation, such numbers are:

- 56 for Haswell-nodes (14 physical cores x 2 CPU per node x 2 hyperthreading);
- 80 for Broadwell-nodes (20 physical cores x 2 CPU per node x 2 hyperthreading);
- 240 for KNC-node (60 physical cores x 1 CPU per node x 4 hyperthreading);
- 272 for KNL-node (68 physical cores x 1 CPU per node x 4 hyperthreading);
- 16 for Elbrus-node (4 physical cores x 4 CPU per node).

Table 6. Test runtimes for the application wemig

Testbench	Number of MPI ranks			
	Min		Min	
	N MPI	time	N MPI	Time
Haswell_64 GB	2	12 min 21 s	56	56 s
Haswell_128 GB	2	12 min 15 s	56	50 s
Broadwell_64 GB	2	16 min 24 s	80	55 s
Broadwell_128 GB	2	10 min 38 s	80	36 s
KNC	2	550 min 21 s	240	12 min 58 s
KNL	2	246 min 34 s	272	2 min 59 s
Elbrus	2	280 min 11 s	16	35 min 47 s

It is interesting to note that the doubling of memory capacity leads to the significant performance increase on the Broadwell testbench, while on the Haswell testbench productivity gains are not substantial. It seems the Broadwell cores with 64 GB RAM configuration were stalled due to memory demands, and the Haswell_64 GB results are more balanced.

Similarly the low test performance observed on the KNC testbench with small memory capacity points out to the severe memory requirements for processing seismic data.

So it is recommended to use RAM more than 3 GB per core for computational node with the Broadwell CPU and no less 2 GB per core for the other types of nodes.

As it was mentioned above the application `cazmig` has the computation pattern similar to the application `wemig`, but the parallel programming model of the application `cazmig` includes MPI and OpenMP levels.

The testing results of the application `cazmig` are listed in the Table 7. The number of MPI-processes and OMP-threads represented in the table reflects the optimal execution options.

Table 7. The minimal execution time T_{run} for the application `cazmig`

Testbench	T_{run}	N_{MPI}	N_{OMP}
Haswell_64 GB	57 min 35 s	14	14
Haswell_128 GB	55 min 39 s	14	14
Broadwell_64 GB	56 min 13 s	4	16
Broadwell_128 GB	43 min 2 s	16	16
KNL	70 min 30 s	34	68

5.2   Module `tNavigator`

The benchmarking results for the Haswell_64 GB/Haswell_128 GB, Broadwell_64 GB/Broadwell_128 GB testbenches are listed in the Fig. 5. The test runs were done using two different input data models. First one, Model A has larger size and includes more active blocks, while Model B is more complex and differs in increased number of components and connections (Table 8).

Table 8. TNavigator data models characteristics

Feature	Model A	Model B
Number of components (including water)	3	9
Total active grid blocks	2418989	36097
Number of connections	853	1882

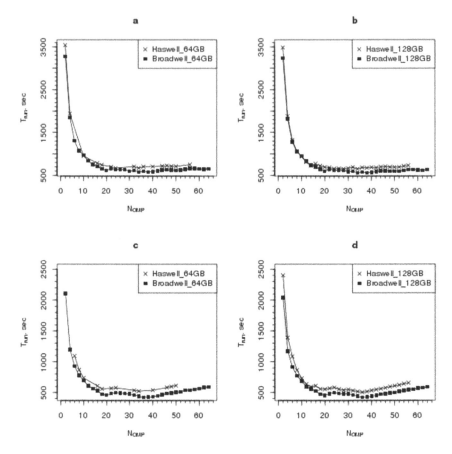

Fig. 5. The execution time T_{run} ratio to the thread number for the tNavigator simulator. (a)–(b) Input data model A (c)–(d) Input data model B

6   Conclusion

This work includes tracing and profiling analysis, as well as workload and time rates results for the 2D and 3D real seismic data processing.

The Intel Xeon E5 testbenches with x86 architecture and 128 GB RAM (the Haswell_128 and Broadwell_128) show the best result in absolute values. Moreover, time and efforts spent on source code compilation, and tracing/profiling were minimal for these testbenches because of the abundance of x86 architectures and documentations.

For the new generation of Intel Xeon Phi processor, KNL testbench, the benchmarking results were also satisfactory; but there was still room for source code improvements. Additional benefit of this CPU is energy efficiency. Besides the code migration time was minimal; while the migration of real seismic module to the previous generation Intel Xeon Phi co-processor, KNC testbench, involved considerable efforts because of cross-compilation issues of the dependence code. The performance level of KNC testbench was also lower than of KNL testbench.

Although for Elbrus testbench overall absolute values were lower than values for the Intel Xeon; the relative performance, i.e. real to peak performance ratio, was high. For some micro-benchmarks, e.g. Fourier transformation benchmark, it was higher than for Intel Xeon architectures. Thus the Elbrus-based computational nodes might be effective for seismic data pre-processing, where the Fourier transformation methods are actively used. It is worth mention that code migration issues could require significant efforts.

In this paper two types of real applications for solving two different groups of tasks were considered. Two different traits of behavior for them were revealed by tracing, profiling, and benchmarking procedures.

Therefore, more effective utilization of computational resources of modern supercomputers for solving problems of the first group can be obtained by massive-parallel execution of one application instance on the large number of computational nodes.

Independent execution of several application instances with different input data on the relatively moderate number of computational nodes can be recommended for the problems of the second group.

Acknowledgments. This research was supported by the Common State Scientific and Technological Programme "SKIF-Nedra" with funding from Ministry of Education and Science of the Russian Federation.

We thank the JSCC RAS (Joint Supercomputer Center of the Russian Academy of Sciences) for the provided computational resources.

References

1. Rechistov, G., et al.: Simulation and performance study of large scale computer cluster configuration: combined multi-level approach. In: International Conference on Computational Science, Omaha, Nebraska, USA, pp. 1–10 (2012)
2. Popovici, A.M.: Prestack migration by split-step DSR. Geophysics **61**, 1412–1416 (1996)
3. Gazdag, J.: Wave equation migration with equation migration with phase shift method. Geophysics **43**, 1342–1351 (1978)
4. tNavigator Technical Description. http://rfdyn.com/technology/technical/. Accessed 22 Nov 2016
5. Paraver: Performance analysis tools: details and intelligence. http://www.bsc.es/computer-sciences/performance-tools/paraver. Accessed 18 Nov 2016
6. Paraver: Trace-generation package. https://www.bsc.es/computer-sciences/extrae. Accessed 22 Sept 2016
7. Intel MPI Performance Snapshot. Home Page. https://software.intel.com/sites/products/snapshots/mpi-snapshot/. Accessed 21 Nov 2016
8. Intel Xeon Processor E5–2697 v3 Technical Specification Page. http://ark.intel.com/ru/products/81059/Intel-Xeon-Processor-E5-2697-v3-35M-Cache-2_60-GHz. Accessed 21 Nov 2016

9. Intel Xeon Processor E5–2698 v4 Technical Specification Page. http://ark. intel.com/ru/products/91753/Intel-Xeon-Processor-E5-2698-v4-50M-Cache-2_ 20-GHz. Accessed 21 Nov 2016

10. Intel Xeon Phi Coprocessor 7120A Technical Specification Page. http:// ark.intel.com/ru/products/80555/Intel-Xeon-Phi-Coprocessor-7120A-16GB-1_ 238-GHz-61-core. Accessed 21 Nov 2016

11. Intel Xeon Phi Processor 7250 Technical Specification Page. http://ark.intel.com/ ru/products/94035/Intel-Xeon-Phi-Processor-7250-16GB-1_40-GHz-68-core. Accessed 21 Nov 2016

12. Matteo, F., Steven, J.G.: The design and implementation of FFTW3. Proc. IEEE **93**(2), 216–231 (2005). Special issue on Program Generation, Optimization, and Platform Adaptation

13. Intel MKL Home Page. https://software.intel.com/en-us/intel-mkl. Accessed 29 Nov 2016

14. Intel Hyper-Threading Technology. http://www.intel.ru/content/www/ru/ru/ architecture-and-technology/hyper-threading/hyper-threading-technology.html. Accessed 29 Nov 2016

15. Wright, N.J., Pfeiffer, W., Snavely, A.: Characterizing parallel scaling of scientific applications using IPM. In: The 10th LCI International Conference on High-Performance Clustered Computing. Boulder, CO., 10–12 March 2009

Simulation of Global Seismicity: New Computing Experiments with the Use of Scientific Visualization Software

Lidiya Melnikova[1], Igor Mikhailov[2], and Valeriy Rozenberg[1(✉)]

[1] Krasovskii Institute of Mathematics and Mechanics of the Ural Branch of the Russian Academy of Sciences, 16, Kovalevskoi Street, 620990 Ekaterinburg, Russia
`{meln,rozen}@imm.uran.ru`
[2] Applied Technologies, Ltd., 81, Lunacharskogo Street, 620075 Ekaterinburg, Russia
`igormich88@gmail.com`

Abstract. The paper briefly describes a modification of the spherical model of lithosphere seismicity taking into account random factors that influence the dynamics of parameters of interacting block structure elements. We present the results of numerical experiments that confirm the benefits of adopting stochastic procedures in the model. Also, we designed a special software to study the process of model stress propagation in the system of tectonic plates in detail. This software allows for the visualization of both the instantaneous stress distribution along a fault and the temporal migration of critical values. New possibilities are used for testing different interconnections between model characteristics and comparing them with real patterns. Finally, we investigate the role of exogenous and endogenous parameters in the model calibration.

Keywords: Block-and-fault models of lithosphere dynamics and seismicity · Earthquake catalogs · Scientific visualization software

1 Introduction: The Necessity of Seismicity Simulation

The statistical analysis of seismicity as the spatial-temporal sequence of earthquakes in a given area on the basis of real catalogs is heavily hindered by the short history of reliable observation data. The patterns of the earthquake occurrence that are identifiable in real catalogs may be only apparent and not be repeated in the future. At the same time, synthetic catalogs obtained from numerical simulations can cover very long time intervals; this allows us to acquire more reliable estimates of parameters of a seismic flow and search for premonitory patterns preceding large events. Such a possibility may be needed in expert systems for global/regional seismic risk monitoring [1,2]. The main result of modeling is a synthetic earthquake catalog; each of its events is characterized by a time moment, epicenter coordinates, a depth, a magnitude, and,

The work was supported by the Program of the Ural Branch of the Russian Academy of Sciences, project no. 15-7-1-13.

L. Sokolinsky and M. Zymbler (Eds.): PCT 2017, CCIS 753, pp. 215–232, 2017.
DOI: 10.1007/978-3-319-67035-5_16

for some models taking into account the geology of a region, an intensity. The simulation of lithosphere dynamics provides the field of velocities at different depths, acting forces, induced displacements, and the character of interaction between structural elements. So far, there is no adequate theory of seismotectonic processes, but, using the available data, one can assume that various features of the lithosphere (e.g., the spatial heterogeneity, hierarchical block structure, and different types of nonlinear rheology) are related to properties of earthquake sequences. The stability of these properties at a quantitative level in different regions allows us to conclude that it is possible to consider the lithosphere as a large dissipative system whose behavior does not essentially depend on particular details of specific processes progressing in a geological system. There exist many different approaches to modeling lithospheric processes (see, for example, [1] and its bibliography); nevertheless, we can mark out the two main directions. The first (traditional) one relies on the detailed investigation of a specific tectonic fault, or, rather often, of a strong earthquake in order to reproduce certain pre- and/or post-seismic phenomena (relevant to this fault or event). In contrast, models of the second direction developed relatively recently treat the seismotectonic process in a rather abstract way; the main goal of simulations is to reproduce general universal properties of observed seismicity (primarily, the power law for earthquake "size" distribution, namely the Gutenberg–Richter law on frequency-magnitude (FM) relation, clustering, migration of events, seismic cycle, and so on). However, it seems that an adequate model designed in the framework of the second direction should reflect both some universal features of nonlinear systems and the specific geometry of interacting faults.

The block models of lithosphere dynamics and seismicity [2] have been developed taking into account both requirements. The approach to modeling is based on the concept of the hierarchical block structure of the lithosphere (see [2]. Tectonic plates are represented as a system of perfectly rigid blocks being in a quasi-static equilibrium state. A model event is a stress drop at a fault separating two blocks occurring under the action of outer forces. In the model, the two main mechanisms are involved in the seismotectonic process: the tectonic loading, with characteristic rate of a few cm/yr, and the elastic stress accumulation and redistribution, with characteristic rate of a few km/sec. They are considered over time as a uniform motion and an instantaneous stress drop, respectively. Wave processes are beyond the scope of existing block models. The plane model [2], where a structure is restricted by two horizontal planes, has been the most extensively studied. Models approximating the dynamics of lithosphere blocks for real seismic regions have been built on its basis [2,3]. However, significant distortions have been revealed when simulating the motion of large tectonic plates; therefore, the spherical geometry has been involved. The computational realization of the spherical modification requires much more expenditures of memory and processor time than the plane model, the use of multiprocessor machines and parallel computing technologies is preferable.

The present paper actually continues the investigations of the problem of lithosphere dynamics and seismicity simulation by means of the spherical block model developed by the authors [4–6]. The last modification has adopted additional possibilities of the mathematical model associated with the introduction of random factors. On the one hand, this allows us to improve some properties of synthetic seismicity [6], and, on the other hand, this creates a need for studying the model stress propagation in the system of tectonic plates for the purpose of feasibility verification. Such a study makes the visualization of both the instantaneous stress distribution along a fault and the temporal migration of critical values along the system of faults very useful. The big data and the necessity of its graphic representation have impeded the realization of the options in question, thus decelerating the process of model development. To overcome these difficulties, the program package has been supplied by scientific visualization tools to verify simulation results. The novelty of the present paper lies in the description of the "fresh" software and the discussion of some results of new computing experiments.

2   Spherical Block Model: A Stochastic Version

Let us briefly describe the last version of the spherical block-and-fault model of lithosphere dynamics and seismicity. The detailed description of all the modifications of the model can be found in [5]; in this paper, we restrict ourselves to a summary of the basic ideas and principles with an emphasis on new constructions. A block structure is a limited and simply connected part of a spherical layer of depth H bounded by two concentric spheres. The outer sphere represents the Earth's surface and the inner one represents the boundary between the lithosphere and the mantle. The partition of the structure into blocks is defined by infinitely thin faults intersecting the layer. Each fault is a part of a cone surface having the same value of the dip angle with the outer sphere at all its points. Faults intersect along curves that meet the outer and inner spheres at points called vertices. A part of such a curve between two respective vertices is called a rib. Fragments of faults limited by two adjacent ribs are called segments. The common parts of blocks with the limiting spheres are spherical polygons; those on the inner sphere are called bottoms. A block structure may be a part of the spherical shell and be bordered by boundary blocks that are adjacent to boundary segments. Another possibility (impossible in plane modifications) is to consider the structure covering the whole surface of the Earth without boundary blocks. The blocks are assumed to be perfectly rigid. All block displacements are considered as negligible compared with block sizes. Therefore, the geometry of the block structure does not change during the simulation, and the structure does not move as a whole. The gravitation forces remain essentially unchanged by the block displacements and, because the structure is in a quasi-static equilibrium state at the initial time moment, gravity does not cause a motion of the blocks. All blocks (both internal and boundary if specified) have six degrees of freedom and can leave the spherical surface. The displacement of each block consists of translation and rotation components. The motions of the boundary blocks as well as those of the underlying medium, considered as an external action on the structure, are assumed to be known. As a rule, they are described as rotations on the sphere, i.e. axes of rotation and angular velocities are given.

The current modification admits the specification of different depths for different blocks (in the range of H) and the changes of fault parameters depending on the depth. Note that this is the first attempt to take into account the inhomogeneity of the lithosphere in block models. In this modification, we consider two ways of introducing stochasticity into the procedures for the calculation of the forces acting on a block and actually determining model earthquakes. This stochasticity consists of (i) adding a noise to the differential equations describing the dynamics of forces and displacements, and (ii) using random variables when specifying strength thresholds for the medium of tectonic faults.

Since the blocks are perfectly rigid, all the deformations take place at the fault zones and block bottoms; the forces arise at the bottoms due to the relative displacements of the blocks with respect to the underlying medium, and at the faults due to the displacements of the neighboring blocks or their underlying medium. Let us present the formulas for the elastic force (f_t, f_l, f_n) acting on a fault per unit area:

$$f_t(\tau) = K_t(\Delta_t(\tau) - \delta_t(\tau)), \quad f_l(\tau) = K_l(\Delta_l(\tau) - \delta_l(\tau)),$$

$$f_n(\tau) = K_n(\Delta_n(\tau) - \delta_n(\tau)). \tag{1}$$

Here τ is the time; (t, l, n) is the rectangular coordinate system with origin at the point of application of the force (the axes t and l lie in the plane tangent to the fault's surface, the axis n is perpendicular to this plane); $\Delta_t, \Delta_l, \Delta_n$ are the components of the relative displacement in the system (t, l, n) of neighboring elements of the structure; $\delta_t, \delta_l, \delta_n$ are corresponding inelastic displacements, the evolution of which is described by the linear stochastic differential equations

$$d\delta_t(\tau) = W_t K_t(\Delta_t(\tau) - \delta_t(\tau))d\tau + \lambda_t d\xi_t(\tau),$$

$$d\delta_l(\tau) = W_l K_l(\Delta_l(\tau) - \delta_l(\tau))d\tau + \lambda_l d\xi_l(\tau),$$

$$d\delta_n(\tau) = W_n K_n(\Delta_n(\tau) - \delta_n(\tau))d\tau + \lambda_n d\xi_n(\tau). \tag{2}$$

We denote by ξ_t, ξ_l and ξ_n standard independent scalar Wiener processes (i.e. processes starting from zero with zero mathematical expectation and dispersion equal to τ); the coefficients λ_t, λ_l and λ_n characterize the amplitude of random noises. There exists a unique solution (in Ito's sense) to each equation of (2); this is a normal Markov random process with continuous realizations [7]. The coefficients K_t, K_l, K_n (1) and (2) characterizing the elastic properties of the faults, the coefficients W_t, W_l, W_n (2) characterizing the viscous properties, and the coefficients λ_t, λ_l, λ_n (2) may be different for different faults and, in addition, may depend on the depth. The formulas for calculating the elastic forces and inelastic displacements at the block bottoms are similar to (1) and (2). At the initial time $\tau = 0$, all the forces and displacements are equal to zero; actually, the source of motion of the model structure is the motion of the underlying medium and the boundary blocks (if specified) determining the dynamics of relative displacements (like $\Delta_t, \Delta_l, \Delta_n$) at both the faults and block bottoms.

The translation vectors of the inner blocks and the angles of their rotation can be found from the condition that the total force and the total moment of the forces acting on each block are equal to zero. This is the condition of quasi-static equilibrium of the system, and at the same time the condition of energy minimum. It is important that the dependence of forces and moments on displacements and rotations of blocks is linear (explicit cumbersome formulas are omitted). Therefore, the system of equations for determining these values is also linear:

$$\mathbf{A}\mathbf{w} = \mathbf{b}. \tag{3}$$

Here the components of the unknown vector $\mathbf{w} = (w_1, w_2, ..., w_{6n})$ are the translation vectors of the inner blocks and the angles of their rotation (n is the number of blocks). The elements of the matrix $\mathbf{A}$ (of dimension $6n \times 6n$) do not depend on the time and can be calculated only once, at the beginning of the process. For realistic values of the model parameters, the matrix $\mathbf{A}$ is non-degenerate; system (3) has a unique solution. To calculate various curvilinear integrals, one should discretize (divide into cells) the spherical surfaces of the block bottoms and fault segments. The values of the forces and inelastic displacements are assumed to be equal for all points of a cell. System (3) is solved at discrete times τ_i.

At every time step τ_i, when computing the force acting on the fault, we find the value of a dimensionless quantity κ (a model stress) by the formula

$$\kappa = \frac{\sqrt{f_t^2 + f_l^2}}{P - f_n}. \tag{4}$$

Here P is a parameter, which may be interpreted as the difference between the lithostatic and the hydrostatic pressure. Thus, the value of κ is actually the ratio of the modulus of the force tending to shift the blocks along the fault to the modulus of the force connecting the blocks to each other. For each fault, three strength levels are specified. In general, they depend on time:

$$B > H_f \geq H_s, \quad B = B(\tau_i) = B_0(\tau_i) + \sigma X(\tau_i),$$

$$H_f = H_f(\tau_i) = aB(\tau_i), \quad H_s = H_s(\tau_i) = bB(\tau_i). \tag{5}$$

For each i, we assume that $0 < B_0(\tau_i) < 1$, $0 < \sigma \ll 1$, $X(\tau_i)$ is a normally distributed random value $N(0; 1)$, $0 < a < 1$, $0 < b \leq a$. The initial conditions are such that the inequality $\kappa < B$ is valid for all the cells of the structure.

The interaction between the blocks (between the block and neighboring underlying medium) is visco-elastic (a "normal state") as long as the value of κ (4) at the part of the fault separating the structural elements remains below the strength level B. When this level is exceeded at some part of the fault (a "critical state"), a stress-drop (a "failure") occurs in accordance with the dry friction model (such failures represent earthquakes). By a failure we mean a slippage by which the inelastic displacements δ_t, δ_l and δ_n in the cells change abruptly to reduce the model stress according to the formulas

$$\delta_t^e = \delta_t + \gamma^e \xi_t f_t, \quad \delta_l^e = \delta_l + \gamma^e f_l, \quad \delta_n^e = \delta_n + \gamma^e \xi_n f_n, \tag{6}$$

where δ_t, f_t, δ_l, f_l, δ_n and f_n are the inelastic displacements and the elastic force per unit area just before the failure. The coefficients $\xi_t = K_l/K_t$ ($\xi_t = 0$ if $K_t = 0$) and $\xi_n = K_l/K_n$ ($\xi_n = 0$ if $K_n = 0$) account for inhomogeneities of the displacements in different directions. The coefficient γ^e is given by

$$\gamma^e = \frac{\sqrt{f_t^2 + f_l^2} - H_f(P - f_n)}{K_l\sqrt{f_t^2 + f_l^2} + K_n H_f \xi_n f_n}. \tag{7}$$

It follows from (1) and (4)–(7) that, after recalculating the new values of the inelastic displacements and elastic forces, the value of κ is equal to H_f. Then, the right-hand part of the system (3) is computed, and the translation vectors and angles of rotation for the blocks are found again. If for some cell(s) $\kappa \geq B$, then the entire procedure is repeated. When $\kappa < B$ for all the cells of the structure, the calculations are continued according to the standard scheme. Immediately after the earthquake, it is assumed that the cells in which the failure occurred are creeping (or are in the creep state). This implies that, for these cells, the parameters W_t^s ($W_t^s \gg W_t$), W_l^s ($W_l^s \gg W_l$), and W_n^s ($W_n^s \gg W_n$) are used instead of W_t, W_l and W_n in equations (2). Such new values provide a faster growth (compared to the normal state) of the inelastic displacements and, consequently, a decrease of the stress. The cells remain in the creep state so long as $\kappa > H_s$ (we may call this process "healing"); when $\kappa \leq H_s$, the cells return to the normal state with the use of W_t, W_l and W_n in (2). The process of stress accumulation starts again. It should be noted that this process is far from periodic, since the values of strength levels (5) depend on time in a special way.

We consider the geometry of the block structure and the characteristics of the outer motions as exogenous model parameters, whereas the coefficients, strength levels and discretization steps described above as endogenous ones.

A synthetic earthquake catalog is produced as a main result of the numerical simulation. All the cells of the same fault, in which the failure occurs at the time τ_i, are considered as a single event. Its epicentral coordinates and depth are the weighted sums of the coordinates and depths of the cells involved in the earthquake. The weighted sum of the vectors $(\gamma^e \xi_t f_t, \gamma^e f_l)$ added to the inelastic displacements δ_t and δ_l computed according to (6) approximates the shift of the blocks along the fault and allows us to determine the mechanism of a synthetic event. Such a mechanism is an important feature of an earthquake; it informs on the process of propagation of different seismic waves from the earthquake source. Depending on the direction of the shift and the dip angle of the fault, the following basic mechanisms are commonly considered: strike-slip, normal faulting, and thrust faulting [8]. In the current version of the model, the magnitude of an earthquake is calculated taking into account its mechanism according to the well-known in seismology empirical formulas [9]

$$M = D \lg S + E, \tag{8}$$

where S is the total area of the cells (in km^2), $D = 1.02$, $E = 3.98$ for strike-slip, $D = 1.02$, $E = 3.93$ for normal faulting, and $D = 0.90$, $E = 4.33$ for

thrust faulting. In addition to the aforesaid, the model informs on the instantaneous kinematics of the blocks and the character of their interaction along the
boundaries.

3   Organization of Numerical Experiments

The spherical block model is most actively applied to investigating the dynamics and seismicity of the global system of largest tectonic plates covering the
whole surface of the Earth [4–6]. The geometry of the structure is assumed to be
invariable: 15 plates (North America, South America, Nazca, Africa, Caribbean,
Cocos, Pacific, Somalia, Arabia, Eurasia, India, Antarctica, Australia, Philip-
pines, and Juan de Fuca), 186 vertices, 199 faults. Boundary blocks are absent,
since the structure occupies the entire spherical shell; the motion of the underlying medium is defined as rotations on the sphere according to the model HS3-
NUVEL1 [10]. Numerical values of different coefficients are varied in order to
obtain the best correspondence of the simulation results to real seismic data.
As main characteristics for the comparative analysis of the modeling quality, we
take the spatial distribution of strong events, the Gutenberg–Richter law on FM
relation, clustering, migration of events, seismic cycle, and so on [1,2].

   In a previous research [5,6], it was established that the spherical block
model admits an effective parallelization based on the standard scheme "master-
worker" with a unique loading MPI-module. At every moment, the most time-
taking procedures are the preservation of information on model events and, substantially, the calculation of forces, inelastic displacements, and stresses (1), (2),
(4)–(6) in all the cells of the structure (so, in a typical variant for the global
system of tectonic plates, we have got more than 235000 cells at the block bottoms and about 3500000 cells at the fault segments). The main calculations are
performed independently from each other; therefore, they are shared uniformly
among processors. The informational exchange between processors involves only
values of small dimension. The detailed description of the parallelization scheme,
the procedure for the dynamical redistribution of loading and the analysis of
some characteristics of the parallelization quality are presented in [6].

   The introduction of the stochastic component in equations (2) and the random temporal fluctuations of strength levels (5) destined to reflect the fault
medium strength variability (not amenable to a precise analytical description
and actually non-predictable) were tested in experiments described in [6]. It was
revealed that the addition of random factors to the procedure of model event
identification provides the time exfoliation concerning the transition of cells from
the same fault into the critical state. Due to this fact, the number of small events
increases and clustering is more clearly detected (clusters consist of foreshocks,
main shocks and aftershocks; as a rule, they are present in the real seismicity
[1,2]). This essentially improves the properties of synthetic seismicity compared
to the previous versions of the model [4,5]. However, we have met serious difficulties when studying the instantaneous model stress distribution along the
faults and the new, randomly influenced, character of temporal stress dynamics.

The matter is that it is useful to compare the states of faults at subsequent times for a better analysis of synthetic seismicity properties. Toward this aim, a high-speed graphic interpretation of 2 to 3 Gb of data is needed. Thus, the need has arisen to provide the existing software with a new program to clearly visualize simulation results.

4   Visualization Program for the Stress Dynamics

We describe the new software that allows us to visualize both the instantaneous stress distribution along a fault and the migration of critical values over time.

Specialized systems are one of the directions for developing scientific visualization tools providing a good quality of mathematical modeling. Due to the specific character of the data and their analysis, universal visualization programs can not be used without significant modifications, comparable in complexity to writing a separate module. The advantages of a specialized system are its independence and easy adaptation to the task, including an appropriate interface.

The Java language was chosen to write the program. Due to its cross-platform feature, it is possible to run the program on either Windows or Linux computers without recompiling. During the development phase of the software, third-party libraries were not involved in order to prevent additional dependencies. To improve the program's performance when working with large amounts of data, its parallelization by standard Java tools (the Thread class) was used; this increased the rendering speed. This optimization did not require significant changes in the program code, since the input file is used as read-only, and the visualized image areas do not intersect at different flows. In connection with the problem of integration, when one view type or one model is insufficient for the verification of visualized data, the program applies the so-called multiple views formalized in [11]. This concept involves the use, for the data interpretation, of several separated views with special relationships established. An important result of the application of multiple views is the reduction of the amount of information displayed simultaneously. In the described program, a binary view is actually used. This view includes the mini-map ensuring the integral (but inaccurate) perception of the structure geometry and the main image showing a detailed and precise (but incomplete) distribution of numerical data. Thus, while working with the visualization program, the user constantly has access to the map of the Earth's surface (see Fig. 1). Its black-and-white schematic representation serves as an auxiliary tool for the orientation in tectonic fault locations in the block structure. The parts of the structure with cells in the critical state are highlighted in red. After selecting a specific fault, a new window opens (also shown in Fig. 1). In this window, all the data on the fault loaded from the input file are graphically represented as a frame vertical scanning, where each point in time corresponds to a separate frame; frames not shown on the screen are available via a scroll bar.

To visualize the state of the cells, gray (from black to white) has been selected as basic color; it marks the cells with model stress below a subcritical level

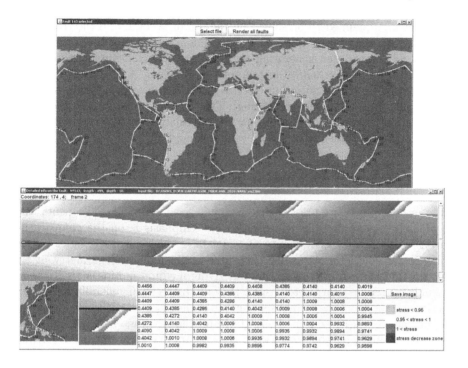

Fig. 1. Binary view of visualized data (Color figure online)

(typically 95% of the level B; the normalization with respect to B is performed providing output values from 0 to 1). To characterize the cells with values in the range from 0.95 to 1 (i.e. up to the critical level), the yellow color is used. The cells in which the stress is greater than the critical level (they form a model earthquake) are colored in bright red, because specifically these cells are of most interest. In addition, the parts of the faults where the model stress decreases compared to the previous frame are colored in blue; these cells were red at the preceding point in time. Thus, the area of red/blue dots on the image shows the earthquake spread dynamics along the system of tectonic faults. In the lower window of Fig. 1, a typical pattern of migration of "quaked" cells along a fault is shown. Then, this window contains a fragment of the global map, the enlarged selected part of the fault and a table with the numerical values of the model stress in the cells of this part. Note that when working with a large number of frames, if it is impossible to place all of them on the screen at the same time at a satisfactory resolution, you can switch on the comparison mode for some selected frames. For example, this option allows us to analyze the initial and final frames of a long time interval.

As one of the most promising directions of development of the visualization program, we can mention a hypothetical online service; the main obstacle here is the security policy of browsers, which makes it difficult to work with local files. To a lesser extent, this approach is limited by the performance of javascript.

The problem is planned to be solved by the use of WebGL and the transfer of calculations to a graphics card. It is also possible to implement the transmission of data to a server, then generating images on it and transmitting them back to the client (remote visualization). Another promising development direction is a version with 3D visualization of data from all the tectonic faults at some selected point in time (volume rendering). This version is restricted by the screen resolution, but it can be very useful for a rapid analysis of the structure's state to correct some parameters in the process of simulation.

5   Discussion of Simulation Results

In the present paper, we present the results of new numerical experiments with the block structure described in Sect. 3. The emphasis is on the comparative analysis of the degree of influence of two sets of model characteristics on the properties of synthetic seismicity. The first set includes the geometry of the structure and the parameters of the underlying medium motion actually being external (exogenous) parameters, since they are taken from generally recognized theories and models. The second set consists of the visco-elasticity coefficients of the fault and block bottom medium; they are treated as internal (endogenous), subjective enough, parameters. Two series of variants were organized: (i) with the same values of coefficients for all the faults and blocks, and (ii) with different values chosen on the basis of the real seismic activity of the corresponding regions. The main idea of the experiment is to verify the hypothesis that adequate results can be obtained not due to the so-called internal calibration of the model but due to the development of its algorithmic part. Let us discuss some simulation results for basic variants of each series.

Variants I and II (from the first series) use the following parameter values for all the faults: $K_t = K_l = K_n = 5$, $W_t = W_l = W_n = 0.01$, and $W_t^s = W_l^s = W_n^s = 10$. As to variants III and IV (from the second series), these values are varied from 10% to 1000% of the basic levels (in variants I and II); the changes are based on the observed seismicity: the coefficients K_t, K_l and K_n are increased, whereas the coefficients W_t, W_l and W_n are decreased for faults with a high level of seismic activity (see [2]). Variants I and III do not engage any random disturbances; they are characterized by the constant strength levels $B = 0.1$, $H_f = 0.085$ and $H_s = 0.7$, and all zero coefficients λ_t, λ_l and λ_n in equations (2). In variants II and IV, the levels B, H_f, H_s are perturbed according to (5) in the unit of model time with the parameters $B(\tau_i) = 0.1$ for each i, $\sigma = 0.005/3$, $a = 0.85$, and $b = 0.7$ (in this case, by the "3σ rule", we assume $B(\tau_i) \in [0.095; 0.105]$); in addition, equations (2) are also disturbed: $\lambda_t = \lambda_l = \lambda_n = 0.1$. In the variants with random values, results averaged by some number of runs are analyzed.

The comparative analysis of the spatial distributions of the epicenters of strong events recorded in the real catalog [12], including events for the period 01.01.1900–04.07.2016 without any restrictions of depth and area (Fig. 2), and those in the model catalogs (variants I–IV give rather similar distributions;

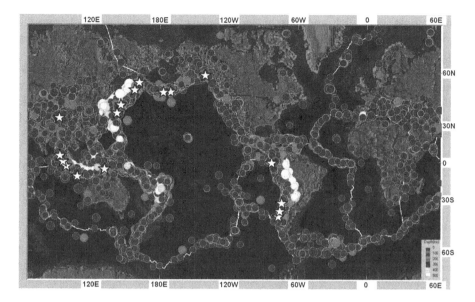

Fig. 2. Registered seismicity: epicenters of strong earthquakes with $M \geq 6.0$, NEIC, 01.01.1900–04.07.2016 [12], 17 extreme events with $M \geq 8.5$ are marked with stars

Fig. 3 corresponds to variants II and IV) shows that a number of common features revealed earlier [6] is preserved. In particular, the most important patterns of the global seismicity should be noted: two main seismic belts, the Circum-Pacific and Alpine-Himalayan, where most of the strong earthquakes occur, and the increased seismic activity associated with triple junctions of plate boundaries. It seems very promising that the strongest events in the model occur approximately at the same places as in reality. The distinctions between the synthetic and real seismicity are also obvious: the absence of model events inside the plates (in the model, earthquakes occur only at faults) and at some boundaries

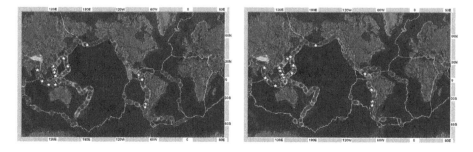

Fig. 3. Synthetic seismicity: epicenters of strong earthquakes with $M \geq 6.0$, 100 units of model time, variant II (*left*) and variant IV (*right*), 17 strongest events are marked with stars

(for example, the north and east of African plate). The epicenter distributions in variants II and IV do not reveal qualitative distinctions, which one can expect taking in mind their essential differences in parameter values.

As to the distribution in depth (see Table 1), we observe (i) the noticeable similarity of the shares of shallow earthquakes in the real and synthetic seismicity and (ii) the essential increase in the shares of model events for average depths. The last fact is mainly explained by the specified values of block depths (most of them range from 30 to 50 km). It is evident that, in order to redistribute events, we need to get more specific information on the rule of change of the model parameters depending on depth. On the other side, the block model is intended for modeling events only in the surface layer of the Earth's crust. Again, we emphasize the absence of serious distinctions between variants of the first and second series.

Table 1. Distribution of earthquakes in depth (in percent with respect to the total number of events with $M \geq 4.0$): NEIC and model (variants I, II, III and IV) catalogs for 100 time units

Depth	NEIC	I	II	III	IV
Up to 10 km	15.98	13.99	14.18	14.65	14.70
[10, 40 km]	46.81	65.45	70.18	68.83	74.50
Over 40 km	37.21	20.56	15.64	16.52	10.80

We analyze parameters of the Gutenberg–Richter law characterizing the power distribution of earthquakes in magnitude. The FM plots in a logarithmic scale for the real and synthetic seismicity are presented in Fig. 4; also, some quantitative computing results are shown in Table 2.

The linearity of the model plots in the range of average and large magnitudes testifies to the "rightness" of the distribution law for model events in the given interval and, consequently, to the possibility of studying real patterns using a synthetic catalog. This quality index has been essentially improved, compared to previous versions of the model [4–6]. On the other hand, the characteristic feature of the model seismicity consists in an insufficient (with respect to the real seismicity) relative number of small events (in this context, with magnitude 4.0–4.5). In the graphic interpretation, this feature corresponds to rather extensive, almost horizontal, parts of the model plots in the range of small magnitudes. The share of small events increases when disturbing the equations (2) in variants II and IV. All the FM plots are approximated by the linear least-squares regression $\lg N = c - kM$ constructed by the least square method. The value of k serves as an estimate of the slope of the plot. The average distance between the points of the plot and the line constructed is treated as an approximation error. For all the model variants, the magnitude interval $[5.3, 8.3]$ of "sufficient" linearity is considered; for the NEIC catalog, the whole range $[4.0, 9.0]$. The FM plot for the registered global seismicity is almost linear and its slope is very close to 1. The slope

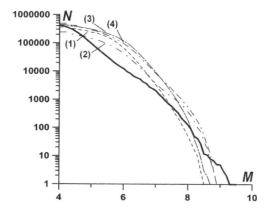

Fig. 4. The FM plots constructed for the real (NEIC, events with $M \geq 4.0$, 01.01.1900–04.07.2016, (*solid line*)) and for four model catalogs of 100 time units (variants I (*1*), II (*2*), III (*3*), and IV (*4*)); N is the accumulated number of earthquakes, M is the magnitude

estimates and approximation errors are presented in Table 2; there are no essential deviations for the two series of variants. At the same time, the approximation is obviously better for variants II and IV with random disturbances.

It seems that a key factor for increasing the interval of linearity of an FM plot and changing the slope is the internal calibration of the model (the coefficients in the equations of the form (1) and (2) for faults and block bottoms, the steps of temporal and spatial discretizations and so on). This aspect has been taken into account when constructing variants III and IV. Let us describe the characteristics that have been considerably improved by means of the model parameter fitting.

The presence of clustering is an explicit plus of variants III and IV. The possibility of clustering events (in the sense of extracting groups of earthquakes consisting of main shocks, foreshocks and aftershocks [1,2,8]) is noticeable in Fig. 5, showing the model seismicity (variant IV) for a part of the year at a small sector of the Philippine plate margin. The magnitude-time dependence has an irregular character: isolated groups of earthquakes concentrated nearby local maximums can be marked out. Note that clustering is an important component of many algorithms for searching premonitory patterns preceding extreme events [2];

Table 2. Simulation results: NEIC and model (variants I, II, III and IV) catalogs of 100 time units with $M \geq 4.0$

	NEIC	I	II	III	IV
Number of events	427950	359750	240976	432698	492018
Slope estimate	1.0	1.43	1.0	1.30	1.05
Approximation error	0.32	0.31	0.14	0.65	0.12

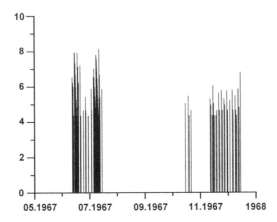

Fig. 5. The magnitude-time dependence for model events (variant IV) at a small sector of the Philippine plate margin

the possibility of investigating this phenomenon in the model should be considered as a positive fact.

The properties of the model seismicity that are characteristic for variants III and IV are tested with the use of the new software described in Sect. 4. In particular, this program allows us to trace the migration of model events along either a tectonic fault or a system of faults over time and to study the influence of model parameters on the stress distribution in a structure. In Fig. 6, the states of a fault located at the boundary South America/Nazca (its length along the Earth's surface is 977 km (977 cells), its depth is 30 km (15 layers)) are shown at subsequent times. A narrow horizontal bar with a color interpretation of the normalized stress distribution corresponds to every time moment.

There are 36 moments on the screen; the moment for which the zoomed distribution pattern with numerical values is presented in the lower window is marked out. The migration (from right to left along the fault) of the two groups of "quaked" cells composing a model earthquake at every time moment (red-colored) and that of stress decrease zones (blue-colored) are clearly noticeable. Actually, the spatial-temporal dynamics of a cluster fragment is visualized with the possibility of qualitative estimation of its parameters. In Fig. 7, the states of a fault located at the eastern boundary of the Philippine plate (its length along the Earth's surface is 499 km (499 cells), its depth is 100 km (50 layers)) are shown at the same time for variants III and IV. The presence of two clearly distinguishable layers is explained by the fact that the fault separates blocks of different depths (30 and 100 km, respectively) and, consequently, on the one side, we have the interaction between two blocks and, on the other side, between the block and the neighboring underlying medium. The rather large size of the fault and its specific form are the reasons of longitudinal inhomogeneities. In variant III (without any random disturbances), we observe a sufficiently regular pattern

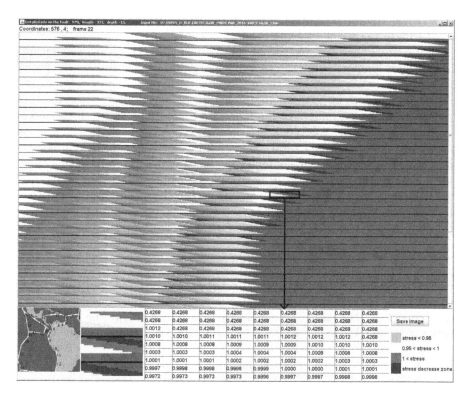

Fig. 6. Migration of model events along a tectonic fault (Color figure online)

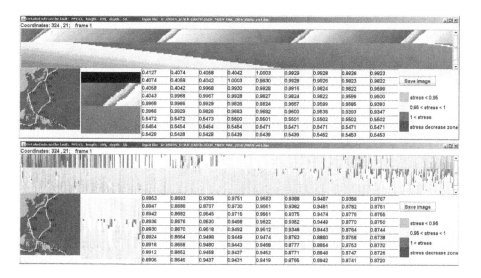

Fig. 7. Model stress distribution at a fault: variant III (*at the top*) and variant IV (*at the bottom*) (Color figure online)

of the model stress distribution along fault's layers, whereas a similar pattern becomes fuzzy in variant IV (with random disturbances).

Summarizing the brief analysis of the conducted experiments, we conclude that the main properties of the synthetic seismicity are justified by the geometry of the block structure and the underlying medium motion, i.e. the exogenous model parameters. It seems that a key role is played by the hypothesis of "geometrical incompatibility" of the faults [13]), according to which the places of possible earthquake epicenters are determined, first of all, by the relative position and motions of structural elements. The endogenous model parameters (the visco-elasticity coefficients of faults and block bottoms, discretization steps, and random factors) can be used for a "subtle" model adjustment to increase the similarity between the synthetic and real seismicity; the danger of incorrect parameter fitting aimed at a specific result should be taken into account.

6   Parallelization Quality

To test the parallelization quality characteristics, we chose the most time-taking variant IV and considered a unit of dimensionless time with a considerable number of model earthquakes. The simulation was performed at the Institute of Mathematics and Mechanics of UB RAS by means of a hybrid machine of cluster type named "Uran" (1864 Intel Xeon CPUs (used in experiments) and 352 NVIDIA Tesla GPUs; the operative memory is 6976 Gb, and the peak performance is about 215 TFlops). The results of the test are presented in Table 3.

Table 3. Computing time, speedup and efficiency for p processors

p	T_p, sec	S_p	E_p
1	2464.33	—	—
2	1232.84	2.00	1.00
4	634.97	3.88	0.97
8	357.76	6.88	0.86
16	175.12	14.07	0.88
32	87.84	28.05	0.88
64	48.94	50.35	0.79

Here T_1 is the performance time for the sequential algorithm, T_p is the performance time for the parallel algorithm on p processors, $S_p = T_1/T_p$ and $E_p = S_p/p$ are speedup and efficiency, respectively [14]. Recall that the speedup is equal to the number of engaged processors for an "ideal" parallel algorithm; in this case, we have the unit efficiency. It follows from Table 3 that the parallelization efficiency is rather high and does not fall below an acceptable level if the number of engaged processors increases. For "large" calculations, we used, as a rule, 64 processors. For a model variant up to 100 units of dimensionless time, it took about 80 min instead of 68 h for calculations on one processor.

7   Conclusive Remarks

The program for visualizing the state of faults provides the possibility of qualitative and quantitative analysis of the instantaneous stress distribution and temporal migration of critical values along the structure. It serves as an effective tool for both the internal calibration of the model and the verification of simulation results. The constructed set of endogenous parameters allowed us to obtain additional, compared to previous experiments, properties of synthetic seismicity similar to real ones. At the same time, the exogenous parameters taken from generally recognized theories and models determine the main characteristics of the model earthquake flow. This fact, in our opinion, testifies to a sufficient degree of adequacy of the spherical block model.

References

1. Gabrielov, A.M., Newman, W.I.: Seismicity Modeling and Earthquake Prediction: A Review. Geophysical Monograph 83, vol. 18, pp. 7–13. IUGG, Washington (1994)
2. Keilis-Borok, V.I., Soloviev, A.A. (eds.): Nonlinear Dynamics of the Lithosphere and Earthquake Prediction. Springer, Heidelberg (2003). doi:10.1007/978-3-662-05298
3. Peresan, A., Vorobieva, I.A., Soloviev, A.A., Panza, G.F.: Simulation of seismicity in the block-structure model of Italy and its surroundings. Pure Appl. Geophys. **164**, 2193–2234 (2007). doi:10.1007/s00024-007-0273-9
4. Rozenberg, V.L., Sobolev, P.O., Soloviev, A.A., Melnikova, L.A.: The spherical block model: dynamics of the global system of tectonic plates and seismicity. Pure Appl. Geophys. **162**, 145–164 (2005). doi:10.1007/s00024-004-2584-4
5. Melnikova, L.A., Rozenberg, V.L.: Spherical block model of lithosphere dynamics and seismicity: different modifications and numerical experiments. In: Proceedings of IMM UB RAS, Ekaterinburg, vol. 13, no. 3, pp. 95–120 (2007). (in Russian)
6. Melnikova, L.A., Rozenberg, V.L.: A stochastic modification of the spherical block-and-fault model of lithosphere dynamics and seismicity. Numer. Methods Program. **16**, 112–122 (2015). (in Russian)
7. Oksendal, B.: Stochastic Differential Equations: An Introduction with Application. Springer, New York (1998). doi:10.1007/978-3-662-03620-4. Mir, Moscow (2003)
8. Aki, K., Richards, P.G.: Quantitative Seismology: Theory and Methods. Freeman, San Francisco (1980). doi:10.1002/gj.3350160110. Mir, Moscow (1983)
9. Wells, D.L., Coppersmith, K.L.: New empirical relationships among magnitude, rupture length, rupture width, rupture area, and surface displacement. Bull. Seism. Soc. Am. **84**(4), 974–1002 (1994)
10. Gripp, A.E., Gordon, R.G.: Young tracks of hotspots and current plate velocities. Geophys. J. Int. **150**, 321–361 (2002). doi:10.1046/j.1365-246x.2002.01627.x
11. Manakov, D.V., Averbukh, V.L.: Verification of visualization. Sci. Visual. **8**(1), 58–94 (2016). (in Russian)

12. Global Hypocenters Data Base, NEIC/USGS, Denver, CO. http://earthquake.usgs.gov/regional/neic/
13. Gabrielov, A.M., Keilis-Borok, V.I., Jackson, D.D.: Geometric incompatibility in a fault system. Proc. Natl. Acad. Sci. USA **93**(9), 3838–3842 (1996). doi:10.1073/pnas.93.9.3838
14. Gergel', V.P.: Theory and Practice of Parallel Computing. Binom, Moscow (2007). (in Russian)

Parallel Implementation of a Monte Carlo Algorithm for Simulation of Cathodoluminescence Contrast Maps

Karl K. Sabelfeld⑩ and Anastasiya E. Kireeva$^{(\boxtimes)}$

Institute of Computational Mathematics and Mathematical Geophysics,
6, Prospekt Lavrentjeva, Novosibirsk 630090, Russia
`karl@osmf.sscc.ru`, `kireeva@ssd.sscc.ru`

Abstract. We suggest a parallel implementation of a Monte Carlo method for cathodoluminescence contrast maps simulation based on a random walk on spheres algorithm developed by K. K. Sabelfeld for solving drift-diffusion problems. The method for cathodoluminescence imaging in the vicinity of external forces is based on the explicit representation of the exit point probability density. This makes it possible to simulate exciton trajectories governed by drift-diffusion-reaction equations with a recombination condition on the surface of dislocations or other defects in crystals. In this study, we apply the developed stochastic algorithm to construct a parallel implementation that uses the OpenMP and MPI standards and is based on a distribution of simulated exciton trajectories starting at a given source. The number of self-annihilated excitons is evaluated as a function of the distance between the exciton source and the dislocation. The algorithm is tested against exact results.

Keywords: Cathodoluminescence · Drift-diffusion problem · Random walk on spheres algorithm · Monte Carlo algorithm · Parallel implementation

1 Introduction

Many physical phenomena where particles, heat, charges and other physical quantities are transferred as a result of two processes, diffusion and convection, are described by the drift-diffusion equation [1,2]. For instance, the drift-diffusion equation may be used to describe semiconductor devices, where a current of charge carriers take place owing to both its diffusion (movement from a place of higher concentration to a place of lower concentration) and the force of the electric field. Another important phenomenon which may be described by the drift-diffusion equation is cathodoluminescence. An electron beam causes the flow of electrons from the valence band to the conduction band and formation of

---

The support of the Russian Science Foundation under grant No. 14-11-00083 is kindly acknowledged.

L. Sokolinsky and M. Zymbler (Eds.): PCT 2017, CCIS 753, pp. 233–246, 2017.
DOI: 10.1007/978-3-319-67035-5_17

excitons [3,4]. An exciton is a bound state of an electron and an electron hole that are attracted to each other by the electrostatic Coulomb force. The electron-hole and exciton recombination is accompanied by a photon emission, which is called cathodoluminescence. Cathodoluminescence intensity depends on the material, its lattice structure, purity and defects. This is what makes the cathodolumines-cence microscope such a powerful technique to reveal morphological structures in materials, and determine the presence and location of impurities and defects.

A random walk on spheres algorithm for solving the drift-diffusion equation has been developed in [5]. The random walk on spheres method is known as a probabilistic algorithm for solving the Laplace's equation [6]. The probability density of the exit point for the drift-diffusion process starting at the center of the sphere was derived in [5]. Simultaneous drift and diffusion of excitons can be simulated using this probability density. During their motion, the excitons can either recombine on dislocations or self-annihilate with photon emission (luminescence). According to the Monte Carlo method [7], many independent exciton trajectories must be simulated to obtain results with sufficient accuracy. Moreover, the construction of cathodoluminescence contrast maps used for the analysis of materials morphology requires the cathodoluminescence intensity to be calculated at various positions of the exciton source.

A parallel implementation of the random walk on spheres algorithm for the drift-diffusion process is developed in the present paper, which allows for the reduction of computation time. The main approaches to the parallel implemen-tation of Monte Carlo algorithms have been described in [8,9]. These papers discuss difficulties and peculiarities of parallel Monte Carlo algorithms. The main approach to the parallel implementation of Monte Carlo algorithms is the distribution of independent tasks among the processors of a supercomputer, computing some number of samples of the simulated random variables. In the present paper, we provide a parallel implementation of the drift-diffusion process simulation using the OpenMP and MPI standards. The calculations of cathodo-luminescence intensity at different positions of the exciton source are distributed among MPI processes. For each MPI process, the exciton trajectories starting at the given source are distributed among OpenMP threads. Also, the efficiency of the parallel implementation is estimated. Using the parallel code, the cathodo-luminescence intensity is investigated regarding diffusion length, drift velocity and recombination coefficients on dislocations.

2 Simulation Algorithm of the Drift-Diffusion Process

2.1 The Drift-Diffusion Equation

According to [5], the drift-diffusion process is described by the following equation:

$$D\Delta u(\mathbf{r}) + \mathbf{v} \cdot \nabla u - \frac{1}{\tau} u = 0 , \quad \mathbf{r} \in G , \qquad (1)$$

where u is particle concentration, D is a constant diffusion coefficient, $\mathbf{v}$ is the drift velocity, τ is the mean lifetime, and $\mathbf{r}$ is a space coordinate.

Following [5], we transform the Eq. (1) into the Eq. (2) by the change of variables $\mathbf{x} = \mathbf{r}/L$, where $L = \sqrt{D\tau}$ is the diffusion length:

$$\Delta w(\mathbf{x}) + s\alpha \cdot \nabla w - w = 0 , \tag{2}$$

where $\alpha = \text{Pe} \cdot \frac{\mathbf{v}}{|\mathbf{v}|}$ is a new expression for the velocity, and $\text{Pe} = \frac{|\mathbf{v}| \cdot L}{D}$ is the Péclet number.

We will try to find the solution of Eq. (2) in the center $\mathbf{x}_0$ of the sphere $S(\mathbf{x}_0, R)$ of radius R. The solution satisfies the spherical integral relation

$$w(\mathbf{x}_0) = \frac{\mu R}{\sinh(\mu R)} \frac{\sinh \kappa}{\kappa} \int_0^\pi \int_0^{2\pi} p(\theta, \varphi; \gamma, \beta) w(\mathbf{x}_0 + R\zeta) \, d\theta \, d\varphi , \tag{3}$$

where $\kappa = \frac{\text{Pe} \cdot R}{2}$ and $\mu = \sqrt{1 + \frac{1}{4}\text{Pe}^2}$. Here γ and β denote the zenith and azimuthal angles of the unit direction vector α of the drift in the polar spherical coordinate system. The zenith angle is counted from the direction of the Z-axis, while the azimuthal angle is counted in the XY-plane horizontal direction. Analogously, θ and φ are the zenith and azimuthal angles of an arbitrary unit direction vector ζ.

According to the probabilistic interpretation of this relation, a particle starting at the center of the sphere $S(\mathbf{x}_0, R)$ survives with probability

$$P_{\text{surv}} = \frac{\mu R}{\sinh(\mu R)} \frac{\sinh \kappa}{\kappa} , \tag{4}$$

and reaches the surface of the sphere at a random exit point whose distribution density $p(\theta, \varphi; \gamma, \beta)$ is

$$p(\theta, \varphi; \gamma, \beta) = \frac{\kappa}{4\pi \sinh \kappa} \cdot \exp\{\kappa[\sin \theta \sin \gamma \cos(\varphi - \gamma) + \cos \theta \cos \gamma]\} \cdot \sin \theta ,$$
$$0 \le \theta \le \pi , \ 0 \le \varphi \le 2\pi . \tag{5}$$

The distribution is axially symmetric with respect to the velocity direction α. Therefore, the coordinate system is chosen in such a manner that the Z-axis coincides with the velocity vector, and $\gamma = 0$. Then the probability density can be simplified as follows:

$$p(\theta, \varphi; \gamma, \beta) = \frac{\kappa \cdot \exp\{\kappa \cos \theta\} \cdot \sin \theta}{4\pi \sinh \kappa} , \quad 0 \le \theta \le \pi, \ 0 \le \varphi \le 2\pi . \tag{6}$$

According to this distribution, the azimuthal angle φ is uniformly distributed on $[0, 2\pi]$, i.e. $\varphi = 2\pi \cdot rand$, where $rand$ stands for a random number uniformly distributed on $[0, 1]$. The cosine of the zenith angle θ in the formula (5) is sampled by the following formula:

$$\cos \theta = 1 + \frac{1}{\kappa} \cdot \log[1 - (1 - e^{-2\kappa}) \cdot rand] . \tag{7}$$

Thus, if one rotates the coordinate system in such a way that the Z-axis coincides with the velocity vector $\boldsymbol{\alpha}$, then the coordinates of the random exit point on the sphere $S(0, R)$ are expressed as

$$
\begin{aligned}
x &= R \cdot \sin \theta \cos \varphi \,, \\
y &= R \cdot \sin \theta \sin \varphi \,, \\
z &= R \cdot \cos \theta \,.
\end{aligned} \tag{8}
$$

To rotate the Cartesian system back to the initial position, the X-axis should be rotated so that it coincides with the velocity vector $\boldsymbol{\alpha} = (\alpha_1, \alpha_2, \alpha_3)$, i.e. by the angle ψ. Let us consider the case when the velocity vector is directed along the vector $\boldsymbol{\alpha} = (\alpha_1, \alpha_2, 0)$. Then the coordinates of the random exit point on the sphere $S(0, R)$ are calculated as follows:

$$
\begin{aligned}
x &= R \cdot \cos \theta \cos \varphi - R \cdot \sin \psi \sin \theta \cos \varphi \,, \\
y &= R \cdot \cos \theta \sin \psi + R \cdot \cos \psi \sin \theta \cos \varphi \,, \\
z &= R \cdot \sin \theta \sin \varphi \,,
\end{aligned} \tag{9}
$$

where $\cos \psi = \dfrac{\alpha_1}{\sqrt{\alpha_1^2 + \alpha_2^2}}$, $\sin \psi = \dfrac{\alpha_2}{\sqrt{\alpha_1^2 + \alpha_2^2}}$.

In the general case, when the drift vector is directed along the vector $\boldsymbol{\alpha} = (\alpha_1, \alpha_2, \alpha_3)$, the coordinates of the random exit point are obtained by three successive rotations of the original coordinate system [5].

2.2   Random Walk on Spheres Algorithm for the Drift-Diffusion Process

A random walk on spheres algorithm was developed in [5] for solving the drift-diffusion equation in the following form:

$$
\Delta w(\mathbf{x}) + \boldsymbol{\alpha} \cdot \nabla w - w + \delta(\mathbf{x} - \mathbf{x}_0) = 0 \,, \quad x \in G \,, \tag{10}
$$

where $\delta(\mathbf{x} - \mathbf{x}_0)$ is a point source of particles placed at a point $\mathbf{x}_0$ of the domain G.

In the case of cathodoluminescence simulation, the domain is considered as a half-space containing a dislocation in the form of a half-cylinder normal to the top surface [10, 11] (Fig. 1). The exciton generated in the source diffuses in the domain, can be recombined either with rate S_1 on the top plane boundary or with rate S_2 on the surface of the dislocation-cylinder, and can self-annihilate with photon emission.

According to [5, 11], the concentration density of excitons at a point x in G is governed by Eq. (10) with Robin boundary conditions:

$$
\begin{aligned}
\frac{\partial w(\mathbf{x})}{\partial \mathbf{n}} + \frac{S_1 \cdot L}{D} \cdot w = 0 \,, \quad \mathbf{x} \in \Gamma_1 \,, \\
\frac{\partial w(\mathbf{x})}{\partial \mathbf{n}} + \frac{S_2 \cdot L}{D} \cdot w = 0 \,, \quad \mathbf{x} \in \Gamma_2 \,,
\end{aligned}
$$

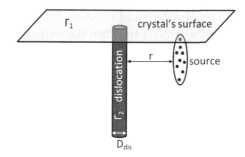

Fig. 1. Schematic depiction of the domain in which cathodoluminescence is simulated.

where $\mathbf{n}$ is the unit exterior normal vector to the boundary, Γ_1 denotes the top plane boundary with recombination rate S_1, and Γ_2 denotes the surface of the cylinder boundary with recombination rate S_2.

The random walk on spheres algorithm for simulation of cathodolumines-cence conforming to Eq. (10) with boundary condition (11) can be described as follows [5]:

1. Generate an exciton in the source point $\mathbf{x} = \mathbf{x}_0$.
2. Construct a sphere centered at $\mathbf{x}$ with radius $d(\mathbf{x})$ equal to the minimal distance from $\mathbf{x}$ to the boundaries Γ_1 and Γ_2.
3. Calculate the survival probability P_{surv} by formula (4). With probability $P_{\text{absorb}} = 1 - P_{\text{surv}}$, the exciton self-annihilates, the trajectory terminates, and the cathodoluminescence intensity I_{CL} is incremented by 1. Go to step 1.
4. If the trajectory does not terminate, then the exciton jumps to a random position $\mathbf{x}_1$ on the surface of the sphere. The coordinates of $\mathbf{x}_1$ are calculated by formula (9) according to the algorithm given in Sect. 2.1.
5. If $\mathbf{x}_1$ hits the ϵ-neighborhood of the boundary Γ_1, then the exciton is recom-bined on the top plane, the trajectory terminates, and the flux value F_1 to the top plane is incremented by 1. Go to Step 1.
6. If $\mathbf{x}_1$ hits the ϵ-neighborhood of the boundary Γ_2, then the exciton is recom-bined on the cylinder-dislocation, the trajectory terminates, and the flux value F_2 to the cylinder-dislocation is incremented by 1. Go to Step 1.
7. If $\mathbf{x}_1$ is not adsorbed on neither of the boundaries Γ_1 and Γ_2, then go to Step 2.

The ϵ-neighborhood of the boundary Γ_i, $i = 1, 2$, is the set $\Gamma_{i\epsilon} = \{x \in G : \rho(x, y) \leq \epsilon, \ x \in G, \ y \in \Gamma_i\}$, where $\rho(x, y)$ is the minimal distance between the points x and y. The value of ϵ is set equal to 0.01 nm.

To calculate the cathodoluminescence intensity I_{CL} and flux to the bound-aries Γ_1 and Γ_2, it is necessary to simulate N trajectories; after that, I_{CL}, F_1 and F_2 are calculated as the arithmetic means of the relevant scores.

Moreover, in [5], the exact solution of the drift-diffusion Eq. (1) with Robin boundary condition is derived in the case when the recombination rate S_1 of the top plane boundary is zero:

$$u(\mathbf{r}) = \frac{e^{[-\mathbf{v}\cdot(\mathbf{r}-\mathbf{R})]/2D} \cdot K_0(\mu\mathbf{r})}{\left(1 + \dfrac{|\mathbf{v}|\cos(\varphi - \psi)}{2S_2D}\right)K_0(\mu\mathbf{R}) + \dfrac{\mu}{S_2}K_1(\mu\mathbf{R})}, \tag{11}$$

where $\mu = \sqrt{\left(\dfrac{|\mathbf{v}|}{2D}\right)^2 + \dfrac{1}{D\tau}}$. Here $\mathbf{R}$ is a point on the dislocation's surface, φ is the polar coordinate of $\mathbf{r}$, ψ is the direction of the drift vector $\mathbf{v}$, K_i, $i = 0, 1$, is the i-order Bessel function of the second kind. The exact solution will be used below to test the results of the cathodoluminescence simulation.

3   Parallel Implementation of the Random Walk on Spheres Algorithm for Simulation of Cathodoluminescence

To obtain reliable values for the cathodoluminescence intensity I_{CL} and the fluxes F_1 and F_2 to the boundaries, we need to compute the average over a sufficiently large number of trajectories. In addition, the construction of cathodoluminescence contrast maps involves the calculation of the cathodoluminescence intensity for various positions of the exciton source. Thus, a large amount of computer time is required to calculate the cathodoluminescence contrast map with sufficient accuracy. A parallel implementation of this task will allow for an essential reduction of computational time.

The parallel implementation uses the OpenMP and MPI standards. The calculations of the characteristics I_{CL}, F_1 and F_2 for different positions of the exciton source are distributed among n_{mpi} MPI processes. For each MPI process, the N exciton trajectories starting at each given position are distributed among n_{omp} OpenMP threads. The OpenMP threads simultaneously calculate the values of the characteristics and summarize the obtained values in the global shared variable using the reduction clause. After computing all the trajectories for all the source positions, the root MPI process gathers the obtained values of the characteristics I_{CL}, F_1 and F_2, normalizes them and save the computed cathodoluminescence contrast map. The characteristic values are normalized to unity at large distances from the dislocation.

The efficiency of the parallel implementation of the cathodoluminescence simulation code is estimated on a computing experiment with the following parameter values: the exciton lifetime $\tau = 1$ ns, the diffusion length $L = 370$ nm, the drift velocity $\alpha = (10, 20)$ nm/ns, the recombination rates $S_1 = 1500$ nm/ns and $S_2 = 48\,000$ nm/ns, the dislocation radius $R_{dis} = 1$ nm, the size of the

Table 1. Characteristics of the parallel implementation of the cathodoluminescence simulation code executed on a single computational node.

$n_{\mathrm{mpi}}; n_{\mathrm{omp}}$	1;1	1;4	1;8	1;16	1;32	**2;16**	4;8	8;4	16;2	32;1
T, minutes	306	82	44	24	16	**15**	18	22	28	89
S		1	3.7	7	13	19.2	**20.3**	17.1	13.7	10.8	3.4

neighborhood for both boundaries $\epsilon = 0.01$ nm, and the number of trajectories $N = 10^5$. The simulations are performed on the "MVS-10P" cluster of the Joint Supercomputer Center of the Russian Academy of Sciences (JSCC RAS)[1]. The "MVS-10P" cluster consists of 207 computational nodes, each featuring two Xeon E5-2690 processors.

At first, the optimal ratio between the numbers of MPI processes and OpenMP threads for a single computational node is determined. Table 1 presents computational times $T(n_{\mathrm{mpi}}, n_{\mathrm{omp}})$ and speedups $S(n_{\mathrm{mpi}}, n_{\mathrm{omp}}) = T(1,1)/T(n_{\mathrm{mpi}}, n_{\mathrm{omp}})$ for different numbers of MPI processes n_{mpi} and OpenMP threads n_{omp}. As it can be seen from the table, the minimal computational time is attained for the parallel code implementation using 2 MPI processes and 16 OpenMP threads. This result corresponds to a computational node architecture featuring two processors.

Further, the efficiency of the parallel code is investigated regarding the number of MPI processes, in the case when two MPI processes are executed on a single node, and each MPI process features 16 OpenMP threads. Table 2 contains computational times $T(n_{\mathrm{mpi}}, 16)$ and speedups $S(n_{\mathrm{mpi}}, 16) = T(1,1)/T(n_{\mathrm{mpi}}, 16)$ for various numbers of MPI processes n_{mpi}. The speedup of the parallel implementation of the code is increased until the number of MPI processes reaches 16. To increase the number of MPI processes even more, the load on the computational nodes is insufficient, then the speedup is reduced, and the computing time grows. According to this result, the best parameters for the execution of the parallel code implementing the cathodoluminescence simulation algorithm was found to be 16 MPI processes, each featuring 16 OpenMP threads.

Table 2. Characteristics of the parallel implementation of the cathodoluminescence simulation code depending on the number of MPI processes.

n_{mpi}	2	4	8	16	32	64
T, minutes	15	9	5.8	3.6	6.2	5.7
S		20.3	33.7	52.8	83.9	49.3	53.2

---

[1] JSCC RAS website: http://www.jscc.ru/.

4   Results of the Cathodoluminescence Simulation

Cathodoluminescence is simulated by a parallel implementation of the random walk on spheres algorithm for different values of the modeling parameters: the exciton lifetime τ (ns), the diffusion length L (nm), the drift velocity α (nm/ns), the recombination rates S_1 (nm/ns) and S_2 (nm/ns), the dislocation radius R_{dis} (nm), the depth of the source location z (nm), and the value of the ϵ-neighborhood of the boundaries Γ_1 and Γ_2. In addition, the accuracy of calculations depends on the number of trajectories N. The cathodoluminescence contrast map is calculated by computing the cathodoluminescence intensity for a point source of excitons located at a different distance r from the dislocation surface and at the same depth z from the top surface.

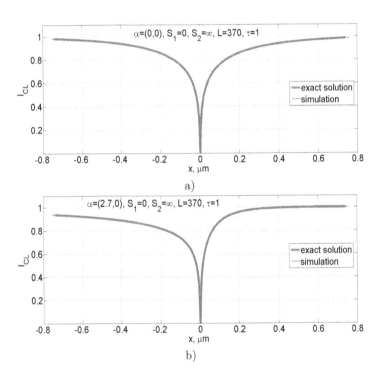

Fig. 2. Cathodoluminescence contrast map obtained by computer simulation, and analytical solution for: (a) pure diffusion case, and (b) diffusion with drift case.

To verify the cathodoluminescence algorithm implementation, a comparison of the simulated results against the exact solution (11) of the drift-diffusion-reaction equation derived in [5] is made for the rates $S_1 = 0$ and $S_2 = \infty$ in two cases: pure diffusion without drift ($\alpha = 0$), and diffusion with drift ($\alpha = (2.7, 0)$ nm/ns). The values of the other modeling parameters are set as follows: $\tau = 1$ ns, $L = 370$ nm, $R_{dis} = 1$ nm, $z = 1$ nm, $\epsilon = 0.01$ nm, and $N = 10^5$. The semi-infinite cylinder imitating the dislocation is placed in the domain center. The point of intersection of the cylinder axis with the upper plane is considered as the origin of a coordinate right-hand coordinate system. The exciton source is located at different positions $(x, 0, 1)$, where x is a coordinate on the X-axis.

The obtained cathodoluminescence contrast map and the exact solution for the same parameters are shown in Fig. 2. The values of the cathodoluminescence intensity obtained by computer simulation are in good agreement with those of the analytical solution both in the pure diffusion case and in the diffusion with drift case. The presence of drift yields the asymmetry of the cathodoluminescence contrast map. This result will be explained below.

In subsequent computer experiments, we studied the influence of the basic parameters (namely recombination rates S_1 and S_2, diffusion length L and drift velocity α) on the cathodoluminescence intensity for the following values of the other modeling parameters: $\tau = 1$ ns, $R_{\text{dis}} = 1$ nm, $z = 1$ nm, $\epsilon = 0.01$, and $N = 10^5$.

Figures 3 and 4 present the cathodoluminescence contrast map obtained in the pure diffusion case ($\alpha = 0$) with diffusion length $L = 370$ nm for two calculation experiments:

1. with different values of the recombination rate S_1 of the top plane boundary and the same recombination rate of the dislocation, $S_2 = 48\,000$ nm/ns;
2. with different values of the recombination rate of the dislocation and the same recombination rate of the top plane boundary, $S_1 = 1500$ nm/ns.

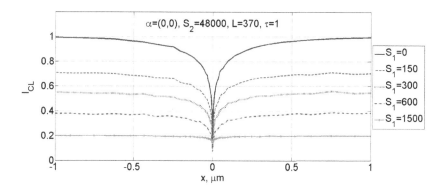

Fig. 3. Cathodoluminescence contrast map for different values of the recombination rate S_1 of the top plane and the same value of the recombination rate S_2 of the dislocation.

As mentioned before, an exciton can recombine on either the top plane boundary or the dislocation surface, and can self-annihilate through radiative recombination. Thus, the values of the recombination rates essentially influence cathodoluminescence intensity. Since the dislocation surface is considerably less

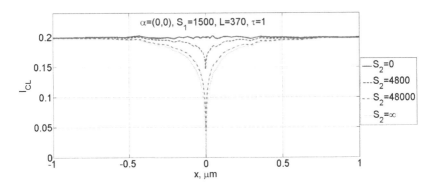

Fig. 4. Cathodoluminescence contrast map for different values of the recombination rate S_2 of the dislocation and the same value of the recombination rate S_1 of the top plane.

than the involved surface of the top plane boundary, the cathodoluminescence intensity maximum value is defined by the recombination rate of the top plane. As it can be seen in Fig. 3, the higher the recombination rate S_1 is, the less the maximum value of the cathodoluminescence intensity I_{CL} is. In addition, the intensity I_{CL} reaches the equilibrium maximum state faster when the values of S_1 are large.

When the recombination rate of the dislocation changes while the recombination rate of the top plane remains the same, the cathodoluminescence intensity maximum value remains constant, and the value of I_{CL} changes only near the dislocation (Fig. 4). The higher the recombination rate S_2 is, the more excitons recombine on the dislocation, and the less the minimal value of the cathodoluminescence intensity I_{CL} is. In addition, large values of S_2 influence the intensity values I_{CL} at greater distances, and I_{CL} reaches the equilibrium state more slowly.

The dependence of the cathodoluminescence contrast map on the diffusion length L is investigated for the pure diffusion case ($\alpha = 0$) at recombination rates $S_1 = 1500$ nm/ns and $S_2 = 48\,000$ nm/ns. Figure 5 shows that the larger the value of the diffusion length is, the larger the maximum value of the cathodoluminescence intensity is. This result can be explained as follows. The larger the diffusion length is, the larger the distance the exciton moves during the same time. The diffusion moving direction is random. Therefore, for larger diffusion lengths, the exciton is more likely to go away from the top plane, bypass the dislocation and self-annihilate. Moreover, the larger the diffusion length is, the larger the distance at which the dislocation influences the cathodoluminescence intensity, since an exciton can reach the dislocation faster.

The influence of the drift velocity on the cathodoluminescence contrast map is analyzed regarding two velocity parameters: value and direction. Figure 6 shows the cathodoluminescence intensity obtained for various values $\alpha = (\alpha_1, 0)$ of the drift velocity, directed along the X-axis, recombination rates $S_1 = 1500$ nm/ns

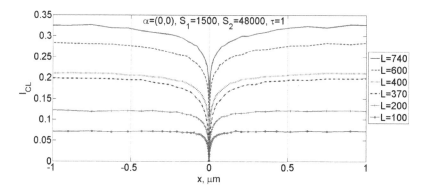

Fig. 5. Cathodoluminescence contrast map for different values of the diffusion length L.

and $S_2 = 48\,000$ nm/ns, and diffusion length $L = 370$ nm. The presence of drift yields the asymmetry of the cathodoluminescence contrast map. This is due to the fact that, on the one hand, the drift causes the transfer of an exciton located to the left of the dislocation to its surface, and consequently, the flux on the dislocation increases, whereas the cathodoluminescence intensity decreases. On the other hand, the drift moves an exciton located to the right of the dislocation away from it, decreasing the flux on the dislocation and increasing the cathodoluminescence intensity. The larger the velocity value is, the steeper the cathodoluminescence intensity curve is.

Figure 7 shows the cathodoluminescence contrast map for different directions of the drift velocity α and the same velocity value $\alpha = (|2.7|, |2.7|)$. The cathodoluminescence contrast maps obtained for oppositely directed along the X-axis velocities ($\alpha = (2.7, 0)$ and $\alpha = (-2.7, 0)$) are the mirror images of each other with respect to the Y-axis. Adding the drift velocity along the Y-axis does not affect the values of the cathodoluminescence intensity to the right of the dislocation, and leads to an increase in intensity to the left of the dislocation. Moreover, the direction of the velocity along the Y-axis has absolutely no effect on the result. The cathodoluminescence contrast maps obtained for oppositely directed along the Y-axis velocities ($\alpha = (2.7, 2.7)$ and $\alpha = (2.7, -2.7)$) totally coincide with each other. This result can be explained as follows. As is usual, the source is located on the same line as the dislocation along the X-axis. The drift along the X-axis takes an exciton to the dislocation, whereas the drift along the Y-axis removes an exciton from the dislocation. Therefore, the flux on the dislocation decreases and the cathodoluminescence intensity increases. At the same time, it does not matter in which direction the exciton is shifted: along the Y-axis or in the opposite direction.

Let us consider the model case of a variable drift velocity depending on the exciton position. In this case, according to [5], in the random walk on spheres algorithm, the radius $d(\mathbf{x})$ of the sphere should be small, since inside a small sphere, the value of the velocity may be considered as a constant. For example,

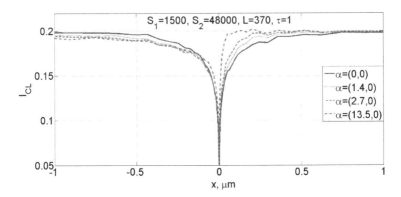

Fig. 6. Cathodoluminescence contrast map for different values of the drift velocity α directed along the X-axis.

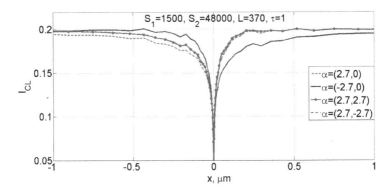

Fig. 7. Cathodoluminescence contrast map for different directions of the drift velocity α.

let us suppose that the velocity vector is directed to the cylinder located as mentioned above (Fig. 1). The drift velocity is defined as a function of the minimal distance r between the cylinder axis $\mathbf{c}_0 = (x_0, y_0, z)$ and the exciton position $\mathbf{c} = (x, y, z)$:

$$\boldsymbol{\alpha} = (\alpha_1, \alpha_2, 0) = \left(f(r) \cdot \frac{x_0 - x}{r}, \ f(r) \cdot \frac{y_0 - y}{r}, \ 0 \right),$$

$$f(r) = \begin{cases} k \cdot (r_{\max}^2 - r^2) \text{if } r \leq r_{\max}, \\ k \cdot \exp(-r) \text{otherwise}, \end{cases} \tag{12}$$

where $r_{\max}$ is the distance at which the cylinder influences the velocity, and k is a velocity coefficient. The cathodoluminescence intensity is calculated in the same way as before.

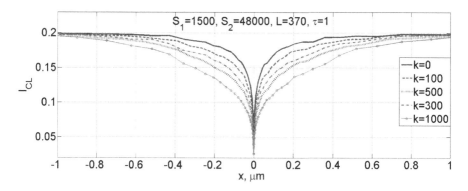

Fig. 8. Cathodoluminescence contrast map for the model case of a variable drift velocity α with different values of the velocity coefficient k.

Figure 8 shows the cathodoluminescence contrast map computed for the drift velocity defined by formula (12) and different values of the velocity coefficient k. The distance $r_{\max}$ of the cylinder influence is taken equal to the diffusion length L. The modeling parameters are set as in the previous experiments: $L = 370$ nm, $S_1 = 1500$ nm/ns and $S_2 = 48\,000$ nm/ns, $\tau = 1$ ns, $R_{\mathrm{dis}} = 1$ nm, $z = 1$ nm, $\epsilon = 0.01$, and $N = 10^5$. Compared to the basic case without drift velocity, $\alpha = (0,0)$, the cathodoluminescence intensity obtained for a variable drift velocity decreases faster near the cylinder. The larger the velocity coefficient k is, the wider the profile of the cathodoluminescence contrast map around the cylinder axis is. This behavior corresponds to the choice of a velocity function defined in such a manner that, at any position, the velocity vector is directed to the cylinder, where the exciton is annihilated. In addition, the larger the velocity coefficient k is, the faster the exciton reaches the cylinder.

This example shows the ability of the cathodoluminescence model to take into account a variable drift velocity depending on space position or on some variable physical parameters.

5   Conclusions

We presented in this paper a parallel implementation of the Monte Carlo method for cathodoluminescence contrast maps simulation based on a random walk on spheres algorithm developed in [5] for solving drift-diffusion problems. The parallel implementation is made using the OpenMP and MPI standards, and is based on a distribution of simulated exciton trajectories starting at various source positions. We determined the optimal ratio between the number of MPI processes and that of OpenMP threads on the "MVS-10P" cluster of the JSCC RAS.

This parallel implementation of the random walk on spheres algorithm has allowed to study cathodoluminescence contrast maps regarding various parameters, namely the recombination rates S_1 and S_2, the diffusion length L and the drift velocity α. The results computed by simulation were compared with the exact solution of the drift-diffusion equation, and a good agreement was observed between both approaches. It was shown in a test problem that the random walk on spheres algorithm for the simulation of cathodoluminescence contrast maps is able to take into account a variable drift velocity.

References

1. Selberherr, S.: Analysis and Simulation of Semiconductor Devices. Springer, Vienna (1984). doi:10.1007/978-3-7091-8752-4
2. Sijerčić, E., Mueller, K., Pejčinović, B.: Simulation of InSb devices using drift-diffusion equations. Solid-State Electron. **49**(8), 1414–1421 (2005). doi:10.1016/j.sse.2005.05.012
3. Boggs, S., Krinsley, D.: Application of Cathodoluminescence Imaging to the Study of Sedimentary Rocks. Cambridge University Press, New York (2006). doi:10.1017/cbo9780511535475.008
4. Knox, R.S.: Theory of excitons. In: Seitz, F., Turnbul, D. (eds.) Solid State Physics. Academic Press, New York (1963)
5. Sabelfeld, K.K.: Random walk on spheres method for solving drift-diffusion problems. Monte Carlo Methods Appl. **22**(4), 265–275 (2016). doi:10.1515/mcma-2016-0118
6. Müller, M.E.: Some continuous Monte Carlo methods for the Dirichlet problem. Ann. Math. Statist. **27**(3), 569–589 (1956). doi:10.1214/aoms/1177728169
7. Fishman, G.: Monte Carlo: Algorithms and Applications. Operations Research and Financial Engineering, Concepts. Springer, New York (1996)
8. Rosenthal, J.S.: Parallel computing and Monte Carlo algorithms. Far East J. Theor. Stat. **4**, 207–236 (2000)
9. Esselink, K., Loyens, L.D.J.C., Smit, B.: Parallel Monte Carlo Simulations. Phys. Rev. E **51**(2), 1560–1568 (1995). doi:10.1103/physreve.51.1560
10. Sabelfeld, K.K., Kaganer, V.M., Pfüller, C., Brandt, O.: Dislocation contrast in cathodoluminescence and electron-beam induced current maps on GaN(0001). Cornell University Library. Materials Science arxiv:1611.06895 [cond-mat.mtrl-sci]
11. Sabelfeld, K.K.: Splitting and survival probabilities in stochastic random walk methods and applications. Monte Carlo Methods Appl. **22**(1), 55–72 (2016). doi:10.1515/mcma-2016-0103

Supercomputer Modeling of Generation of Electromagnetic Radiation by Beam–Plasma Interaction

Evgeny Berendeev[1], Marina Boronina[1], Galina Dudnikova[2],
Anna Efimova[1(✉)], and Vitaly Vshivkov[1]

[1] Institute of Computational Mathematics and Mathematical Geophysics,
Siberian Branch of the Russian Academy of Sciences,
Prospect Akademika Lavrentieva, 6, Novosibirsk 630090, Russia
{berendeev,boronina,efimova,vsh}@ssd.sscc.ru
[2] Institute of Computational Technologies,
Siberian Branch of the Russian Academy of Sciences, Novosibirsk, Russia
dudn@ict.nsc.ru
http://www.icmmg.nsc.ru, http://www.ict.nsc.ru

Abstract. We construct in this article a two-dimensional particle-in-cell (PIC) numerical model of electron beam - plasma interaction based on the kinetic description of both ion and electron components with a continuously injected electron beam, and develop the corresponding parallel code. In this model, an electron beam, entering the plasma along magnetic field lines through one boundary and leaving it through the other, provides a continuous pumping of plasma oscillations. Such a problem statement requires that the model be constructed in a sufficiently long plasma region, where the time is long enough for the beam to be captured by the exciting wave field. The parallel algorithm was successfully applied to the solution of resource-intensive problem by efficiently using large numbers of computational cores.

Keywords: Particle-in-cell methods · Maxwell's equations · Vlasov equation · Open plasma trap · Generation of electromagnetic radiation

1 Introduction

We present a 2D particle-in-cell (PIC) model and the corresponding parallel code for the computer-aided simulation of the generation of electromagnetic radiation in a "relativistic electron beam – magnetized plasma" system under conditions of laboratory experiments. The solution of the problem has an important meaning for both computational plasma physics and thermonuclear research in the open plasma traps. The development of new schemes of generation of electromagnetic

The computational experiments were supported by the Russian Science Foundation, project no. 16-11-10028. The development of numerical algorithms was supported by the Russian Foundation for Basic Research, project no. 16-31-00304, 16-31-00301.

© Springer International Publishing AG 2017
L. Sokolinsky and M. Zymbler (Eds.): PCT 2017, CCIS 753, pp. 247–260, 2017.
DOI: 10.1007/978-3-319-67035-5_18

radiation in the terahertz frequency range allows one to understand the turbulent processes in the system "beam–plasma". These studies are relevant to various physical problems, such as generation of radiation in solar radio splashes, fast ignition in inertial confinement fusion and formation of collisionless shocks in astrophysics.

A solution to the problem of electron beam–plasma interaction with periodic boundary conditions is given in [1,2]. This result allowed us to estimate the level of peak radiation power at the stage of dynamic development of beam instability. However, to study the electromagnetic emission in a quasi-stationary fully developed turbulence regime, it is necessary to consider the possibility of continuous injection of an electron beam into the plasma with open boundary conditions.

We have developed a two-dimensional numerical model based on the kinetic description of both ion and electron components plasma and a continuously injected relativistic electron beam. In this model, the electron beam, entering the plasma along the magnetic field lines through one boundary and leaving it through the other, provides a continuous pumping of plasma oscillations. Such a problem statement requires that the model be constructed in a sufficiently long plasma region, where the time is long enough for the electron beam to be captured by the field of an exciting wave. The plasma is adjoined across the magnetic field lines with a vacuum regions and absorbing walls. This allows us to compare the radiation flows leaving the plasma with the radiation flows leaving the plasma in the laboratory experiments [3].

The effective use of supercomputer systems requires the development of a mathematical model appropriate to investigate the physical processes, algorithms and software environment. With the advent of the new generation of exaflop computing systems, the role of high-performance modeling in solving different fundamental astrophysical problems and building energy sources based on controlled thermonuclear fusion is acquiring a growing importance. The numerical model is based on a particle-in-cell method. We also have developed parallel computation algorithms and software for performing a series of calculations on supercomputers. Numerical experiments with different magnetic fields, beam and plasma parameters were carried out on the Novosibirsk State University supercomputer and the "Lomonosov" supercomputer of the Supercomputing Center at Lomonosov Moscow State University.

2   The Problem Statement

The physical process can be described by the Vlasov equation for the distribution functions of ions and electrons (1),

$$\frac{\partial f_\alpha}{\partial t} + (\overrightarrow{v}, \overrightarrow{\nabla}) f_\alpha + q_\alpha \left(\overrightarrow{E} + \frac{1}{c} [\overrightarrow{v} \times \overrightarrow{B}] \right) \frac{\partial f_\alpha}{\partial \overrightarrow{p}} = 0, \tag{1}$$

and Maxwell's equations for the electromagnetic fields (2)–(4),

$$\mathrm{rot}\,\overrightarrow{B} = \frac{4\pi}{c}\overrightarrow{j} + \frac{1}{c}\frac{\partial \overrightarrow{E}}{\partial t}, \quad \mathrm{rot}\,\overrightarrow{E} = -\frac{1}{c}\frac{\partial \overrightarrow{B}}{\partial t}, \tag{2}$$

$$\text{div } \overrightarrow{E} = 4\pi\rho, \tag{3}$$

$$\text{div } \overrightarrow{B} = 0, \tag{4}$$

$$\overrightarrow{j} = \sum_\alpha q_\alpha \int \overrightarrow{v} f_\alpha(\overrightarrow{p}, \overrightarrow{r}, t)\, d\overrightarrow{p}, \quad \rho = \sum_\alpha q_\alpha \int f_\alpha(\overrightarrow{p}, \overrightarrow{r}, t)\, d\overrightarrow{p}, \tag{5}$$

where f_α is the particle distribution function of the species α (the beam electrons, the plasma electrons and the ions), $\overrightarrow{B}$ is the magnetic field, $\overrightarrow{E}$ is the electric field, c is the speed of light, $\overrightarrow{v}$ is the velocity of the particles, $\overrightarrow{p}$ is the momentum of the particles, ρ is the electric charge density, j is the electric current density, and q_α is the charge of the particle of the species α.

The computation area represented in Fig. 1 has a rectangular shape, the plasma in the area is bounded by vacuum from both sides. There are two buffers inside the area in the plasma; they are bounded from the other two sides (at the boundaries $X = 0$ and $X = L_x$), and constantly maintain the initial plasma distribution. The magnetic field is $\overrightarrow{B} = (B_x, 0, 0)$.

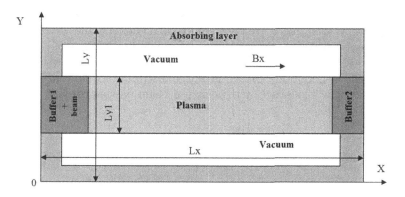

Fig. 1. Computation area

The electron beam enters the plasma along the magnetic field lines through the left boundary and leaves it through the the right one; the beam injection is persistent.

Special attention should be paid to the boundary conditions. It is necessary to set consistent boundary conditions both for particles and for electromagnetic fields. It is assumed that at the boundaries $x = 0$ and $x = L_x$ the plasma is weakly disturbed. Therefore, the following boundary condition should be introduced for the electromagnetic fields in the plasma region (6),

$$\left.\frac{\partial Ex}{\partial x}\right|_{x=0, x=L_x} = 0. \tag{6}$$

The second-order absorbing boundary conditions (ABC) based on the one-way wave equations (OWWE) [4,5] are set on the other boundary.

A special absorbing layer is used for radiation diagnostics. The absorption of electromagnetic waves is specified by multiplying at each step the electric and magnetic field values by the absorption coefficient depending on the distance to the border. The quadratic dependence of the absorption coefficient is used to smooth the wave damping (7),

$$k = \begin{cases} \frac{a-1}{l^2}x^2 - 2\frac{a-1}{l}x + a, \ x < l, \\ 1, \ x \geq l, \end{cases} \tag{7}$$

where l is the width of the absorbing layer, x is the distance to the boundary, $0 < a < 1$ is the value characterizing the absorption distribution within the absorption layer. It is difficult to estimate the energy left by the electromagnetic wave in the ABC case. We introduce an artificial coefficient to account for the absorbed energy. This coefficient allows to remove part of the electromagnetic field energy at each time step. The absence of artifacts has been proven experimentally by comparing the absorption with the coefficient k and using the Mur's ABC.

A special buffer is used for setting boundary conditions on the velocity of the particles (Fig. 2). The particles penetrate from the plasma to the buffer, pass through it freely, then they leave the computational domain and are removed. At the same time, a particle with parameters corresponding to those of the removed particle is added to the buffer through the opposite boundary. This allows keeping the particle distribution close to the initial distribution, and causes a natural plasma counterflow generated by the electron beam passing through it.

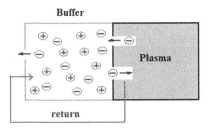

Fig. 2. Scheme of the boundary buffer

3   Solutions of the Main Equations

A PIC method (see [7–9]) is used to solve the Vlasov equation. In this method, the plasma is simulated by a set of separate particles, each characterizing the motion of many physical particles. The characteristics of the Vlasov equation describe the trajectories of the particles. The equations for these characteristics can be written as

$$\frac{d\vec{p}_\alpha}{dt} = q_\alpha(\vec{E} + [\vec{v_\alpha}, \vec{B}]), \quad \frac{d\vec{r}_\alpha}{dt} = \vec{v}_\alpha, \quad \vec{p}_\alpha = \frac{\vec{v}_\alpha m_\alpha}{\sqrt{1 - (\frac{\vec{v}_\alpha}{c})^2}}. \tag{8}$$

The Boris scheme [10] is used to solve these equations. The Langdon–Lazinsky scheme (9), described by A. Langdon and B. Lasinski in [11], is used to solve Maxwell's equations,

$$\frac{B^{m+1/2} - B^{m-1/2}}{\Delta t} = -c\,\mathrm{rot}_h\,E^m, \quad \frac{E^{m+1} - E^m}{\Delta t} = -4\pi j^{m+1/2} + c\,\mathrm{rot}_h\,B^{m+1/2}, \quad (9)$$

where rot_h is the operator of the central difference scheme defined on the Yee cell [6]. In this scheme, the electric and magnetic fields are calculated on staggered grids. This yields second-order accuracy with respect to space and time. The value of B is calculated at the fractional time step $m + 1/2$, while E is calculated at the integer time step $m + 1$. Mur has developed special ABC [5] for discretization and application of OWWE to the Yee algorithm.

In this scheme, the electric and magnetic fields are calculated on staggered grids. This yields second-order accuracy with respect to space and time.

Charge densities and current densities are defined by particle velocities and coordinates:

$$\rho(r,t) = \sum_{k=1}^{K} q_k R(r, r_k(t)), \quad j(r,t) = \sum_{k=1}^{K} q_k v_k(t) R(r, r_k(t)). \quad (10)$$

Here q_k is the charge of the k-th particle, and the function $R(r, r_k(t))$ is the form factor of the PIC method. It characterizes the shape and size of the particle, as well as the charge distribution [8].

We calculate the current density according to [12] and our previous works [13,14]; the continuity Eq. (11) is automatically satisfied:

$$\frac{\partial \rho}{\partial t} + \mathrm{div}\,\vec{j} = 0. \quad (11)$$

By using this method, we avoid having to solve Poisson's equation at each time step and find the correct grid values of the electric and magnetic fields.

4   Parallel Algorithm

To correctly describe the electromagnetic wave caused by the interaction of the electron beam with the plasma, we used no less than 20 grid nodes per wavelength. When the wavelength is equal to c/w_{pe} and the size of the area is $100 \times 100 \; (c/w_{pe})^2$, we need a 2000×2000 grid. Even if the number of particles per cell is small ($\sim$100), at least 10^9 particles are required for the calculations, making necessary to use high-performance computers and efficient parallel algorithms.

There are several approaches to the parallel implementation of the particle-in-cell method. Since the trajectories of the model particles can be computed independently, the easiest way is to distribute the particles evenly among all processors regardless of their coordinates. In this case, each processor solves the Maxwell's equations in the whole computation area. The time for computation of the electromagnetic fields is much smaller than that of the trajectories: there are up to 1000 particles of each sort in each grid cell and, for example, a grid with 1000×1000 nodes needs 3×10^6 exchanges for the values of the currents, it is only ~ 20 Mb. Thus, the parallelization is highly effective. Another possible way is to decompose the computational domain in one direction, for instance, perpendicular to the direction of the beam motion. We use this approach of parallelization, called the Euler–Lagrange domain decomposition [1,2]. In this case, the computational domain is divided into several sub-domains. A group of processors is associated with each sub-domain [15]. The groups of processors communicate with each other by the boundary values of the fields and particles. The particles are distributed among the processor groups within each sub-domain, independently of their location. Within each group, there is a collective exchange (all-to-all) by the grid values of the current density at each step. Thus, Eqs. (3)–(4) are solved with the same input parameters by each processor, and there is no need to send the electromagnetic field values to the group of processors. There are some limitations on the number of processor groups in this case. The number of groups cannot exceed the number of nodes in the direction of the domain decomposition. Moreover, the number of particles in each sub-domain may vary significantly, therefore, a large number of groups may lead to an uneven loading of the processors. A comparison of different ways of decomposition in the PIC method is made in [16].

The scheme of the parallel algorithm for the one-dimensional domain decomposition is shown in Fig. 3.

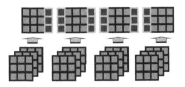

Fig. 3. Scheme of the parallel algorithm

As the length of the plasma area is significantly larger than its width, the most natural partitioning of the area is its division in the x direction. The solution of the equations for the particles is distributed among the group of processors. Thus, the gain in acceleration depends on both the number of particles and the area size.

In addition, at the initial stage of calculations, when the beam has not occupied the whole area yet, there is an imbalance in computational load among the sub-domains where the beam has already entered and the sub-domains that the beam has not reached yet.

The speed of calculation of the problem is important. However, the practical implication of the algorithm lies on the possibility of solving this large-scale problem, and improving the accuracy by increasing the number of particles and the number of grid nodes. Thus, we want to evaluate how the computation time depends on the number of available processors assuming that the volume of calculations for each processor is constant.

The collective exchanges of current density within a group of processors is the bottleneck of the algorithm, therefore, we will increase the number of processors within a group keeping the grid size constant for each processor.

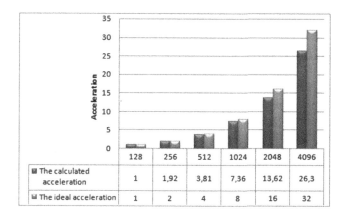

Fig. 4. Acceleration of computations

The Fig. 4 shows the acceleration of computations relative to 128 processor cores for various numbers of processor cores, namely 256, 512, 1024, 2048, and 4096. In these calculations, we used a 1024×1024 grid, the total number of particles is 10^9, and eight sub-domains are used.

Table 1. Computation time (sec) of one time step for different numbers of grid nodes and various numbers of cores

Number of cores	512×512	512×1024	1024×1024	1024×2048
128	0,027	0,076	0,139	0,426
256	0,036	0,077	0,146	0,431
512	0,038	0,081	0,147	0,375
1024	0,039	0,082	0,153	0,492
2048	0,049	0,113	0,157	0,437
4096	0,054	0,119	0,197	0,428

Table 1 shows the computation time of one step of the algorithm for different numbers of processors. We used grids with 512×512, 512×1024, 1024×1024 and 1024×2048 nodes. At the same time, the numbers of particles per processor were 2×10^6, 4×10^6, 8×10^6, 16×10^6. These computations were performed on the

"Lomonosov" supercomputer (Moscow State University, Intel Xeon processors 5570 2932 MHz). Table 1 demonstrates that the computation time increases for a grid of small size as the number of cores increases, but it remains almost unchanged for a grid of large size. Thus, both the volume of data to be sent and the number of processors among which data exchange occurs play a significant role. However, for large grids, increasing the number of processors within a group does not yield additional computing costs. Thus, the parallel algorithm can be successfully applied to the solution of resource-intensive problems by efficiently using large numbers of computational cores.

5   Simulation Results

The grid size, the number of particles and the decomposition of the computational domain, required in the solution of the problem of generation of electromagnetic radiation by interaction of an electron beam with plasma, were investigated in computational experiments.

The simulation parameters correspond to those of real physical experiments performed at the GOL-3 facility (BINP SB RAS) [3]: plasma density $n_p = 10^{15}$; density ratio $\frac{n_b}{n_p} = 0.02$; beam temperature $T_b = 100eV$; ratio of the beam velocity to the speed of light $\frac{v_b}{c} = 0.9$; plasma electron temperature $T_e = 14eV$; and plasma ion temperature $T_i = 0$. We assume that the characteristic size of the area and the characteristic wavelengths of the generated electromagnetic radiation for these parameters are the following: area length $Lx = 86 \ c/w_{pe}$, and plasma layer width $Ly1 = 6.4 \ c/w_{pe}$. Here and below, all the lengths are expressed in c/w_{pe}.

The numerical experiments were performed on the NSU supercomputer (Intel Xeon X5670, 2932 MGz). The standard parameters for the calculations were: 128 processor cores, 360×384 grid, number of particles per core 8×10^6. The total time of computation was about 34 h (500 000 time steps).

Figure 5 shows the time history of the electron beam density and the corresponding x-p phase space. The graph shows the dynamics of the entered beam and a forming shock wave with wavelength ~ 5.

For the simulation of the beam injection throughout a long time, a large area is required, and the grid step size is rather small. However, to reduce these high computation costs, it is sufficient to add a new group of processors. The scalability of the algorithm (in the sense of weak scalability) remains high, since the exchanges among the groups are simultaneously executed and the transferred data size does not increase. The tests have demonstrated that an eight-fold increase of the computational domain with the corresponding eight-fold increase of the number of processors requires only 0.27 s instead of 0.24 s, i.e. the difference is about 12%.

We analyzed the electromagnetic waves generated by the beam–plasma interaction. Figure 6 shows the dynamics of the x- and z-components of the electric field. We can observe in Fig. 6 the radial propagation of the electromagnetic

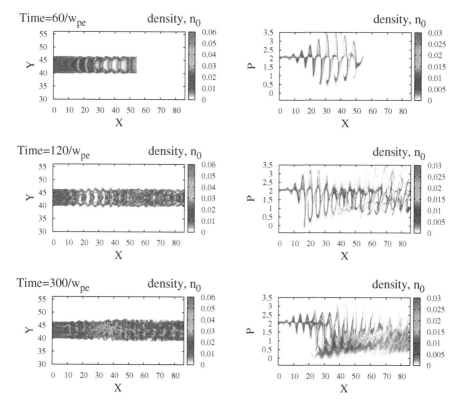

Fig. 5. Time history of the electron beam density, and the corresponding x-p phase space

waves in the vacuum. The amplitudes of the waves propagated in the vacuum reach 20% of the amplitudes of those generated in the plasma.

A Fourier analysis was made at every time step for the accurate study of the frequency of the radiation generated by the electromagnetic field. Figure 7 shows the results of the Fourier analysis for the electric field at the following points:

1. $X = 30, Y = 7$ (inside the absorbent layer);
2. $X = 15, Y = 43$ (the central part of the area).

Figure 7 demonstrates the dominant plasma frequency of the electromagnetic waves generated in the plasma. If the plasma density $n_e = 10^{15}$ cm^{-3}, then the frequency is equal to 0.28 THz. The radiation propagating to the vacuum has both the dominant plasma frequency and the dominant double plasma frequency, while the generated electromagnetic radiation is located in the subterahertz frequency range.

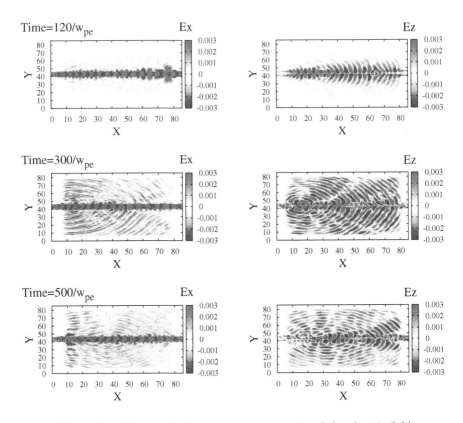

Fig. 6. Time history of the x- and z-components of the electric field

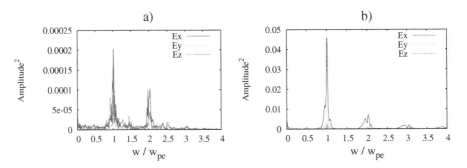

Fig. 7. Fourier analysis of the electric field at the points (a) $X = 30, Y = 7$, (b) $X = 15, Y = 43$

The low amplitude of the electromagnetic waves in the vacuum requires a small spatial step. Since the domain decomposition is carried out only in the x direction, the load of each processor increases as the spatial step decreases. The exchanges of the large-scale grids within each group lead to a decrease in computing costs. The greatest part of the computational domain in the y direction is situated in the vacuum; for this reason, a two-dimensional decomposition is not efficient in this case. The optimal solution, however, consists in preprocessing the current density array before carrying out the exchanges, so as to remove the zero values corresponding to the zero currents in the vacuum. The total time costs will not increase, since removing the zero elements takes significantly less time than the transfer itself. It was confirmed by numerical experiments that an increase of the grid along the y direction from 1024 to 8196 nodes yields 0.26 s (instead of 0.24) for one step of computation with the same number of processors. The standard transfer method, according to Table 1, yields a significantly higher increase in computation time.

Let us consider the dynamics of the ion density. Figure 8 shows the ion density and the average ion density in the y direction. From Fig. 8, one can see that the ion density modulation appears in the process of continuous injection of the beam into the plasma at the time instant $530/w_{pe}$, and the amplitude of modulation is up to 40% of the average ion density. It is supposed that this effect provides an efficient generation of radiation in the vacuum cavities. For a more accurate description of the size and amplitude cavities formed, a further decrease of the grid space in both x and y directions is required. In this case, the amount of data transferred within a group increases, because the number of nodes with non-zero

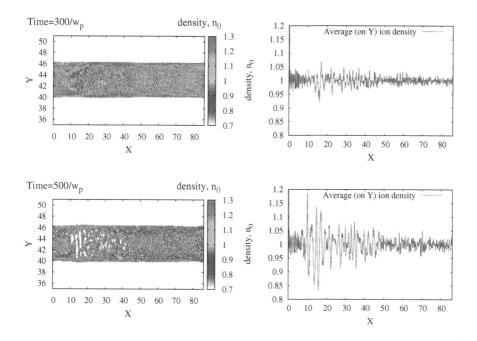

Fig. 8. Ion density and averaged ion density in the Y direction

current values increases. The development of an irregular two-dimensional domain decomposition will improve the scalability.

We have studied the dependence between the radiation and the entrance angle of the beam. Since the plasma is confined by an homogeneous magnetic field parallel to the axis x, the particles of the entering beam continue in spiral motion along the magnetic field. The angle of the entering beam should not be large, otherwise the beam will extend out of the plasma very quickly. The computational experiments show that an increase of the angle of the entering beam leads to an increase in radiation intensity. The propagation of the electromagnetic waves takes place along a sector, and the width of this sector increases as the injection angle increases. The center of the radiation does not move in the course of time. Figure 9 shows the calculation results when the injection angle is 30°. The results of the numerical experiments demonstrate the existence of a wave structure in the beam density, the density waves are directed at a certain angle to the axis x. This fact confirms the assumption that density modulation is a reason for the efficient generation of electromagnetic radiation.

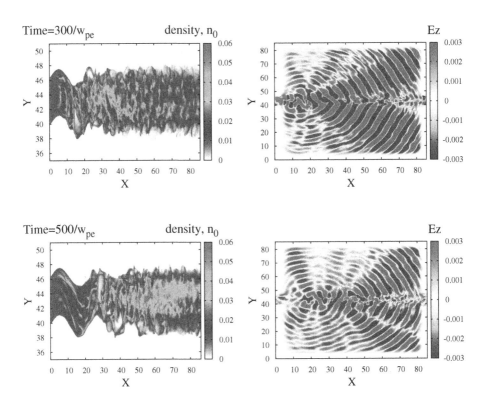

Fig. 9. Beam density and electric field when the angle of the entering beam is 20°

6 Conclusions

We presented in this article the numerical model of the continuous injection of an electron beam in a plasma with open boundary conditions. This model is based on a particle-in-cell method (PIC), and allows to perform numerical experiments on generation of electromagnetic radiation by beam–plasma interaction. The developed parallel algorithm uses the Euler–Lagrange domain decomposition for high-performance computations.

We made an analysis of the weak scalability of the parallel algorithm. It has shown that an increase in the number of processors from 128 to 4096 and a proportional increase in computational load yield less than a two-fold increase in computation time.

Computational experiments have been performed on the basis of the developed models and programs. The simulation parameters correspond to those of physical experiments performed at the GOL-3 facility (BINP SB RAS). The beam density, the plasma and the electromagnetic waves in the plasma and vacuum were defined.

Also, the characteristic domain sizes, the length and width of the plasma layer for the most efficient generation of electromagnetic radiation were determined.

The formation of the electromagnetic spectrum was analyzed by Fourier analysis. It was proven that the resulting radiation is located in the subterahertz frequency range. We discovered the dominant harmonics of the electromagnetic fields in the beam–plasma interaction in the vacuum.

It was shown that the radiation intensity increases when the injection of the beam occurs at a certain angle to the magnetic field.

The calculations performed have demonstrated that the parallel algorithm allows to solve successfully problems of continuous beam injection into a plasma and explain the results obtained in laboratory experiments.

References

1. Berendeev, E.A., Efimova, A.A.: Effective parallel calculations realization for modeling big problems of the plasma physics by the PIC-method. Vestnik UGATU **17**(2(55)), 112–116 (2013)
2. Efimova, A.A., Berendeev, E.A., Dudnikova, G.I., Vshivkov, V.A.: Numerical simulation of nonlinear processes in a beam-plasma system. In: AIP Conference Proceedings, vol. 1684, p. 100001 (2015)
3. Burdakov, A.V., Avrorov, A.P., Arzhannikov, A.V., Astrelin, V.T., et al.: Development of extended heating pulse operation mode at GOL-3. Fusion Sci. Technol. **63**(1T), 29–34 (2013)
4. Engquist, B., Majda, A.: Absorbing boundary conditions for the numerical simulation of waves. Math. Comp. **31**(139), 629–651 (1977)
5. Mur, G.: Absorbing boundary conditions for the finite-difference approximation of the time-domain electromagnetic-field equations. IEEE Trans. Electromagnet. Compatib. **23**(4), 377–382 (1981)

6. Yee, K.S.: Numerical solution of initial boundary value problems involving Maxwell's equations in isotropic media. IEEE Trans. Antenn. Propagat. **AP–14**(3), 302–307 (1966)
7. Birdsall, C.K., Langdon, A.B.: Plasma Physics via Computer Simulation. Institute of Physics Publishing, Bristol (1991)
8. Berezin, Y.A., Vshivkov, V.A.: Particles Method in the Dynamics of a Rarefied Plasma. Nauka, Novosibirsk (1980)
9. Hockney, R.W., Eastwood, J.W.: Computer Simulation Using Particles. CRC Press, Boca Raton (1988)
10. Boris, J.P.: Relativistic plasma simulation - optimization of a hybrid code. In: Fourth Conference on Numerical Simulation of Plasmas, Washington, pp. 3–67 (1970)
11. Langdon, A.B., Lasinski, B.F.: Electromagnetic and relativistic plasma simulation models. Meth. Comput. Phys. **16**, 327–366 (1976)
12. Villasenor, J., Buneman, O.: Rigorous charge conservation for local electromagnetic field solver. Comput. Phys. **69**, 306–316 (1992)
13. Berendeev, E., Dudnikova, G., Efimova, A., Vshivkov, V.: Computer simulation of plasma dynamics in open plasma trap. In: Dimov, I., Faragó, I., Vulkov, L. (eds.) NAA 2016. LNCS, vol. 10187, pp. 227–234. Springer, Cham (2017). doi:10.1007/978-3-319-57099-0_23
14. Berendeev, E.A., Dudnikova, G.I., Efimova, A.A., Ivanov, A.V., Vshivkov, V.A.: Computer simulation of cylindrical plasma target trap with inverse magnetic mirrors. In: AIP Conference Proceedings, vol. 1771, p. 030009 (2016)
15. Boronina, M.A., Vshivkov, V.A.: Parallel 3-D particle-in-cell modelling of charged ultrarelativistic beam dynamics. J. Plasma Phys. **81**(6) (2015)
16. Andrianov, A.N., Efimkin, K.N.: Approach to parallel implementation of the particles in cell method. Preprints Keldysh Institute of Applied Mathematics, No. 9 (2009)

The Hybrid-Cluster Multilevel Approach to Solving the Elastic Wave Propagation Problem

Boris Glinskiy, Anna Sapetina, Valeriy Martynov, Dmitry Weins,
and Igor Chernykh[(⊠)]

Institute of Computational Mathematics and Mathematical Geophysics
of the Siberian Branch of the Russian Academy of Sciences,
6, Prospect Akademika Lavrentjeva, Novosibirsk 630090, Russia
gbm@opg.sscc.ru, afsapetina@gmail.com, vnm@nmsf.sscc.ru,
wns.dmitry@gmail.com , chernykh@parbz.sscc.ru

Abstract. We propose in this paper an integrated approach to creating algorithms and software to simulate seismic problems. Our extended co-design concept considers supercomputer architecture at all stages, beginning with the physical and mathematical formulation of the problem, and then developing an algorithm and code. In particular, we compare the efficiency of the parallel implementation (on a supercomputer equipped with GPU) of algorithms for solving two mathematical statements of dynamic elasticity problem for the numerical modeling of seismic wave fields in 3D media, which is typical of volcanic structures.

The scalability of the algorithms is investigated using the multi-agent system AGNES to simulate the behavior of computational nodes based on the current state of computer equipment characteristics. We present the results obtained for the efficiency of the implementation of the algorithms when using millions of cores. Also, we assess the energy efficiency of these algorithms.

Keywords: Elastic waves · 3D modeling · Finite-difference schemes · Hybrid cluster · GPU · Co-design · Agent simulation · Energy efficiency of algorithms

1 Introduction

Numerical modeling of elastic wave propagation in heterogeneous 3D media with complex subsurface geometries is a complex problem in terms of computation, thereby demanding the use of efficient methods for the parallelization and scalability of algorithms. This class of problems includes studying special aspects of propagation of seismic waves in media that are typical of the magmatic structures of active (or dormant) volcanoes. Both active and sleeping volcanoes are

This work was supported by the Russian Foundation for Basic Research, grants No. 16-29-15120, No. 16-07-00434, No. 16-01-00455, and a grant of the Russian Ministry of Education No. MK-152.2017.5.

L. Sokolinsky and M. Zymbler (Eds.): PCT 2017, CCIS 753, pp. 261–274, 2017.
DOI: 10.1007/978-3-319-67035-5_19

potentially dangerous to the environment because sudden catastrophic eruptions are possible. At the present state of scientific and technological achievements, preventing such eruptions is impossible. However, predicting a probable time of eruption using methods of active vibroseismic monitoring of magmatic structures has become feasible.

Scientists at the Institute of Computational Mathematics and Mathematical Geophysics of the Siberian Branch of the Russian Academy of Sciences have developed a method of monitoring earthquake-prone zones and volcanic structures with the use of powerful vibrators [1,2]. The main idea of the method, as applied to the monitoring of volcanoes, is as follows. On the surface of a volcanic structure, a vibrator and a monitoring system recording incoming seismic waves are placed at specific points. From a change in the characteristics of the observed signal, one can make conclusions about the behavior dynamics of the magmatic structures of volcanoes. The method proposed allows to use the main advantage of the modern technology of seismic sounding aimed at detecting and tracking changes in the state of seismic-acoustic parameters of volcanic vents and their environs. This advantage is based on the high long-range stability of the emission parameters of vibration signals. This is achieved by means of modern systems of computer-aided control of the vibrators. This technique has been tested on the mud volcanoes of the Taman province, Russia [1].

The relief of many real geophysical objects does not allow one to install an observational system. Therefore, the construction of 3D models of such objects requires solving the inverse problem by means of a set of direct problems: for different values of the elastic parameters of a heterogeneous medium and various geometries of the objects composing a model.

In the process of obtaining the solutions, we need to collect synthetic data: seismograms and snapshots of the wave field. One of the approaches to solving the problem is to use a finite-difference method and, correspondingly, a 3D mesh, which requires storing and processing large amounts of data [3]. Thus, we need solutions to a set of direct problems to solve an inverse problem requiring a large volume of memory, and all this must be done rapidly.

To solve this class of problems, the authors suggest the approach described earlier in [4]. It consists of three stages. The first stage is the co-design, which is based on the development of a parallel computational technology taking into account all the aspects of parallelism. The second is the development of anticipated algorithms and software for the exascale supercomputers. This stage is based on the simulation of the algorithm behavior within a certain supercomputer architecture. The third stage is to estimate the energy efficiency of the algorithm with different implementations for either a concrete architecture or different supercomputer architectures [4,5].

We apply the approach discussed above to the solution of the problem of seismic wave propagation in a heterogeneous medium typical of magmatic volcanoes.

2   The Co-design Application in Modeling Seismic Wave Propagation in an Elastic Medium

The co-design of parallel methods for solving large-scale problems is a sufficiently complex process and one that is difficult to formalize. This is primarily due to the fact that each individual problem, even within one and the same area of knowledge, has its own features. Consideration of these features may seriously change not only the parallel computing method but also the mathematical model and, of course, the program implementation and its efficiency.

It is impossible, indeed, to "collect recipes" for the efficient solution of all large-scale problems. Nonetheless, it is possible to propose a general approach. The concept of the co-design can be formulated as follows:

1. Physical formulation of the problem.
2. Mathematical formulation of the physical problem.
3. Development of the numerical method for solving the mathematical formulation of the physical problem.
4. Selection of a data structure and a parallel algorithm.
5. Consideration of a supercomputer architecture.
6. Choice and use of the development tools.

In the approach discussed here, we use the extended definition of the co-design, in contrast to the approaches described, for instance, in [6]. In the context of the numerical simulation, the co-design is the design of a physical model and a mathematical model, a numerical method, a parallel algorithm and an implementation with the efficient use of a hybrid supercomputer architecture. In the above-discussed approach, not only the comparison of the methods used to solve the problem is relevant but also the efficiencies obtained by using various physical and mathematical statements.

The co-design has been successfully applied to the development of software for solving problems of plasma physics and astrophysics [4,5]. In the next section, we will consider the use of the co-design for modeling seismic wave propagation in an elastic medium.

2.1   Description of Mathematical Statements of Dynamic Elasticity

The domain of simulation is assumed to be an isotropic 3D inhomogeneous elastic complex medium in the form of a parallelepiped in which one of the faces is a free surface (the plane $z = 0$). Let us introduce a Cartesian rectangular coordinate system so that the axis Oz is directed vertically downwards, and the axes Ox and Oy are directed along the free surface. The system of elasticity equations with appropriate initial and boundary conditions can be written down in terms of the displacement velocity vector $\mathbf{u} = (u, v, w)^T$ and the stress tensor $\boldsymbol{\sigma} = (\sigma_{xx}, \sigma_{yy}, \sigma_{zz}, \sigma_{xy}, \sigma_{xz}, \sigma_{yz})^T$:

$$\rho \frac{\partial \mathbf{u}}{\partial t} = [A]\boldsymbol{\sigma} + \mathbf{F}(t, x, y, z), \tag{1}$$

$$\frac{\partial \boldsymbol{\sigma}}{\partial t} = [B]\mathbf{u}, \tag{2}$$

$$A = \begin{bmatrix} \frac{\partial}{\partial x} & 0 & 0 & \frac{\partial}{\partial y} & \frac{\partial}{\partial z} & 0 \\ 0 & \frac{\partial}{\partial y} & 0 & \frac{\partial}{\partial x} & 0 & \frac{\partial}{\partial z} \\ 0 & 0 & \frac{\partial}{\partial z} & 0 & \frac{\partial}{\partial x} & \frac{\partial}{\partial y} \end{bmatrix}, B = \begin{bmatrix} (\lambda + 2\mu)\frac{\partial}{\partial x} & \lambda\frac{\partial}{\partial y} & \lambda\frac{\partial}{\partial z} \\ \lambda\frac{\partial}{\partial x} & (\lambda + 2\mu)\frac{\partial}{\partial y} & \lambda\frac{\partial}{\partial z} \\ \lambda\frac{\partial}{\partial x} & \lambda\frac{\partial}{\partial y} & (\lambda + 2\mu)\frac{\partial}{\partial z} \\ \mu\frac{\partial}{\partial y} & \mu\frac{\partial}{\partial x} & 0 \\ \mu\frac{\partial}{\partial z} & 0 & \mu\frac{\partial}{\partial x} \\ 0 & \mu\frac{\partial}{\partial z} & \mu\frac{\partial}{\partial y} \end{bmatrix},$$

where t is time, $\rho(x, y, z)$ is the density, and $\lambda(x, y, z)$ and $\mu(x, y, z)$ are the Lame coefficients.

The initial conditions are

$$\sigma_{xz}|_{t=0} = 0, \ \sigma_{yz}|_{t=0} = 0, \ \sigma_{xy}|_{t=0} = 0, \ \sigma_{xx}|_{t=0} = 0, \ \sigma_{yy}|_{t=0} = 0, \ \sigma_{zz}|_{t=0} = 0,$$
$$u|_{t=0} = 0, \ v|_{t=0} = 0, \ w|_{t=0} = 0. \tag{3}$$

The boundary conditions at the free surface are

$$\sigma_{xz}|_{z=0} = 0, \ \sigma_{yz}|_{z=0} = 0, \ \sigma_{zz}|_{z=0} = 0. \tag{4}$$

The system of elasticity equations for the displacement vector $\mathbf{U} = (U, V, W)^T$ is

$$\rho \frac{\partial^2 \mathbf{U}}{\partial t^2} = [C]\mathbf{U} + \mathbf{F}(t, x, y, z), \tag{5}$$

$$C = \begin{bmatrix} C_1 \\ C_2 \\ C_3 \end{bmatrix},$$

$$C_1 = \left[(\lambda + 2\mu)\frac{\partial^2}{\partial x^2} + \mu\left(\frac{\partial^2}{\partial y^2} + \frac{\partial^2}{\partial z^2}\right) \ (\lambda + \mu)\frac{\partial^2}{\partial x \partial y} \ (\lambda + \mu)\frac{\partial^2}{\partial x \partial z} \right],$$

$$C_2 = \left[(\lambda + \mu)\frac{\partial^2}{\partial y \partial x} \ (\lambda + 2\mu)\frac{\partial^2}{\partial y^2} + \mu\left(\frac{\partial^2}{\partial x^2} + \frac{\partial^2}{\partial z^2}\right) \ (\lambda + \mu)\frac{\partial^2}{\partial y \partial z} \right],$$

$$C_3 = \left[(\lambda + \mu)\frac{\partial^2}{\partial z \partial x} \ (\lambda + \mu)\frac{\partial^2}{\partial z \partial y} \ (\lambda + 2\mu)\frac{\partial^2}{\partial z^2} + \mu\left(\frac{\partial^2}{\partial x^2} + \frac{\partial^2}{\partial y^2}\right) \right].$$

The initial conditions are

$$U|_{t=0} = 0, \ V|_{t=0} = 0, \ W|_{t=0} = 0. \tag{6}$$

The boundary conditions at the free surface are

$$\sigma_{xz}|_{z=0} = 0, \ \sigma_{yz}|_{z=0} = 0, \ \sigma_{zz}|_{z=0} = 0. \tag{7}$$

In both cases, it is assumed that the right-hand side term (i.e. the mass force) is expressed as $\mathbf{F}(t, x, y, z) = F_x\mathbf{i} + F_y\mathbf{j} + F_z\mathbf{k}$, where $\mathbf{i}, \mathbf{j}, \mathbf{k}$ are the unit direction vectors of the coordinate axes.

2.2   Finite-Difference Method for Solving the Problem

In the case of a three-dimensional elastodynamic problem, the most "flexible" and widespread method is a finite-difference method. It is important to note that explicit finite-difference schemes fit the architecture of graphics accelerators, since they are directly mapped onto the topology of the GPU architecture, and involve independent computations of values at the next step in each cell of the computational domain.

To numerically solve Eqs. (1)–(4), we apply the well-known Verrier finite-difference scheme on a staggered grid [7,8]. The calculation of its difference coefficients uses integral conservation laws. The method is of second order of approximation with respect to time and space [7]. Here we consider only uniform grids. Let us present, as an example, a few finite-difference equations of the scheme used:

$$\frac{\rho_{i,j,k} + \rho_{i-1,j,k}}{2} \frac{u_{i-\frac{1}{2},j,k}^{n+1} - u_{i-\frac{1}{2},j,k}^{n}}{\tau} = \frac{\sigma_{xxi,j,k}^{n+\frac{1}{2}} - \sigma_{xxi-1,j,k}^{n+\frac{1}{2}}}{h}$$
$$+ \frac{\sigma_{xyi-\frac{1}{2},j+\frac{1}{2},k}^{n+\frac{1}{2}} - \sigma_{xyi-\frac{1}{2},j-\frac{1}{2},k}^{n+\frac{1}{2}}}{h} + \frac{\sigma_{xzi-\frac{1}{2},j,k+\frac{1}{2}}^{n+\frac{1}{2}} - \sigma_{xzi-\frac{1}{2},j,k-\frac{1}{2}}^{n+\frac{1}{2}}}{h} + f_{xi,j,k}^{n},$$

$$\frac{\sigma_{xzi-\frac{1}{2},j,k-\frac{1}{2}}^{n+\frac{1}{2}} - \sigma_{xzi-\frac{1}{2},j,k-\frac{1}{2}}^{n-\frac{1}{2}}}{\tau}$$
$$= M1_{i-\frac{1}{2},j,k-\frac{1}{2}} \left(\frac{u_{i-\frac{1}{2},j,k}^{n} - u_{i-\frac{1}{2},j,k-1}^{n}}{h} + \frac{w_{i,j,k-\frac{1}{2}}^{n} - w_{i-1,j,k-\frac{1}{2}}^{n}}{h} \right).$$

$$\text{where } M1_{i-\frac{1}{2},j,k-\frac{1}{2}} = \left(\frac{1}{4} \left(\frac{1}{\mu_{i,j,k}} + \frac{1}{\mu_{i-1,j,k}} + \frac{1}{\mu_{i,j,k-1}} + \frac{1}{\mu_{i-1,j,k-1}} \right) \right)^{-1}.$$

To solve the problem in terms of displacements, we use a similar finite-difference scheme. The finite-difference equation for the component U is

$$\frac{\rho_{i,j,k} + \rho_{i-1,j,k}}{2} \frac{U^{n+1}_{i-\frac{1}{2},j,k} - 2U^n_{i-\frac{1}{2},j,k} + U^{n-1}_{i-\frac{1}{2},j,k}}{\tau^2}$$

$$= (\lambda_{i,j,k} + 2\mu_{i,j,k}) \frac{U^n_{i+\frac{1}{2},j,k} - U^n_{i-\frac{1}{2},j,k}}{h^2}$$

$$+ \lambda_{i,j,k} \left(\frac{V^n_{i,j+\frac{1}{2},k} - V^n_{i,j-\frac{1}{2},k}}{h^2} + \frac{W^n_{i,j,k+\frac{1}{2}} - W^n_{i,j,k-\frac{1}{2}}}{h^2} \right)$$

$$- (\lambda_{i-1,j,k} + 2\mu_{i-1,j,k}) \frac{U^n_{i-\frac{1}{2},j,k} - U^n_{i-\frac{3}{2},j,k}}{h^2}$$

$$- \lambda_{i-1,j,k} \left(\frac{V^n_{i-1,j+\frac{1}{2},k} - V^n_{i-1,j-\frac{1}{2},k}}{h^2} + \frac{W^n_{i-1,j,k+\frac{1}{2}} - W^n_{i-1,j,k-\frac{1}{2}}}{h^2} \right)$$

$$+ M2_{i-\frac{1}{2},j+\frac{1}{2},k} \left(\frac{U^n_{i-\frac{1}{2},j+1,k} - U^n_{i-\frac{1}{2},j,k}}{h^2} + \frac{V^n_{i,j+\frac{1}{2},k} - V^n_{i-1,j+\frac{1}{2},k}}{h^2} \right)$$

$$- M2_{i-\frac{1}{2},j-\frac{1}{2},k} \left(\frac{U^n_{i-\frac{1}{2},j,k} - U^n_{i-\frac{1}{2},j-1,k}}{h^2} + \frac{V^n_{i,j-\frac{1}{2},k} - V^n_{i-1,j-\frac{1}{2},k}}{h^2} \right)$$

$$+ M1_{i-\frac{1}{2},j,k+\frac{1}{2}} \left(\frac{U^n_{i-\frac{1}{2},j,k+1} - U^n_{i-\frac{1}{2},j,k}}{h^2} + \frac{W^n_{i,j,k+\frac{1}{2}} - W^n_{i-1,j,k+\frac{1}{2}}}{h^2} \right)$$

$$- M1_{i-\frac{1}{2},j,k-\frac{1}{2}} \left(\frac{U^n_{i-\frac{1}{2},j,k} - U^n_{i-\frac{1}{2},j,k-1}}{h^2} + \frac{W^n_{i,j,k-\frac{1}{2}} - W^n_{i-1,j,k-\frac{1}{2}}}{h^2} \right)$$

where $M2_{i-\frac{1}{2},j-\frac{1}{2},k} = \left(\frac{1}{4} \left(\frac{1}{\mu_{i,j,k}} + \frac{1}{\mu_{i-1,j,k}} + \frac{1}{\mu_{i,j-1,k}} + \frac{1}{\mu_{i-1,j-1,k}} \right) \right)^{-1}$.

Let us note the main difference between the algorithms that can be constructed with the above-mentioned finite-difference schemes. The calculation of the velocities of displacement and stress requires a larger memory size (at least, nine 3D arrays of unknowns must be stored), but requires a smaller number of floating-point operations in total. On the contrary, the calculation of the displacements requires a smaller memory size (at least, three 3D arrays of unknowns) but a larger number of operations.

2.3 Parallel Implementation

Graphic accelerators are well suited for solving finite-difference equations because of their massively parallel architecture and quick access to the device memory. Many computer systems in Russia and abroad are equipped with GPUs, including the NKS-30T+GPU (Siberian Supercomputer Center) which consists of forty HP SL390s G7 computational nodes, each equipped with two six-core Xeon X5670 CPUs and three NVIDIA Tesla M2090 graphics cards with Fermi architecture. Each card contains one GPU with 512 cores and 6 Gb of GDDR5 RAM. The peak performance of this system is 85 teraflops.

The development of a program code for such hybrid systems requires additional knowledge and time but provides a significant increase in performance. The efficient use of a hybrid architecture requires the parallelization and optimization of the simulation algorithms. Since the methods for solving the considered statements of the elastodynamic problem are similar in their nature, the comparison of the correctness, adaptation and optimization of these methods is carried out in similar ways. As for the software tools, we chose CUDA and MPI, which make possible the simultaneous use of the largest number of parallel processes and, ultimately, the attainment of the maximum efficiency.

To carry out the parallelization, we decompose the computational domain into layers along one of the coordinate axes. Each layer is calculated at a separate node, where, in turn, it is sub-divided into sub-layers along another coordinate axis (to achieve better scalability) according to the number of graphics accelerators available at a node. Thus, we come to a truncated 2D decomposition whose scalability in one of the directions is limited by the number of graphics cards present at a computational node. In such an implementation, each graphics card calculates its own grid domain inside the sub-layer at each time step independently of the other cards, except for the points at the interface between two adjacent domains. These points are common to each of two domains, and to continue the calculation, it is necessary to exchange information on the required values among "neighbors".

In order to minimize the time of data exchange, data are transferred among nodes using appropriate non-blocking asynchronous functions of MPI, and exploiting the asynchronous copying function of CUDA for exchanges among the graphics cards. Let us note that the data for exchange have the same size in both approaches. The effective use of graphics cards requires handling and optimizing different memory types. To reduce the time necessary for global memory access, we have made the same optimal arrangement of all three-dimensional arrays used and an appropriate load distribution among threads. All the basic constants used at each time step are specially selected and stored in the constant memory of a graphics card. Also, we can try to use the faster shared memory of a graphics card [9] to reduce the number of readings from the GPU global memory. In the future, such optimization will be implemented and presumably will bring about a greater acceleration when computing the displacement, as in this case one re-uses a large amount of data and, correspondingly, a smaller number of copies in the shared memory is involved.

Finally, calculating the velocities of displacement and stress requires the allocation of memory to store 26 three-dimensional arrays, while calculating the displacement requires only 14 arrays (the source grid coefficients $\lambda_{i,j,k}$, $\mu_{i,j,k}$ and $\rho_{i,j,k}$ are not stored in the GPU memory, but their modifications used in the calculation scheme at each time step are stored to avoid re-calculation). This reduces the amount of used memory to almost a half, which is very important for calculations with computer systems that contain a small number of nodes.

When using CUDA it is necessary to divide all parallel-running threads into equal blocks. The dependence of software performance on the dimension and size of a thread block has been studied for the Tesla M2090 graphics card.

Thus, the calculation domain at each GPU is split into a three-dimensional grid of blocks, their size for the component x must be a multiple of the length of the warp (the number of threads that are simultaneously executed on GPU). The specific size of each block is empirically selected for each algorithm. Some results of measurements are shown in Table 1. Such an optimization (uncomplicated in the context of a code change) allows to accelerate substantially the program run by effectively using the graphics accelerator global memory.

Table 1. Dependence of the software run speed on the thread block size (a spatial grid of size $500 \times 500 \times 600$ for the displacement velocity problem, a spatial grid of size $600 \times 600 \times 600$ for the displacement problem, and 1000 time steps for both problems)

Block size, $x \times y \times z$	Calculation time in displacement velocity problems, s	Calculation time in displacement problems, s
$4 \times 4 \times 4$	789.4	—
$8 \times 8 \times 8$	506.3	422.8
$2 \times 16 \times 2$	1429.2	—
$16 \times 2 \times 2$	352.7	—
$16 \times 4 \times 4$	358.5	317.3
$4 \times 32 \times 4$	—	629.2
$32 \times 2 \times 2$	290.1	—
$32 \times 4 \times 4$	290.9	287.5
$64 \times 2 \times 2$	272.9	260.1
$64 \times 4 \times 2$	273.8	—
$128 \times 2 \times 2$	264.0	258.4
$128 \times 1 \times 1$	277.2	258.1
$256 \times 1 \times 1$	269.9	257.7
$512 \times 1 \times 1$	300.1	256.9

For the comparison of the running speed of both approaches, we have carried out calculations for the same media with a spatial grid of size $1500 \times 700 \times 2100$ and 1000 time steps. This grid is close to the maximum grid placed in the memory of 45 graphics cards (15 nodes of the NKS-30+GPU cluster) for calculating the velocities of displacement and stress. To calculate the displacements on the same spatial grid, one needs about half of the memory and can use, at least, 8 nodes instead of 15. The measurement results are presented in Table 2. Calculations of the displacements and the velocities of displacement and stress on equal numbers of nodes take roughly the same amount of time (in fact, the calculation of displacements runs a little faster).

The conducted numerical experiments show that the approach based on the calculation of displacements is faster, thereby allowing to carry out calculations for very large grids requiring a smaller number of free nodes. This allows quick access to computer cluster resources in queue conditions. The approach provides a reasonable calculation time (several hours for a full-scale actual problem).

Table 2. Comparison of the running speed of the algorithms

	Number of nodes	Time, s
Calculating velocities of displacement and stress	15	183.1
Calculating displacements	15	174.8
Calculating displacements	8	247.4

3   Study of the Algorithms Scalability Using Agent Simulation

For the simulation of distributed systems, it is best to use a distributed simulation based on message passing. The multi-agent approach [10] is used for the simulation of parallel programs running on a large number of cores owing to such properties as decentralization, self-organization and intelligent behavior [11]. We chose the adaptable distributed simulation system AGNES [12] among many other multi-agent simulation platforms. It has demonstrated high efficiency in the simulation of telecommunication and information networks [13] and systems of operative management of distributed computing systems [14]. Also, it was successfully used in a scalability study of a series of parallel problems when executed on a large number of cores [4, 15, 16].

AGNES is a package based on the Java Agent Development Framework (JADE) providing tools for creating MAC in Java [11]. AGNES takes the advantages provided by JADE, and extends the multi-agent systems to a simulation system. For the simulation of large-scale calculations, it is important that JADE is FIPA-compatible as well as a distributed agent-based platform which can use one or more computers (nodes). In this case, only one virtual JAVA machine must work on each of them.

To simulate the execution of a parallel algorithm, let us represent the computational process as a set of threads that are executed in parallel at an individual node and interact with each other by exchanging values. The main characteristic of the threads is the execution time and the time of data exchange with another computing node. To collect this information, we construct an interaction diagram of computational processes, and investigate the profile of program execution. We draw a general scheme of interaction of computational processes to study the scalability of parallel implementations of the considered problem.

We implemented classes of GeoWGrid functional agents to simulate the execution of the algorithm of numerical modeling of seismic fields. These are agents-calculators, each simulating the calculation of grid methods at a single computational node. When calculating, for instance, the velocities of displacement and stress, this process can be described as follows. At the first step, each agent calculates the stress tensor components in a subdomain for each node, and then sends the results obtained to the neighbors and waits for a signal before taking further actions. Then the agents start calculating the components of the displacement velocity vector in a subdomain for each node after receiving a confirmation that all the agents have received data. Then the agents simultaneously exchange the values obtained. A similar scheme was adopted to calculate the displacements. The characteristics of the calculation and exchange times for each approach mentioned above were taken from the execution profile of appropriate programs. The interaction scheme of the agents corresponds to the scheme of interaction of computational processes.

We designed a model experiment for the execution of algorithms on a small number of computational nodes with the purpose of verifying the models. The verification results can be seen in Fig. 1, where the efficiency means the ratio of the calculation time on n GPUs (cores) for a problem that is n times greater, to the calculation time of the initial problem on a single node. As we can see, a comparison of the actual execution data with the model experiment data indicates that our model corresponds to the execution of the parallel algorithm.

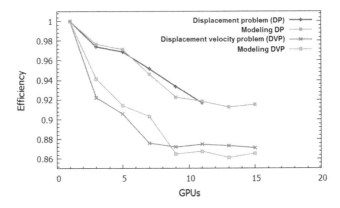

Fig. 1. Verification of the model of execution of the algorithm of numerical modeling (by two approaches) of seismic wave propagation.

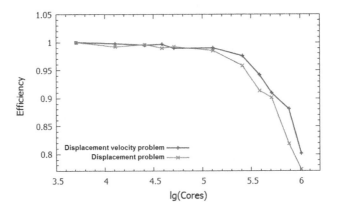

Fig. 2. Algorithm scalability depending on the simulated number of cores (the horizontal axis features a logarithmic scale).

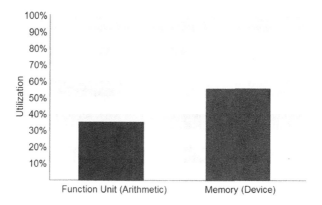

Fig. 3. Arithmetic and memory utilization of the Nvidia K40M GPU in the displacement problem tests.

We investigated the scalability of the parallel algorithms by conducting a model experiment for their execution on a large number of processing cores (up to 10^6). The simulation results are shown in Fig. 2. According to these results, we can conclude that, even though the schemes of interaction for both approaches are similar, their scalability differs slightly; the algorithms proposed are suitable for execution on a large number of processing cores.

4 Assessment of the Energy Efficiency of the Algorithms

In the present paper, the term "energy efficiency" for scientific HPC applications means all the following: the most efficient use of each core, processor or computational accelerator; the minimization of communication among computational nodes; and a good workload balancing of a program. The minimization of communication allows to reduce the idle time for processors and accelerators. A good workload balancing enables us to uniformly load a computational system. In case of a good workload balancing and a stable node balancing, we can carry out a set of program runs to show the relation between the power consumption and the usage of cores. The most energy efficient algorithm produces the best FLOPS per watt (joules/sec) value.

We used the Nvidia Tesla K40M GPU for our energy efficiency tests with CUDA 7.5 toolkit. In the energy efficiency tests, we performed a numerical simulation of the displacement velocity problem and the displacement problem. In the case of the displacement test (Fig. 3), these utilization levels indicate that the performance of the core is most likely limited by the latency of arithmetic or memory operations. In the displacement velocity problem tests (Fig. 4), these utilization levels indicate that the performance of the core is most likely being limited by the memory system of the K40M.

The average GPU power for the displacement problem and the displacement velocity problem tests is 149 and 156 W, respectively. The energy efficiency

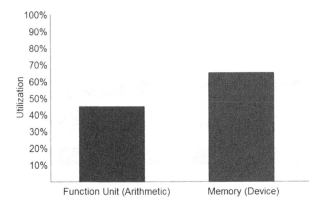

Fig. 4. Arithmetic and memory utilization of the Nvidia K40M GPU in the displacement velocity problem tests.

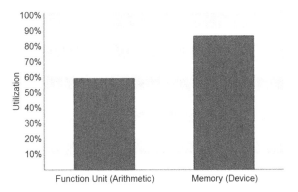

Fig. 5. Arithmetic and memory utilization of the Nvidia Tesla 2090M GPU in the displacement velocity problem tests.

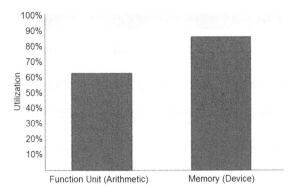

Fig. 6. Arithmetic and memory utilization of the Nvidia Tesla 2090M GPU in the displacement problem tests.

for the displacement problem and the displacement velocity problem tests is 9 GFLOPS/W and 12 GFLOPS/W, respectively.

We have also carried out the same tests on the Nvidia Tesla 2090M GPU. Figures 5 and 6 show the arithmetic and memory utilization of the Nvidia Tesla 2090M GPU in the displacement and displacement velocity problems tests. The energy efficiency in the displacement problem and the displacement velocity problem tests for the Nvidia Tesla 2090M is 4.3 GFLOPS/W and 4.5 GFLOPS/W, respectively. We can see that the Tesla K40M energy efficiency index is three times greater for the same code.

5   Conclusion

This paper suggests an integrated approach to the development of algorithms and software for the numerical simulation of seismic wave propagation in media with complex subsurface geometries, on a hybrid cluster. For the purpose of the co-design, we have compared the developed parallel implementations of the solutions with the elastodynamic problem written both in terms of the velocities of displacement and stress and in terms of displacements, on computational clusters equipped with graphics cards. We have studied the execution time of the created parallel programs, their scalability (including the simulation of program execution on a large number of cores by using the AGNES simulation system) and energy efficiency.

Based on the results obtained, we can conclude that the approach proposed to calculate the displacements should be preferred. In spite of a small gain in time, it allows to solve large 3D dynamic problems of elastodynamic theory using a significantly smaller number of graphics accelerators compared with the approach based on the calculation of the velocities of displacement and stress. The schemes of interaction for both approaches are similar but their scalability differs slightly: about 80% in the simulation of their execution on 106 cores. It has been shown that the energy efficiency is close to 12 GFLOPS/W with the Nvidia Tesla K40M accelerator, which is almost three times higher than that of the Nvidia Tesla 2090M.

Thus, using the approach proposed, we have been able to develop a set of parallel programs allowing to carry out three-dimensional calculations of a required size with necessary computational resources in a reasonable time.

References

1. Glinskii, B.M., Kovalevskii, V.V., Khairetdinov, M.S.: Vibroseismic monitoring of earthquake-prone areas. Volcanol. Seismol. **21**(6), 723–730 (2000)
2. Alekseev, A.S., Glinsky, B.M., Kovalevsky, V.V., Khairetdinov, M.S.: Active vibromonitoring: experimental systems and fieldwork results. In: Handbook of Geophysical Exploration: Seismic Exploration, vol. 40, pp. 105–120 (2010). doi:10. 1016/s0950-1401(10)04011-5
3. Glinskii, B.M., Martynov, V.N., Sapetina, A.F.: 3D modeling of seismic wave fields in a medium specific to volcanic structures. Yakutian Math. J. **22**(3), 84–98 (2015)

4. Glinskiy, B.M., Kulikov, I.M., Snytnikov, A.V., Chernykh, I.G., Weins, D.: Mnogourovnevyj podhod k razrabotke algoritmicheskogo i programmnogo obespechenija jekzaflopsnyh superJeVM (A multilevel approach to algorithm and software design for exaflops supercomputers). Vychislitel'nye metody i programmirovanie (Vychisl. Metody Programm.) **16**, 543–556 (2015)
5. Glinskiy, B., Kulikov, I., Snytnikov, A., Romanenko, A., Chernykh, I., Vshivkov, V.: Co-design of parallel numerical methods for plasma physics and astrophysics. Supercomput. Front. Innov. **1**(3), 88–98 (2014). doi:10.14529/jsfi140305
6. Dosanjh, S.S., et al.: Exascale design space exploration and co-design. Future Gener. Comput. Syst. **30**, 46–58 (2014). doi:10.1016/j.future.2013.04.018
7. Bihn, M., Weiland, T.: A stable discretization scheme for the simulation of elastic waves. In: Proceedings of the 15th IMACS World Congress on Scientific Computation, Modelling and Applied Mathematics (IMACS 1997), Berlin, vol. 2, pp. 75–80 (1997)
8. Karavaev, D.A.: Parallel'naja realizacija metoda chislennogo modelirovanija volnovyh polej v trehmernyh modeljah neodnorodnyh sred (Parallel Implementation of Wave Field Numerical Modeling Method in 3D Models of Inhomogeneous Media). Vestnik Nizhegorodskogo universiteta im. N.I. Lobachevskogo (Vestnik of Lobachevsky State University of Nizhni Novgorod) **6**(1), 203–209 (2009)
9. Nakata, N., Tsuji, T., Matsuoka, T.: Acceleration of computation speed for elastic wave simulation using a graphic processing unit. Explor. Geophys. **42**(1), 98–104 (2011). doi:10.1071/eg10039
10. Wooldridge, M.: Introduction to MultiAgent Systems. Wiley, England (2002)
11. Bellifemine, F.L., Caire, G., Greenwood, D.: Developing Multi-Agent Systems with JADE. Wiley, New York (2007). doi:10.1002/9780470058411
12. Podkorytov, D., Rodionov, A., Sokolova, O., Yurgenson, A.: Using agent-oriented simulation system AGNES for evaluation of sensor networks. In: Vinel, A., Bellalta, B., Sacchi, C., Lyakhov, A., Telek, M., Oliver, M. (eds.) MACOM 2010. LNCS, vol. 6235, pp. 247–250. Springer, Heidelberg (2010). doi:10.1007/978-3-642-15428-7_24
13. Podkorytov, D., Rodionov, A., Choo, H.: Agent-based simulation system AGNES for networks modeling: review and researching. In: Proceedings of the 6th International Conference on Ubiquitous Information Management and Communication (ACM ICUIMC 2012), p. 115. ACM (2012). ISBN 978-1-4503-1172-4. doi:10.1145/2184751.2184883
14. Vins, D.V.: Analiz jeffektivnosti sistemy upravlenija potokom zadanij dlja CKP v mul'tiagentnoj imitacionnoj modeli (Analysis of Effectiveness of Job Stream Management System for the Center of Collective Use in Multi-agent Simulation Model). Vestnik NGU (Vestnik NSU: Information Technologies) **12**(2), 33–41 (2014)
15. Glinsky, B.M., Marchenko, M.A., Mikhailenko, B.G., Rodionov, A.S., Chernykh, I.G., Karavaev, D.A., Podkorytov, D.I., Vins, D.V.: Simulation modeling of parallel algorithms for exaflop supercomputers (in Russian). Inf. Technol. Comput. Syst. **4**, 3–14 (2013)
16. Kulikov, I., Chernykh, I., Glinsky, B., Weins, D., Shmelev, A.: Astrophysics simulation on RSC massively parallel architecture. In: Proceedings of 2015 IEEE/ACM 15th International Symposium on Cluster, Cloud, and Grid Computing, CCGrid, pp. 1131–1134. IEEE Press (2015). doi:10.1109/ccgrid.2015.102

Supercomputer Simulation of Components and Processes in the New Type Li-ion Power Sources

Vadim Volokhov, Dmitry Varlamov$^{(\boxtimes)}$, Tatyana Zyubina, Alexander Zyubin,
Alexander Volokhov, and Elena Amosova

Institute of Problems of Chemical Physics of RAS, Academician Semenov avenue 1,
Chernogolovka, Moscow region 142432, Russian Federation
{vvm,dima,zyubin,vav,aes}@icp.ac.ru
http://www.icp.ac.ru/en/

Abstract. As a result of a large amount of computational experiments (more than two thousand) performed on a number of supercomputing resources, we have developed quantum-chemical models of various constituent components of Li-ion power sources based on silicon-carbon nanostructure composites and promising solid electrolytes.

Transport, structural and energetic processes occurring in Li-ion power sources during numerous charge-discharge cycles have been simulated. Using methods of molecular dynamics, we estimated the effects of different temperature regimes on the nanosystem structure as well as the characteristics of these processes.

Keywords: Computer simulation · Silicon-carbon nanocomposites · Solid electrolytes · Li-ion power sources · VASP applied package · Quantum chemistry · Molecular dynamics

Introduction

This article describes the results of the project "Computer simulation of absorption and transport properties of solid electrolytes and nanostructured electrodes based on carbon and silicon in Li-ion power sources". The aim of this project is the supercomputer simulation of quantum-chemistry and molecular dynamics of new nanocomposite materials (based on silicon and carbon) and solid electrolytes with high ionic conductivity, as well as non-reactive electrode materials during operation of a current source. Also, transport, structural and energetic processes occurring in the modeled nanostructures and at the "interface" between them have been simulated.

The activity is part of the work "Creation of an effective environment of computer modeling of quantum-chemical processes and nanostructures on the basis of a program complex of newest computing services and high-level web- and grid-interfaces to them", supported by the Russian Foundation of Basic Research (project no. 15-07-07867-a).

L. Sokolinsky and M. Zymbler (Eds.): PCT 2017, CCIS 753, pp. 275–287, 2017.
DOI: 10.1007/978-3-319-67035-5_20

Li-ion power sources (LPS) are currently the most promising and common types of power sources and batteries. LPS are based on the transport of Li-ions through a liquid or solid electrolyte from cathode to anode (and back when charging). The design of new types of LPS is needed to improve their efficiency parameters, such as energy capacity, number of charge-discharge cycles, resistance to external conditions (temperatures), safety of their production and utilization from an environmental point of view, and cost (prime cost of materials in main components).

Here is a brief description of the operating principle of Li-ion power sources (Fig. 1).

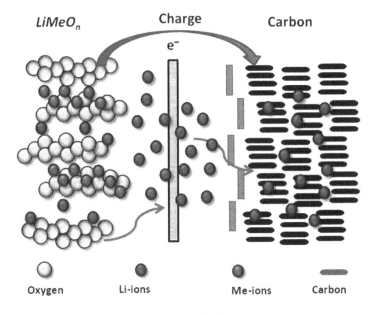

Fig. 1. Schematic diagram of a Li-ion power source.

The following reactions occur in a Li-ion power source during charge:

on the positive plates: on the negative plates:
$$LiMeO_2 \rightarrow Li_{1-x}MeO_2 + xLi^+ + xe^- \quad C + xLi^+ + xe^- \rightarrow CLi_x$$

The reverse reactions occur when discharging. Therefore, both *lithiation* (lithium saturation) and *delithiation* (returning of lithium to electrolyte and cathode) are basic processes.

Simulated materials should be the basis for the design and creation of new types of electrochemical and ecologically safe Li-ion power sources (LPS).

These power sources will be able to operate at low and medium temperatures, provide significantly higher energy densities, and improve operational and cost characteristics.

The synthesis of new nanocomposite materials, the study of their properties and predictable applications are only possible as a result of a detailed computer modeling of crystalline composite structures, elementary processes and different mechanisms of chemical reactions and transport processes at molecular level.

Experimental studies of factors having a major influence on the solution of the issues listed above are complex, expensive, not always possible, and in most cases, do not give clear answers to the following questions: mechanisms of ongoing physical and chemical processes, reasons for their differences depending on the composition of the system and conditions, possible course of reactions, etc.

The experimental (analogue) simulation of the influence of various factors on the properties of the constituent components of Li-ion power sources and the processes occurring in them poses labor-intensive and costly tasks.

Since experiments give only initial and final information about processes (like electromotive force, voltage, resistance, etc.), it is quite difficult to build a genuine analytical model. Such tasks can be solved partially in laboratory conditions, where experiments give incomplete or indirect information about mechanisms and structures of experiment components. However, modern numerical methods of quantum-chemical and molecular-dynamics simulation can provide substantial assistance in determining the characteristics of processes and assessing the impact of individual factors with a high degree of accuracy. These methods allow to obtain new theoretical data on the structure and properties of both nanostructured cathode-anode systems and ion-conducting solid electrolytes, making it possible to subsequently develop new highly effective materials for electrochemical devices.

A detailed simulation of elementary processes as well as mechanisms of lithiation/delithiation and ion-transport processes in Li-ion power sources at the micro level leads to a better control over chemical reactions occurring in them, allowing to design the most appropriate anode materials in terms of efficiency of electricity generation, lithiation processes, stability of materials during numerous charge-discharge cycles, cost of LPS constructive materials and environmental recycling processes.

Also, the created models can be reviewed for adequacy by comparing them (and the properties of materials modeled on their basis) against observable analytical, experimental and theoretical data published in specialized literature.

For this task, the authors carried out a detailed quantum-chemical and molecular-dynamical simulation of various nanosystems based on carbon and silicon, as well as solid electrolytes with high ionic conductivity, both in cluster approximation and for periodic boundary conditions with projector-augmented wave (PAW), using VASP, CPMD and Gaussian application packages on a number of high-performance computing resources.

The objects of the computer simulation are composites based on carbon and silicon, which have the ability to repeatedly absorb Li without damage and are promising materials for Li-ion power sources (nanoclusters, nanotubes, nanowires, nanopapers and active crystal surfaces). Also objects of this computer simulation are solid electrolytes with high ion conductivity based on glasses, salts and polymer composites that do not react with the electrode material during operation of a current source.

Some simulation experiments were conducted using authors' computer system based on up-to-date software packages for quantum-chemistry and molecular dynamics, "hybrid" computing technologies, web services, data storage, visualization of results, etc. Using high-performance resources (supercomputers, problem-oriented clusters and hybrid systems) would greatly improve the details and the quality of the created models of nano-objects and those of the processes accompanying them, and would also allow to solve tasks previously inaccessible due to their computational complexity.

1   Simulation Methods

The models of nanocomposite materials and processes occurring in them were constructed by methods of quantum chemistry computer simulation on the clusters of the Computation Center at the Institute of Problems of Chemical Physics (IPCP), and then at the Supercomputer Center of Moscow State University "Lomonosov" [1], using the applied software packages VASP (Vienna Ab initio Simulation Package, https://www.vasp.at) and CPMD (Car-Parinello Molecular Dynamics, http://www.cpmd.org) for the calculation of complex nanostructures and the dynamics of their behavior depending on time and temperature.

The VASP package has been used by the authors during a long time for modeling materials and components of complex electrochemical objects. This package is applied to the simulation of various processes both in the volume and on the surface of solids (first of all, catalysis and ionic conductivity) within the non-empirical approaches based on the use of density functional theory with periodic boundary conditions and a plane wave basis set. VASP allows to optimize the structures and to model processes within a molecular dynamics framework.

VASP implements effective schemes of iterative matrix diagonalization and the highly efficient Broyden–Pulay electronic charge density mixing. In addition, the MSP processes convergence procedure (self-consistent field) and optimization are significantly improved, which greatly increases the efficiency of calculations. This package provides a good accuracy of description for structural and energy characteristics of systems containing up to several hundred atoms. First of all, we conducted a full optimization of the geometric and energy parameters of molecules under consideration within the established basis and method of calculation.

In this paper, we applied an approach based on density functional theory (DFT) with periodic boundary conditions to simulate learning systems. We applied the projector-augmented wave (PAW) with the corresponding PAW pseudopotentials and PBE functional (Perdew–Burke–Ernzerhof). The limit of energy (E_c) defining the completeness of the basis set was established at 400 eV.

When simulating two-dimensional plates, the vacuum layer between them was not less than 10 Å. To simulate the $Li_{10}GeP_2S_{12}$ electrolyte volume, we used a canned double cell $Li_{20}Ge_2P_4S_{24}$ involving 50 atoms; for the simulation of the surface, four such cells (200 atoms) were used.

To solve the problem of interaction of the surfaces between electrodes and electrolytes, we modeled (with full optimization of geometric parameters) a structure continuously propagating in two directions, solid electrolyte fragments (propagated $Li_{80}Ge_8P_{16}S_{96}$-fragment), a silicon-carbon paper (propagated $Si_{32}C_{38}$-fragment) and the result of their interaction (propagated $[Si_{32}C_{38}]\cdot[Li_{80}Ge_8P_{16}S_{96}]$-fragment).

In the case of polymer electrolytes, we modeled (with full optimization of geometric parameters) the structure of infinite nanowires of Li$Nafion\cdot$nDMSO (n = 0,1,8,16), and also spatially propagated fragments of $Li(C_{15}O_5F_{29}S)\cdot$ $n(H_6C_2OS)$ and $[Li(C_{15}O_5F_{29}S)\cdot n(H_6C_2OS)]_2$ of 51 to 262 atoms.

For the optimization, we applied the Methfessel–Paxton method of electronic state (with blur parameter (σ) 0.2 and energy approximation of the value $\sigma = 0$). This approach allows for the automatic detection of system's multiplicity. The estimate of energy stability of combined systems was determined according to De/n(Li), computed as the difference between the calculated energy of the system and the total energy of isolated lithium atoms divided by the number of atoms of adsorbed lithium, for example, De/n(Li) = -[E((SiC)$_k$Si$_m$Li$_n$) - E((SiC)$_k$Si$_m$) -nE(Li)]/n.

We used two approaches for the simulation of transport processes in the framework of an ab initio non-empirical molecular dynamics with periodic boundary conditions: CPMD (Car-Parrinello approximation), in which the calculated wave function for the starting configuration is approximated by a set of classically-moving low-mass particles, and a more accurate but slower approximation, namely MD-VASP (MD/PBE/PAW), which uses the same algorithms as normal optimization structures, but with rougher calculation criteria.

Generally, the use of MD-VASP allows a substantially faster simulation than CPMD. MD-VASP requires about 6 to 8 times less computation steps to achieve the same penetration depth.

The Gaussian software package (http://gaussian.com) was used for comparison and estimation of the accuracy of the simulation of some nano-objects at DFT/B3LYP level. By comparing different levels of calculation, we noted that the calculated values used in VASP and Gaussian software for average bond energies and distances of identical objects give consistent results with accuracy of 0.02 to 0.04 eV and 0.005 to 0.01 Å, respectively.

It should be noted that the difference of calculation results at B3LYP/6-31G(d,p), PBE/6-31G(d,p) and PBE/PAW levels does not exceed 0–2% for distances and 1–13% for energies. The chosen calculation level provides the following calculation accuracy in computer models: the Si crystal lattice calculated parameters a = b = c are 5.48 Å (experimental: 5.43 Å), the Si–Si distance is 2.37 Å (experimental: 2.34 Å), and the energy of the crystal is 4.44 eV (experimental: 4.52 eV).

The adequacy of the computer models was also evaluated by comparing the values calculated on the basis of their physico-chemical characteristics (optical and X-ray spectra, thermodynamic measurements, energy parameters) with those observed in physical experiments. For example, the calculated structural parameters for crystal electrolytes (a = b = 8.79 Å and c = 12.80) are in good agreement with X-ray experiments (a = b = 8.72 Å and c = 12.63 Å)

2   Computational Complexity and Processing Efficiency

In earlier times, similar computer simulations were hindered by a catastrophic lack of computing resources, since calculating the behavior of small/medium atomic clusters of the Si_{7-126} type, even in a simplified form, required months, and modeling systems as a whole (containing thousands of atoms) required approximately $n \cdot 10^6$ CPU-hours per year.

Only in recent times, the same simulation became feasible using high-performance supercomputing centers and grid polygons. Currently, the use of computing resources with speeds of the order of teraflops and petaflops allows to make sufficiently detailed simulations of geometrical and energy characteristics of modeled nanostructures. It is also possible to study the effects of various factors and processes occurring in these nanostructures for a variety of conditions determining the efficiency of the created LPS.

Let us summarize the computational complexity and use efficiency of computing resources in the process of quantum-chemical simulation of learnt structures. We used the IPCP cluster (176 dual-node HP Proliant, making a total of 1472 cores based on 4- and 6-core Intel Xeon processors 5450 and 5670 3 GHz, 8 and 12 GB of RAM per node; InfiniBand DDR communication network, transport and network management – Gigabit Ethernet; hard drives – no less than 36 GB per node), and the SCC of MSU supercomputing installations "Lomonosov-1,2" having various pools of processors (8 to 128 CPU) with obligatory presence of local drives and no less than 2 GB of RAM per core.

A sufficient effective acceleration of the VASP package for this type of tasks was observed for 40 to 48 CPU (similar to calculations for the simulation of fuel cell catalysts previously carried out by the authors [2]). The further growth of the efficiency of task parallelization is limited (or even reduced) by the rate of data exchange due to a significant increase in the amount of data being transferred between nodes. Thus, increasing the number of CPU over 48 (at least for this task variant) is meaningless for the moment.

The average effective time for calculation of Si_n clusters (n = 2 ÷ 350) and C_nSi_m nanofibers increases as the dimension of the silicon-carbon fragment increases, taking up to 4 days (78 h on a pool based on 4-core Intel Xeon 5450 3 GHz processors) and even more (due to complications of the structure). The calculation time of lithiated large mesostructures of silicon and aggregates reinforced with nanotubes or nanowires took tens of days to complete.

The most critical calculation parameter is the amount of memory per core, with an effect of acceleration of calculations with a decrease in the number of allocated cores by increasing the amount of RAM per core. For MD calculations, we used 14000 steps per calculation (for example, heating up to 400 K for 2000 steps, holding at 400 K for 10000 steps, cooling down to 10 K for 2000 steps, and optimizing the structure in standard mode; the time step model was 1 femtosecond). The calculation of complex structures, such as those described in Sect. 3.6, requires up to 80000 CPU-hours.

In the latest versions of VASP, starting with version 5.4.1 (February 2015), the application package supports CUDA technology for the calculation method of standard and hybrid DFT (Hartree–Fock equation). For most tasks, using DFT on Tesla C2075 accelerators at the IPCP, we achieved (comparing VASP versions with support and without support of GPU acceleration) 1.6- to 6-fold accelerations depending on the dimension of the problem and its type. This gives the prospect of a significant acceleration for VASP calculations on "hybrid" computing nodes (following VASP upgrade to versions above 5.4), including existing CPU-GPU pools on the SC "Lomonosov-1.2" and hybrid IPCP stations. In addition to upgrading VASP, it is necessary to do a further reconfiguration of VASP settings files, and update CUDA library at least to version 7.5 (current version: 8.0).

The total number of computing experiments performed at all stages of the work reached more than 2000.

3   Simulation Results

The results of the multi-step simulation have been described in detail in a number of publications by the authors of the present work [3–11]. Here is a brief description of the most representative results of the computer simulation of nanostructures and processes occurring in them.

3.1   Computer Simulation of Various Types of Porous Nanocomposite Materials Based on Carbon and Silicon

Computer models of the following types of Si–C nanocomposites have been created by the authors [6, 7]:

- pure silicon aggregates with different morphologies (clusters of "snowballs", "core/shell", etc., size up to 3 nm), and a number of silicon atoms ranging from 2 to 350;
- silicon clusters with silicon carbide core (rod-shaped), 1.2 to 2.8 nm in diameter, and nanofibers of Si_nC_m type, n/m = 1÷3;
- carbon nanotubes (CNT) with dimension (6,6) and 0.8 nm in diameter, surrounded by a layer of silicon clusters of various dimensions;
- silicon nanowires with a rod on the basis of silicon carbide and silicon shell;
- infinite carbon nanofibers coated with silicon nanoclusters;
- silicon-carbon "nanopapers".

A conclusion following from our research is that the use of different types of simulated nanocomposites may be a promising opportunity in the construction of new types of electrodes for LPS. Examples of the simulated nanostructures are shown in Fig. 2.

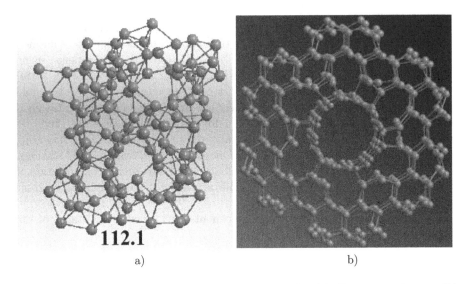

a)   b)

Fig. 2. Examples of nanocomposites based on Si–C: (a) Si_{112} mesocluster; (b) Si_{288}/C_{120} nanotube. The models have been derived from the authors' computer simulation using the VASP package (PBE/PAW level of calculation).

3.2   Quantum-Chemical Simulation of Transport Processes of Lithium Ions in Nanocomposite Materials Based on Carbon and Silicon

On the basis of the constructed models of nanocomposites (see above), we made a quantum-chemical simulation of various processes occurring during charge-discharge cycles of LPS (i.e. processes of lithiation and delithiation on electrodes based on the above-described nanostructures) [8,10]. A majority of characteristics of these processes have been established, including:

– Li-ion transport processes and processes of lithium consistent implementation in Si–C nanostructures of various types and dimensions;
– structural and energetic changes of nano-objects in processes of absorption of lithium atoms;
– possible paths and migration barriers for lithium atoms in the process of nanoparticle saturation;
– construction of models of sequential removal of lithium atoms from lithiated nanoparticles and determination of structural and energetic changes identified in this process;

– determination of the limits of resistance to fracture for nanoparticles during delithiation processes.

3.3 Computer Models of Aggregation Processes of Initial and Lithiated Nanoparticles

In LPS operation, there are aggregation processes of small nanostructures into larger ones and vice versa, affecting greatly the characteristics of components and the whole LPS. We have obtained computer models of aggregation processes of original and lithiated nanoparticles [4,8], including:

– formation of a mesostructure based on original silicon-carbon nanoparticles;
– formation of a mesostructure based on lithiated (saturated lithium atoms) silicon-carbon nanocomposite structures.

3.4 Quantum-Chemical and Molecular Dynamics Simulation of Highly Conductive Solid Electrolytes

This work presents the results of the computer quantum-chemistry and molecular dynamics simulation [5,11] of highly conductive solid electrolytes based on $Li_{10}GeP_2S_{12}$ systems and polymer electrolytes based on $LiNafion^{TM}$. dimethylsulfoxide (LiNafion· 8DMSO) with an ionic conductivity that is higher than that of liquid electrolytes.

We modeled the structures and the contact surface of superionic solid electrolytes. In this connection, the ionic conductivity mechanism was determined during simulation. We also defined the types of surface structure and the nature of the $Li_{10}GeP_2S_{12}$ electrolyte contacts with anode nanocomposite materials (carbon fibers coated with Si_nC_m silicon nanoclusters and silicon-carbon "nanopaper").

We modeled the contact surfaces of superionic solid electrolytes with different Si–C nanocomposites. It was shown that a layer of liquid or plastic polymer electrolyte, such as dimethylsulfoxide (hereinafter DMSO), can be used to enhance the contact between solid surfaces. The simulated structure and its surface are shown in Fig. 3.

We constructed a computer model of interaction of solid and polymer lithium-based electrolytes with composites based on carbon fibers and silicon nanoclusters. Examples of the simulated complicate complexes "electrolyte–Si_nC_m" of different dimensions are shown in Fig. 4.

Lithium transition across the interface "electrode–electrolyte", as well as the determination of the migration channels and the potential barrier were modeled taking as example the interaction of solid and polymer lithium electrolytes with composites based on carbon fibers and silicon nanoclusters.

3.5 Simulation of Lithium Ions Migration in Non-aqueous Polymer Electrolytes

Using methods of quantum chemistry and molecular dynamics, we modeled various aspects of the migration of lithium ions in a complex electrolyte

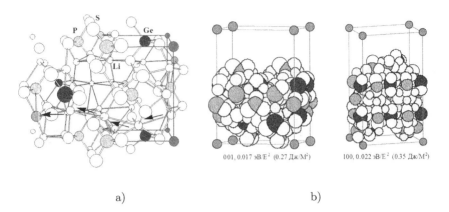

a)   b)

Fig. 3. Simulated structure (a) and (b) two types of modeled surfaces (001) and (100) $Li_{10}GeP_2S_{12}$ solid electrolyte crystal.

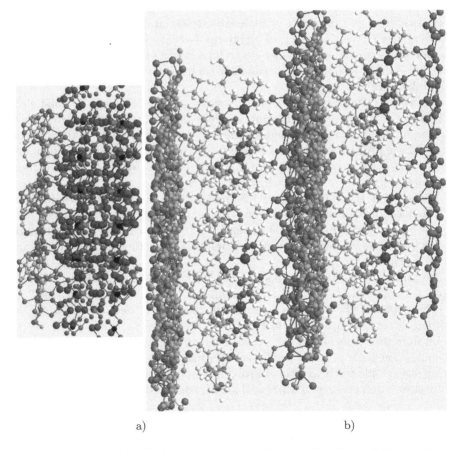

a)   b)

Fig. 4. Structure models of solid electrolyte complexes $Li_{80}Ge_8P_{16}S_{96}$ (a) and polymer electrolyte [LiNafion*8DMSO] (b) with layers of $\mathbf{Si_{32}C_{38}}$ silicon-carbon "paper".

(Li*Nafion·* nDMSO, n = 0 ÷ 18) as well as the structure, stability and electronic properties of membranes based on the electrolyte, including the effect of various parameters, such as the degree of swelling of the electrolyte and a number of physico-chemical properties of the plasticizer containing the molar volume, the viscosity and the coordination number.

On the basis of the transport models, we made some conclusions on the possible paths of lithium migration and energy parameters: (1) four-coordinated lithium transition through three-coordinated state to the next position surrounded by four DMSO molecules, and (2) movement of the Li(DMSO)$^{4+}$ tetrasolvate complex.

3.6 Computer Simulations of Repeated Lithiation/Delithiation Cycles Depending on the Degree of Lithium Saturation and Temperature Conditions

Along with the simulation of individual LPS components (electrodes, electrolytes) and processes occurring at the interfaces between them during operation of the battery, a great importance for the creation of new types of power sources is ascribed to the stability of these elements over time in case of multiple "charge-discharge" cycles, depending on the power capacity of the system (amount of lithium) and temperature conditions.

The molecular dynamics simulation is performed to assess the feasibility of the composite mesostructure return to its original state after repeated lithiation/delithiation cycles depending on the degree of lithium saturation and temperature conditions. MD-VASP (MD/PBE/PAW) approximation is used. For MD calculations, we used 14000 steps per calculation (for example, heating up to 400 K for 2000 steps, holding at 400 K for 10000 steps, cooling down to 10 K for 2000 steps, and optimizing the structure in standard mode; the time step model was 1 femtosecond). We took as initial model the nanocomposite models obtained in the first stages. The models are intended to illustrate the reorganization of the Li / Si layer structure during gradual recovery of lithium from the surface (i.e. discharge): the effect exerted by heating and subsequent cooling on the structure of silicon-carbon delithiated nanosystems and their possible return to the initial state according to the degree of heating of lithium and the saturation level.

The simulation has allowed to determine the most stable mesostructures for electrode materials, the optimal ratio of Li : Si in the Si–C nanocomposites saturation with lithium, and the best energy parameters of charge-discharge cycles. It has been demonstrated that the introduction of lithium into silicon is energetically more favorable than the formation of a metal layer on its surface, but increasing lithium concentrations leads to a reduction of energy difference, i.e. the implementation is less advantageous, the mesh of silicon atoms is broken into smaller pieces, the thickness of the absorbing layer is significantly increased, and its structure becomes amorphous. It is important to note that the energy in the modeled systems does not lead after cooling to stabilization to the substantial structural rearrangement that makes LPS components more resistant.

3.7   Computer Model of Ionic Transport in Lithium-Ion Batteries

We constructed computer models of ion transport using different combinations of the three main components of LPS, namely anode, cathode and membrane (electrolyte)[4,5]. These models allowed to define the basic characteristics of the energy system and evaluate the properties of the target battery based on calculations of the structure and transport properties of the electrode and electrolyte at molecular level. Comparing the results of different simulation options, we could identify the most promising areas of construction of lithium-ion batteries of new type and their characteristics during lithiation / delithiation processes.

4   Conclusions

Thus, on the basis of a large number of numerical experiments regarding computer quantum chemistry and molecular dynamics simulation, we calculated the structures and surfaces of solid and polymeric electrolytes of a new type for LPS, their interaction with various nano-objects based on carbon and silicon with different morphologies, spatial rigidity, power characteristics, saturation potential with lithium ions. We also calculated transport processes of lithium ions (delithiation-lithiation) in nanocomposites, including structural energy characteristics and structures evolving over time (depending on the number of cycles of lithiation).

The model structures calculated, as well as the characteristics of electrolyte and anode materials for LPS and their interaction during charge and discharge were used to simulate a whole picture of lithiation and delithiation processes in Li-ion cells, the interaction of lithium ions with the surfaces of carbon and silicon nanomaterials, the determination of the "container" received by the anode materials, and also to model both components and new LPS types in general.

The simulation results will be used to determine the optimal conditions for the synthesis and production of the most energetically favorable and industrially suitable electrolyte and anode materials for new types of Li-ion power sources.

References

1. Voevodin V.V., Zhumatii S.A., Sobolev S.I., Antonov A.S., Bryzgalov P.A., Nikitenko, D.A., Stefanov, K.S., Voevodin, V.V.: Practice of "Lomonosov" supercomputer. In: Open Systems, vol. 7, pp. 36–39 (2012). (in Russian)
2. Volokhov, V.M., Varlamov, D.A., Zyubina, T.S., Zyubin, A.S., Pokatovich, G.A., Volokhov, A.V.: Quantum-chemical simulation of processes in low-temperature electro-chemical fuel elements. In: Supercomputer Technologies in Science, Education and Industry, no. 5, pp. 172–176. MSU printing, Moscow (2013). (in Russian)
3. Volokhov, V.M., Varlamov, D.A., Zyubina, T.S., Zyubin, A.S., Volokhov, A.V., Pokatovich, G.A.: Supercomputer simulation of nanocomposites on silicon-carbon base for new types of Li-ion power sources. In: Proceedings of International Conference on Russian Supercomputing Days, pp. 453–464. MSU printing, Moscow, 28–29 September 2015. (in Russian)

4. Volokhov, V.M., Varlamov, D.A.,Zyubina, T.S., Zyubin, A.S., Volokhov, A.V., Pokatovich, G.A.: Supercomputer simulation of transport and energetic processes in silicon-carbon based nanocomposites. In: Proceedings of International Conference on Parallel Computing Technologies, PAVT 2016, Arkhangelsk, Russia, pp. 105–117. South Ural State Univ. printing, Chelyabinsk, 28 March–1 April (2016). (in Russian)

5. Volokhov, V.M., Varlamov, D.A., Zyubina, T.S., Zyubin, A.S., Volokhov, A.V.: Supercomputer simulation of interaction processes between Si-C nanostructured electrodes and solid electrolytes in new types of Li-ion power sources. In: Proceedings of International Conference Russian Supercomputing Days, pp. 690–699. MSU printing, Moscow, 26–27 September (2016). (in Russian)

6. Zyubin, A.S., Zyubina, T.S., Dobrovol'skii, Y.A., Volokhov, V.M.: Silicon-and carbon-based anode materials: a quantum-chemical modeling. Russ. J. Inorg. Chem. **61**(1), 48–54 (2016)

7. Zyubina, T.S., Zyubin, A.S., Dobrovolsky, Y.A., Volokhov, V.M.: Quantum chemical modeling of nanostructured silicon Si_n (n = 2–308). The snowball-type structures. Russ. Chem. Bull. **V65**(3), 621–630 (2016)

8. Zyubin, A.S., Zyubina, T.S., Dobrovol'skii, Y.A., Volokhov, V.M.: Quantum-chemical modeling of Lithiation of a Silicon-Silicon Carbide composite. Russ. J. Inorg. Chem. **61**(11), 1423–1429 (2016)

9. Zyubina, T.S., Zyubin, A.S., Dobrovol'skii, Y.A., Volokhov, V.M.: Lithiation-delithiation of infinite nanofibers of the Si_nC_m type - the possible promising anodic materials for lithium-ion batteries. Quantum-chemical modeling. Russ. J. Electrochem. **52**(10), 988–991 (2016)

10. Zyubina, T.S., Zyubina, A.S., Dobrovol'skii, Y.A., Volokhov, V.M.: Quantum-chemical modeling of lithiation-delithiation of infinite fibers $[Si_nC_m]_k$ (k = ∞) for n = 12–16 and m = 8–19 and small silicon clusters. Russ. J. Inorg. Chem. **61**(13), 1677–1687 (2016)

11. Zyubina, T.S., Zyubin, A.S., Dobrovol'skii, Y.A., Volokhov, V.M.: Migration of lithium ions in a nonaqueous nafion-based polymeric electrolyte: quantum-chemical modeling. Russ. J. Inorg. Chem. **61**(12), 1545–1553 (2016)

Development of a High Performance Code for Hydrodynamic Calculations Using Graphics Processor Units

Andrey V. Sentyabov[1,2](✉) ⓘ, Andrey A. Gavrilov[1,2], Maxim A. Krivov[3],
Alexander A. Dekterev[1,2], and Mikhail N. Pritula[4]

[1] Siberian Federal University, Krasnoyarsk, Russia
{sentyabov_a_v,dekterev}@mail.ru, gavand@yandex.ru
[2] Institute of Thermophysics SB RAS, Novosibirsk, Russia
[3] Lomonosov Moscow State University, Moscow, Russia
m_krivov@cs.msu.su
[4] CTP PCP RAS, Moscow, Russia
pritmick@yandex.ru

Abstract. The paper presents the results of the implementation of computational algorithms of hydrodynamics for using graphics processor units. The implementation was carried out on the basis of the in-house CFD code SigmaFlow. Numerical simulations were based on the solution of the Navier-Stokes equations using SIMPLE-like procedure. The discretization of the differential equations was based on the control volume method on unstructured mesh. In the case of multiple CPU/GPU, parallel calculations were performed by means of domain decomposition. In the GPU-version of the code, basic computational functions were implemented as CUDA kernels to perform on GPUs. The code has been verified using several test cases. The computational efficiencies of several GPUs were compared with each other and that of modern CPUs. A modern GPU can increase the calculation performance of CFD problems by more than two times compared to a modern six-core CPU.

Keywords: GPGPU · CUDA · MPI · CFD · Numerical simulation · Control volume method · SIMPLE · Incompressible flow · Domain decomposition

1 Introduction

Modelling of natural phenomena and industrial processes requires a continuous growth of computing performance. In recent years, the performance of graphics processor units (GPU) has increased so much that they have become attractive for scientific and engineering simulations. As a result, techniques of GPGPU

The work was financially supported by the Russian Foundation for Basic Research, the Government of Krasnoyarsk Territory, and the Krasnoyarsk Region Science and Technology Support Fund, research project No16-41-243033.

L. Sokolinsky and M. Zymbler (Eds.): PCT 2017, CCIS 753, pp. 288–300, 2017.
DOI: 10.1007/978-3-319-67035-5_21

(General-Purpose computation on GPUs) have been developed. The high computational efficiency of graphical processor units leads to wide applications of GPUs in supercomputing systems [1]. There are many fields application of GPGPU calculations such as linear algebra [2], molecular dynamics [3], aeromechanics [4,5] and so on. The GPU performance is rapidly increasing and is much higher than the computational performance of central processor units (CPU).

Most algorithms of computational fluid dynamics for incompressible flows are based on the elliptic equation for the pressure correction. The algorithms include the following main steps: a discretization of the pressure correction equation and velocity equations, solving the linear systems, corrections of the pressure and velocity fields. The implementation of only certain algorithm functions in the GPU is not effective. All the data for the calculation are required to be stored in the GPU memory since their transfer between the GPU and the CPU takes a very long time. Consequently, all the main computational operations should be implemented in the GPU. The limited memory of the graphics processor units imposes serious restrictions on the computational problem in the case of a single GPU. Multiple-GPU systems allow computing the complex problem with fine mesh. In this case, a computational domain decomposition can be used for the parallel calculation on the multiple GPU. It is known that the efficiency of the parallel calculations goes down as the number of the mesh cells increases, which is a result of the intensification of the data exchange between the computational units. In the case of graphics processor units, the speed of the data exchange is limited by the PCI-E bus.

The paper considers the realization of the GPU-version of an in-house computational fluid dynamics (CFD) code. The CFD code allows modeling the 3D steady and unsteady viscous incompressible flow in a complex computational domain. The main objectives of this paper are the demonstration of the problems that can be solved by means of GPU, and the comparison of the GPU and CPU performances for the incompressible flow modelling.

2   Mathematical Model

The implementation of GPGPU calculations is based on the in-house CFD code SigmaFlow [6]. A three-dimensional incompressible flow is described by the Navier-Stokes equations

$$\nabla \cdot \mathbf{v} = \mathbf{0}, \tag{1}$$

$$\frac{\partial(\rho\mathbf{v})}{\partial t} + \nabla \cdot (\rho\mathbf{v}\mathbf{v}) = -\nabla p + \nabla \cdot \mathbf{T}, \tag{2}$$

where the viscous stress tensor is

$$\mathbf{T}_{ij} = \mu\left(\frac{\partial v_i}{\partial x_j} + \frac{\partial v_j}{\partial x_i}\right), \tag{3}$$

The model LES WALE (Large Eddy Simulation Wall Adapting Local Eddy-viscosity) is used for the simulation of turbulent flows. This version of the LES model is suitable for the simulation of a turbulent flow near a wall. The equations for the filtered velocity differ from the Navier-Stokes equations by the additional subgrid stress tensor [7]

$$\mathbf{T}_{ij}^{t} = \mu_{\text{SGS}} \left(\frac{\partial v_i}{\partial x_j} + \frac{\partial v_j}{\partial x_i} \right), \tag{4}$$

where μ_{SGS} is the subgrid viscosity. In the LES WALE model, the subgrid viscosity is a function of the flow and is defined as follows [8]:

$$\mu_{\text{SGS}} = (C_W \Delta) \frac{(s_{ij}^d s_{ij}^d)^{3/2}}{(s_{ij} s_{ij})^{5/2} + (s_{ij}^d s_{ij}^d)^{5/4}}, s_{ij} = \frac{1}{2} \left(\frac{\partial v_i}{\partial x_j} + \frac{\partial v_j}{\partial x_i} \right), \tag{5}$$

$$s_{ij}^d = \frac{1}{2}(g_{ij}^2 + g_{ji}^2) - \frac{1}{3}g_{kk}^2 \delta_{ij}, g_{ij}^2 = g_{ik}g_{ki}, g_{ij} = \frac{\partial v_i}{\partial x_j}, \tag{6}$$

where Δ is the cell size, $C_W = 0.325$ is a model constant.

The numerical method is based on the finite volume method on an unstructured mesh. The distributions of a field and its gradient are used for the discretization of a differential equation. The gradient value in the center of the finite volume is determined by means of the least-squares method [9].

Coupling of the pressure and velocity fields is one of the main challenges in the numerical modelling of incompressible flows. A SIMPLE-like procedure on the collocated grids is used. There are many references concerning the SIMPLE approach [10–12]. Collocated grids use the same finite volumes for all variables (both pressure and velocity). This is the most efficient approach. To eliminate the pressure field oscillations, the Rhie-Chow method [13] is applied.

The Quick scheme [14] and Umist TVD scheme [15] are used for the approximation of the convective term of the velocity equations. A second order scheme is used for the viscous term. Unsteady calculations are based on an implicit three-level second-order scheme. The linear systems for velocity equations are solved by means of the incomplete factorization method DILU [16]. A variant of Krylov subspace iterative methods is used for solving the linear system of the pressure correction equation.

3   Software Implementation

Computational domain decomposition is used for parallel computations on multiple CPU or GPU. This method is based on a decomposition of the spatial computational domain into subdomains. Each subdomain is handled by a separate computational process. The connection between the subdomains is realized by the MPI interface. The distribution of the computational load among these processes should be uniform to obtain the highest performance. The partitioning of the computational nodes among the subdomains is performed by means of the MeTiS software [17].

In the GPU-version of the code, all the computational operations are performed on the GPU: calculation of the pressure and velocity gradients, discretization of the velocity equation and pressure correction equation, solving the linear systems, correction of the pressure and velocity according to the SIMPLE procedure and calculation of the turbulent viscosity in the case of turbulent flow modelling. The development of the GPU-versions of all the parts of the package was performed to prevent an excessive data transfer over the PCI-E bus. In this code, the data transfer from the GPU memory to the CPU main memory is performed only in the MPI data exchange or data writing. The algorithms of the CUDA-kernels are identical to the corresponding CPU-method except for the DILU method. Due to the intensive use of atomic operations, the CUDA Compute Capabilities 2.0 architecture is used for the implementation of these functions (CUDA-kernels). In order to achieve enough parallelism, the CUDA-kernels perform an operation on values at each mesh cell or face. The kernels take an array representing the field distribution as an argument and perform many identical operations on the elements of the arrays. Thus, the thread corresponds to the index of the mesh cell or mesh face.

Similarly to the CPU-version of the code, computational domain decomposition and MPI were used for the calculations by means of multiple GPU. In this case, each subdomain was calculated on a separated GPU.

4   Laminar Test Cases

The GPU-version of the code was verified in several test cases. Two laminar test cases were considered in this paper in detail. These are a laminar flow in a cylinder with rotating endwall and an unsteady laminar flow around a circular cylinder.

The swirled flow in the cylindrical container is produced by the endwall rotating with angular velocity Ω (Fig. 1). The endwall rotates the fluid by friction force. The centrifugal force throws the fluid near the rotating endwall to the periphery. Near the opposite endwall, the flow returns to the center. Thus, a concentrated vortex is formed on the axis of the container. The flow is determined by the Reynolds number $\mathrm{Re} = \Omega R^2/\nu$ and the ratio H/R, where ν is the kinematic viscosity, H is the height of the cylinder, R is the radius of the container. Two regimes were considered. In the first one, the Reynolds number was $\mathrm{Re} = 1800$, and the geometrical ratio was $H/R = 1$. In this case, the computational mesh included 800 thousand hexahedral cells. In the second regime, the Reynolds number was $\mathrm{Re} = 2752$, and the geometrical ratio was $H/R = 3.25$. In this case, a fine computational mesh was considered. The mesh included 10 million cells, which were concentrated near the wall.

For the regime $H/R = 1.0$ the radial and tangential velocity components along the lines $r = 0.6R$ and $r = 0.9R$ were compared with experimental data [18]. As Fig. 2 shows, the CPU and GPU codes give the same results, which are close to experimental data. In the regime $H/R = 3.25$, $\mathrm{Re} = 2752$ the experiment shows a three-bubble vortex breakdown [19]. A vortex along the axis of the

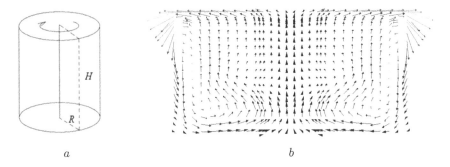

Fig. 1. Flow produced in a cylindrical container by a rotating endwall: (*a*) scheme; (*b*) flow in the central cross-section.

container is directed from the rigid to the rotating endwall. The vortex core transforms (vortex breakdown). As a results, a recirculation zone (a bubble) forms on the axis. Then two more bubbles form near the center of the container. This regime corresponds to a very narrow range of the parameters. However, numerical results agree well with experiment.

The calculation time was considered to compare the computational efficiencies of CPU and GPU. The same number of iterations of the SIMPLEC procedure was fixed in the calculations (2000 iterations in the case $H/R = 1$, Re $= 1800$, and 20 000 iterations in the case $H/R = 3.25$, Re $= 2752$). The Fig. 3 shows that the Titan Black GPU is two times faster than the Core i7-5820k six-core CPU. The calculation time on the Titan Black GPU corresponds to that of a system including two CPUs. The fine mesh including 10 million cells required two Titan Black GPUs due to memory limits.

The second test case is an unsteady flow around a circular cylinder (Fig. 4). The flow depends on the Reynolds number Re $= UD/\nu$, where U is the bulk velocity, D the diameter of the cylinder, and ν is the kinematic viscosity. The Reynolds number is 100. In this case, a Karman vortex street is formed behind the cylinder (Fig. 5). Figure 4a shows a scheme of the computational domain. The external boundary has size $D_{ext} = 40D$. The length of the cylinder is $4D$. A uniform velocity distribution was set as inlet boundary conditions. Non-reflective boundary conditions were used in the outlet. Symmetry was used on the side walls. A number of the O-type meshes included from 0.75 to 1.5 million cells. Mesh cells were concentrated near the wall and in the wake region. The time step is $\tau = 0.04T_{ref}$, where $T_{ref} = D/U_{in}$ is the reference time of the flow. Preliminary calculations showed a Karman vortex street (Fig. 5a). The Strouhal number (dimensionless frequency) of the vortex shedding is St $= fD/U_{in}$, where f is the frequency of the vortex shedding, Hz. The Strouhal number value (see Table 1) agrees with experimental [20,21] and other numerical results [22]. The averaged length of the recirculation zone also agrees well with the results [22].

For comparison of the computational efficiency, unsteady calculations of the flow around the cylinder were performed using a uniform initial velocity field.

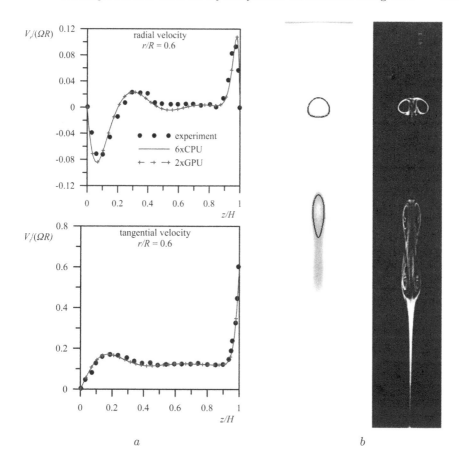

Fig. 2. Flow produced in a cylindrical container by a rotating endwall: (*a*) $H/R = 1$, $\text{Re} = 1800$ (velocity components along the vertical lines, experiment [18]); (*b*) $H/R = 3.25$, $\text{Re} = 2752$ left: iso-line of zero-value of vertical velocity and stagnant zone; right: experimental photo [19]; rotating endwall located on the bottom).

The calculation times of the first $0.6T_{ref}$ (15 time steps) were used to compare the computational efficiencies of the GPUs. The number of iterations of the SIMPLEC procedure per time step was fixed equal 30. Figure 6 shows the computational efficiency of different GPUs. It is shown that the performance of modern GPUs is 2 to 3 times higher than that of six-core CPUs. The highest performance is shown by Titan Black, GeForce 1070 and GeForce 780Ti. The Ge-Force series did not show much lower performance than the Tesla series. On the other hand, Tesla GPGPUs have much more memory (Table 2). Memory is one of the major parameters for computational fluid dynamics, since industrial and scientific problems require very fine computational meshes. Parallel computations on modern GPUs allow to meet these requirements.

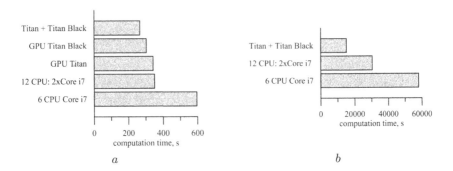

Fig. 3. Calculation time of the flow in a cylindrical container: (*a*) $H/R = 1$, $Re = 1800$ (800 thousand cells); (*b*) $H/R = 3.25$, $Re = 2752$ (10 million cells).

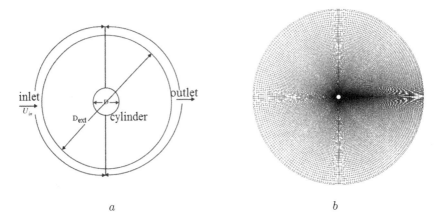

Fig. 4. Flow around a circular cylinder: (*a*) scheme; (*b*) computational mesh.

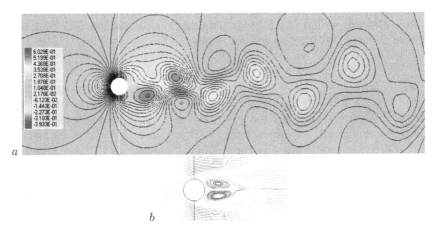

Fig. 5. Unsteady flow around a cylinder: (*a*) Karman vortex street, visualized by instantaneous pressure distribution; (*b*) streamlines of the averaged flow.

Table 1. Flow around a cylinder: mean parameters

	St	L/D
Calculations SigmaFlow-GPU	0.164	1.41
Calculations Shoeybi (2010)	0.168	1.41
Experiment Zdravkovich (1997)	0.165	-

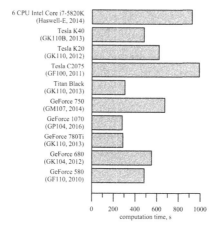

Fig. 6. Flow around a cylinder: computation time (mesh includes 1.5 million cells, single precision).

Table 2. Memory of the considered graphics processor units

GPU	Memory, Gb
Tesla K40	12
Tesla K20	5
Tesla C2075	3
GeForce 750	2
GeForce 1070	8
GeForce 780Ti	3
GeForce 680	2
GeForce 580	1.5
Titan Black	6

Usually, single precision is used in computational fluid dynamics. In some cases, however, single precision is insufficient. At the same time, GPU performance in double precision is lower than in single precision. Therefore, in this test case, performances in single and double precision were compared for different GPUs (Fig. 7). The computational mesh included 0.75 million cells due to

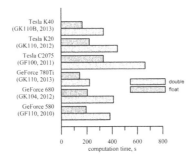

Fig. 7. Flow around a cylinder: computation time (the mesh includes 0.75 million cells, single and double precision).

memory limits of GPUs. Both Tesla and GeForce GPUs show a computational time in double precision that is two times higher than that in single precision. Thus, the performance in double precision is determined primarily by memory bandwidth instead of calculations blocks.

5 Turbulent Test Cases

In addition to laminar test cases, two turbulent problems were solved by means of the LES WALE method. The first turbulent test case was a flow in a cubic cavity with a leading cover wall that moves in the x-direction (Fig. 8a). The Reynolds number, based on the cover wall velocity and the cavity length, was $Re = 10^4$. The computational mesh included 1 million cells ($100 \times 100 \times 100$) concentrated to the walls (Fig. 8b). Figure 9 shows a comparison of the averaged velocity along the central vertical line with experimental data [23]. It has been shown that both the CPU- and the GPU- versions of the code closely agree with the experimental results. A comparison of computational times shows that the performance of two Titan Black GPUs is 4 times higher than that of two Intel Core i7-5820k six-core CPUs (Fig. 10).

The second test case was a turbulent swirled flow in the draft tube of a model hydraulic turbine. The calculations were based on experimental data of the workshop Francis-99 [24] and numerical simulation [25]. The computational domain included the draft tube with a conical inlet (Fig. 11a). The averaged velocity profile in the inlet, obtained in numerical investigations [25], was used as inlet boundary conditions. A part load operation mode was considered. The mesh included 7.8 million hexahedral cells (Fig. 11b). The dimensionless wall distance of the near-wall nodes was $y_+ \approx 2$. The Umist TVD scheme was used for the approximation of the convective terms. It is not sufficient for appropriate LES calculation, although it is reasonable for the assessment of the computational efficiency. Figure 11c shows complex vortex structures in the draft tube. A comparison of the averaged velocity profiles behind the runner shows good

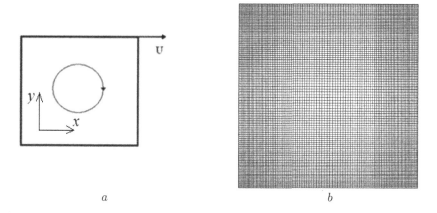

a

b

Fig. 8. Flow in a cavity: (*a*) scheme; (*b*) computational mesh.

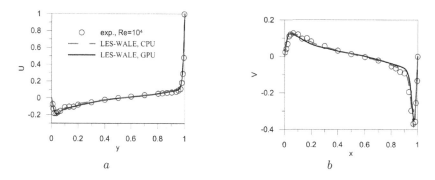

a

b

Fig. 9. Averaged velocity components along the central vertical line: (*a*) horizontal velocity component; (*b*) vertical velocity component.

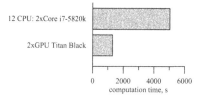

Fig. 10. Cavity: computation time.

agreement with experimental data (Fig. 12). As Fig. 13 shows, the performance of two Titan Black GPUs is six times higher than that of the Intel Core i7-5820k six-core CPU. The performance of four GPUs is 40% higher than that of two GPUs (Fig 13).

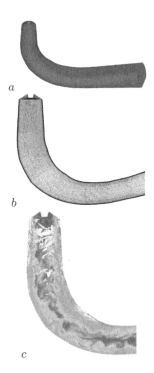

Fig. 11. Flow in the draft tube of the Francis-99 hydraulic turbine: (*a*) computational domain; (*b*) computational mesh in the central longitudinal cross-section; (*c*) vortices visualized by the iso-surface of the *Q*-criterion (second invariant of the velocity gradient) and instantaneous velocity magnitude in the central cross-section.

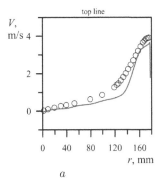
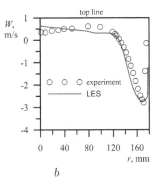

Fig. 12. Averaged velocity components in the Francis-99 draft tube: (*a*) axial velocity; (*b*) tangential velocity.

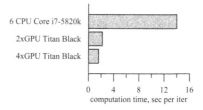

Fig. 13. Flow in the Francis-99 draft tube: the computation time per iteration of the SIMPLE procedure.

6   Conclusions

Thus, a GPU-version of the CFD code was developed for parallel GPGPU simulation of incompressible flows. All test cases calculations showed agreement with experimental and other numerical results. Such problems as steady and unsteady laminar flows, turbulent flow in a cavity and swirled turbulent flow in a part of a hydraulic turbine were considered. Most of the arrays were stored in the GPU memory, and main operations of the SIMPLE procedure were performed on the GPU to provide the highest performance of the code.

The calculations showed a high performance of the graphics processor units. The efficiency of modern GPUs is 2 to 3 times higher than that of a six-core CPU. Parallel calculations in multiple GPU systems allow to overcome the memory limits of a single GPU. In the case of a coarse mesh, parallel calculations on multiple GPUs are not efficient due to data exchange with a GPU. Memory is one of the major parameters for computational fluid dynamics, since industrial and scientific problems require very fine computational meshes. Therefore, a code for parallel computations on modern GPUs can be a useful tool of CFD calculations.

References

1. Top500 list. https://www.top500.org/lists/2016/06/
2. Li, R., Saad, Y.: GPU-accelerated preconditioned iterative linear solvers. J. Supercomputi. **63**(2), 443–466 (2013). doi:10.1007/s11227-012-0825-3
3. Rinaldi, P.R., Dari, E.A., Venere, M.J., Clausse, A.A.: Lattice-Boltzmann solver for 3D Fluid simulation on GPU. Simul. Model. Pract. Theor. **25**, 163–171 (2012). doi:10.1016/j.simpat.2012.03.004
4. Corrigan, A., Camelli, F.F., Lohner, R., Wallin, J.: Running unstructured grid-based CFD solvers on modern graphics hardware. Int. J. Numer. Methods Fluids **66**(2), 221–229 (2011). doi:10.1002/fld.2254
5. Waltz, J.: Performance of a three-dimensional unstructured mesh compressible flow solver on NVIDIA Fermi-class graphics processing unit hardware. Int. J. Numer. Methods Fluids **72**(2), 259–268 (2013). doi:10.1002/fld.3744
6. Gavrilov, A.A., Dekterev, A.A., Sentyabov, A.V.: Modeling of swirling flows with coherent structures using the unsteady reynolds stress transport model. Fluid Dyn. **50**(4), 471–482 (2015). doi:10.1134/S001546281504002X

7. Pope, S.B.: Turbulent Flows. Cambridge University Press, New York (2000)
8. Nicoud, F., Ducros, F.: Subgrid-scale stress modelling based on the square of the velocity gradient tensor. Flow Turbul. Combust. **62**(3), 183–200 (1999). doi:10. 1023/A:1009995426001
9. Mavriplis, J.: Revisiting the least-squares procedure for gradient reconstruction on unstructured meshes. AIAA-Paper 2003-3986, June 2003. doi:10.2514/6.2003-3986
10. Ferziger, J.H., Peric, M.: Computational Methods for Fluid Dynamics. Springer, New York (2002). doi:10.1007/978-3-642-56026-2
11. Moukalled, F., Darwish, M.: A unified formulation of the segregated class of algorithms for fluid flow at all speeds. Numer. Heat Transf. Part B **37**(2), 227–246 (2000). doi:10.1080/104077900275576
12. Patankar, S.: Numerical Heat Transfer and Fluid Flow. Hemisphere Publishing Corporation, New York (1980)
13. Rhie, C.M., Chow, W.L.: A numerical study of the turbulent flow past an isolated airfoil with trailing edge separation. AIAA J. **21**, 1525–1532 (1983)
14. Leonard, B.P.: A stable and accurate convective modeling procedure based on quadratic upstream interpolation. Comput. Math. Appl. Mech. Eng. **19**, 59–98 (1979). doi:10.1016/0045-7825(79)90034-3
15. Leschziner, M.A., Lien, F.S.: Upstream monotonic interpolation for scalar transport with application to complex turbulent flows. Int. J. Numer. Methods Fluids **19**(6), 527–548 (1994). doi:10.1002/fld.1650190606
16. Barrett, R., Berry, M., Chan, T.: Templates for the Solution of Linear Systems: Building Blocks for Iterative Methods. SIAM, Philadelphia (1994). doi:10.1137/1. 9781611971538
17. Karypis, G., Kumar, V.: A fast and highly quality multilevel scheme for partitioning irregular graphs. SIAM J. Sci. Comput. **20**(1), 359–392 (1999). doi:10.1137/ S1064827595287997
18. Michelsen, J.A.: Modeling incompressible rotating fluid flow. Ph.D. thesis, Technical University of Denmark (1986)
19. Escudier, M.P.: Observation of the flow produced in a cylindrical container by a rotating endwall. Exp. Fluids **2**(4), 189–196 (1984). doi:10.1007/BF00571864
20. Zdravkovich, M.M.: Flow Around Circular Cylinders: Fundamentals, vol. 1. Oxford University Press, New York (1997). doi:10.1017/S0022112097227291
21. Rajani, B., Kandasamy, A., Majumdar, S.: Numerical simulation of laminar flow past a circular cylinder. Appl. Math. Model. **33**, 1228–1247 (2009). doi:10.1016/j. apm.2008.01.017
22. Shoeybi, M., Svard, M., Ham, F.E., Moin, P.: An adaptive implicit-explicit scheme for the DNS and LES of compressible flows on unstructured grids. J. Comput. Phys. **229**, 5944–5965 (2010). doi:10.1016/j.jcp.2010.04.027
23. Akula, B., Roy, P., Razi, P., Anderson, S., Girimaji, S.: Partially-Averaged Navier-Stokes (PANS) simulations of lid-driven cavity flow - Part 1: comparison with URANS and LES. In: Girimaji, S., Haase, W., Peng, S.H., Schwamborn, D. (eds.) Progress in Hybrid RANS-LES Modelling. Notes on Numerical Fluid Mechanics and Multidisciplinary Design, vol. 130, pp. 421–430. Springer, Cham (2015). doi:10. 1007/978-3-319-15141-0_29
24. Francis-99 Workshop I (2017). http://www.ltu.se/research/subjects/ Stromningslara/Konferenser/Francis-99
25. Minakov, A.V., Sentyabov, A.V., Platonov, D.V., Dekterev, A.A., Gavrilov, A.A.: Numerical modeling of flow in the Francis-99 turbine with Reynolds stress model and detached eddy simulation method. J. Phys. Conf. Ser. **579**(1) (2015). doi:10. 1088/1742-6596/579/1/012004

Implementation of Implicitly Restarted Arnoldi Method on MultiGPU Architecture with Application to Fluid Dynamics Problems

Nikolay M. Evstigneev[✉]

Federal Research Center "Informatics and Control", Institute for System Analysis,
Russian Academy of Science, Moscow, Russia
evstigneevnm@yandex.ru

Abstract. The parallel CPU+multiGPU implementation of the Implicitly Restarted Arnoldi method (IRA) is presented in the paper. We focus on the problem of implementing an efficient method for large scale non-symmetric eigenvalue problems arising in linear stability and Floquet theory analysis in fluid dynamics problems. We give brief details about the problem in both cases. We use and cross-compare different methods to implement QR shifts with polynomial filtering. Then we conduct a benchmark on standard non-symmetric matrices. The presented results give insight on the best choice of the method of polynomial filters. It turns out that a polynomial filter is essential to accelerate computations in fluid dynamics stability problems as well as to increase the Krylov space dimension. However, some knowledge about the spectral radius of the problem is required. Further, we investigate the parallel efficiency of the method for single-GPU and multi-GPU modes using the problems previously considered. It is shown that the implementation of Givens rotations on one GPU for QR computations (of a Hessenberg matrix) is essential in order to achieve high speed for a considerable Krylov subspace dimension. In the final part we present some results regarding the application of the IRA polynomial filtered method to linear stability and Floquet analysis for large fluid dynamics problem with parallel efficiency comparison on multi-GPU architecture.

Keywords: Arnoldi method · Eigenvalue solvers · Matrix free methods · Krylov methods · Multi GPU · Fluid dynamics · Linear stability analysis · Floquet theory analysis

1 Introduction

The need to calculate a few leading eigenvalues is essential in various fields of research, such as linear stability analysis, structural stability analysis, quantum physics, etc. The systems that are subject to eigenvalue analysis are mostly

The work is supported by the Russian Foundation for Basic Research (grant RFBR 14-07-00123).

L. Sokolinsky and M. Zymbler (Eds.): PCT 2017, CCIS 753, pp. 301–316, 2017.
DOI: 10.1007/978-3-319-67035-5_22

of large scale, hence the matrices under consideration are large. Therefore, it is impractical to use methods such as QR to find the whole spectrum. Such iterative methods include power-type methods, Krylov subspace-type methods, the Jacobi–Davidson method, etc. (for more information, see [1]). Arnoldi-type methods are among the most widely used for the solution of this problem. Suffice it to say that the Implicitly Restarted Arnoldi method in ARPACK [2,3] is included in NumPy (SciPy) Python and standard MATLAB package. For large scale applications, a parallel MPI CPU version exists that is called PARPACK.

Our research focuses on linear stability analysis and the analysis of the monodromy matrix for stationary and time-periodic solutions of fluid dynamics problems, respectively. There are few papers on the subject but the field is growing rapidly. Detailed information concerning bifurcations of the main stationary solution of initial boundary value problems (IBVPs) in partial differential equations (PDEs) is given in the book [4], in the last chapter (including 2D Navier–Stokes bifurcation problems). It is shown that some specific methods are required to achieve convergence in both the space of the discrete problem and the eigenvalue solvers. The author recommends using spatial preconditioners during the eigenvalue problem solution, namely the successful use of the shift-and-invert preconditioner that was initially introduced by Ericsson and Ruhe [5] for symmetric matrices. Parallel implementation of an eigenvalue solver is mandatory due to the problem size. In [6], the authors use the IRA method of ARPACK with both shift-and-invert and Cayley transformations in 2D incompressible Navier–Stokes linear stability problems with finite element discretization. The discontinuous Galerkin and ARPACK IRA methods are used to perform the linear stability analysis of a 2D incompressible flow in a channel with a sudden expansion and also a flow in a pipe with cylindrical symmetry and semi-spherical stenosis (see [7]). The authors applied an *a posteriori* error estimation and adaptive mesh refinement of discontinuous Galerkin approximations to achieve the convergence of the eigenvalues for the stability problem. ARPACK is also successfully used in [8] for a linear stability analysis in the Rayleigh–Benard convection problem using trilinear nodal elements for both two and three dimensional problems. The authors demonstrated that the accuracy of the calculations deteriorates as the flow is advectively dominated due to under-discretization of the problem, but not due to eigensolver difficulties. ARPACK was also successfully applied in [9] to perform a linear stability eigenvalue analysis for the control of the 2D disturbances in the Blasius boundary layer. An alternative to shift-and-invert preconditioning is presented in [10], where a spectral transformation based on the filtering of the linearized equations of motion in a certain frequency range is introduced along with a Krylov subspace iterative solver. The method is also successfully applied to compute least stable eigenvalues in a compressible jet problem. An analysis of bifurcations in a 2D lid driven cavity flow using the Newton–Picard method is carried out in [11] with a linear stability and Floquet theory analysis of some time-periodic solutions.

Some recent papers have been dedicated to the linear stability and Floquet theory analysis of time-periodic solutions for 3D fluid dynamics problems. One of the first such papers was [12] (and some other papers by these authors), where a 3D Floquet stability analysis was applied to a circular cylinder 2D wake stability problem. The fluid flow solver was a spectral element method for 2D equations where Floquet analysis was used through the introduction of infinitesimal three-dimensional perturbations and derivation of a state-transition periodic monodromy matrix. The eigenvalues of the latter are found using Krylov subspace iterations of dimensions between 8 and 20, initialized either from a random starting vector or an eigenmode computed at nearby parameter values. The linear stability of a 3D compressible subsonic flow in a cavity is analyzed in [13] using ARPACK's IRA method and a sixth-order compact finite-difference scheme for spatial discretization. The Krylov subspace iteration method implemented in a matrix-free form is successfully used in [14] to find the eigenvalues and eigenvectors of the monodromy matrix for 3D pipe flows. The authors used the spectral element method to perform the discretization of the problem. Also, the authors of the report [15] successfully apply Krylov subspace iterations to the Floquet stability analysis of the 3D wake flow behind a mechanically oscillated circular cylinder placed in a low Reynolds-number flow, on a GPGPU hardware using CUDA C.

All these problems are solved successfully using Krylov-type methods, mostly the IRA method from ARPACK or some other implementation. In the light of the new computational paradigm based on CPU+multiGPU architecture, it is expedient to adopt an implementation of the IRA method suitable for eigenvalue analysis problems in complex PDEs for that kind of modern architecture. There are only a few implementations of Arnoldi-type methods available for GPU and multiGPU architectures, for example [16] (Explicitly Restarted Arnoldi) and [17] (IRA). The latter IRA method is freely available in source code, it uses multiGPU abstraction through C++ templates and is capable of running efficiently on modern GPU clusters. In our experience, however, we have been unable to achieve a successful convergence of eigenvalues for non-symmetric matrices in flow stability problems using code from [17], whereas ARPACK successfully converges with the same Krylov subspace dimension and for the same problems. Thus, it was decided to develop a CPU and multiGPU IRA method for fluid dynamics eigenvalue problems.

In this paper, we report results on the implementation, benchmarking and some applications of the IRA method for CPU+multiGPU architecture with polynomial filters. The paper is laid out as follows. First, some general information on the linear stability and Floquet theory for fluid dynamics problems is presented. Then we give the IRA method algorithm with polynomial filters. Some key points of the algorithm are indicated that are crucial for the GPU implementation. Further, this is followed by the implementation section, where some variants of implementation are described. Additionally, benchmark problems from standard test cases and fluid dynamics applications are presented, followed by the analysis of the parallel efficiency on GPU/CPU/multiGPU variants. The last two sections consist of examples of application to 3D fluid dynamics problems and conclusions.

2 Fluid Dynamics Eigenvalue Problems

We will focus on an incompressible fluid stability problem. Let us assume that, by some space discretization method, we have derived the following system of Navier–Stokes equations equipped with proper initial-boundary conditions:

$$\frac{d\hat{\mathbf{u}}_{\mathbf{m}}}{dt} + \mathrm{P}\left(\sum_{\mathbf{k}} \hat{\mathbf{u}}_{\mathbf{k}} \mathrm{L}\hat{\mathbf{u}}_{\mathbf{m}-\mathbf{k}}\right) = \frac{1}{R}\mathrm{C}\hat{\mathbf{u}}_{\mathbf{m}} + \hat{\mathbf{f}}_{\mathbf{m}}, \tag{1}$$

where $\mathbf{m}$ and $\mathbf{k}$ are space discretization parameter vectors, $\hat{\mathbf{u}}$ is a discrete velocity vector, P is a pressure projection operator, L is a discrete divergence operator, C is a discrete diffusion operator, R is the Reynolds number and $\hat{\mathbf{f}}$ is a discrete external force vector. If boundary conditions include pressure definition (e.g. Poiseuille flow), then the pressure can be split into dynamic and static; the gradient of static pressure is introduced through the force, whereas the gradient of dynamic pressure is projected onto the divergence free velocity functional subspace to fulfill the solenoidal condition $\mathrm{L}\hat{\mathbf{u}} = 0$. The convolution part can be calculated by the pseudo-spectral finite-element/difference method or directly. We wish to stress that the diffusion part and pressure projection are computed implicitly. This will be essential for further considerations.

We introduce a small magnitude perturbation velocity vector $\hat{\mathbf{v}}$, and substitute it into (1). After subtracting one equation from the other and linearizing, we obtain the following system:

$$\frac{d\hat{\mathbf{v}}_{\mathbf{m}}}{dt} + \mathrm{P}\left(\sum_{\mathbf{k}} (\hat{\mathbf{u}}_{\mathbf{k}} \mathrm{L}\hat{\mathbf{v}}_{\mathbf{m}-\mathbf{k}} + \hat{\mathbf{v}}_{\mathbf{k}} \mathrm{L}\hat{\mathbf{u}}_{\mathbf{m}-\mathbf{k}})\right) - \frac{1}{R}\mathrm{C}\hat{\mathbf{v}}_{\mathbf{m}} = \mathbf{0}. \tag{2}$$

The system (2) is used in the form

$$\frac{d\hat{\mathbf{v}}}{dt} + \mathrm{A}(\hat{\mathbf{u}}, t, R)\hat{\mathbf{v}} = \mathbf{0}, \tag{3}$$

where A is a linearized matrix operator. We will omit the index $\mathbf{m}$ from now on if this does not cause confusion. For the stationary solutions, we have $\frac{d\hat{\mathbf{v}}}{dt} = 0$, and we get the Jacobi matrix of the nonlinear operator (1) in the form

$$\mathrm{A}\hat{\mathbf{v}} = \mathbf{0}. \tag{4}$$

Now, let the system (1) have a periodic solution with period $\mathcal{T}$. Then the system has the form

$$\hat{\mathbf{v}}(t + \mathcal{T}) = \mathrm{M}(\mathcal{T})\hat{\mathbf{v}}(t), \tag{5}$$

where $\mathrm{M}(\mathcal{T})$ is a state-transition operator or a monodromy matrix. We integrate the linearized system over time for a period $\mathcal{T}$:

$$\int_{\hat{\mathbf{v}}_0}^{\hat{\mathbf{v}}_{\mathcal{T}}} d\hat{\mathbf{v}} = -\underbrace{\int_{t}^{t+\mathcal{T}} (\mathrm{A}(\tau, \hat{\mathbf{u}}(\tau)))\, d\tau}_{\mathrm{M}(\mathcal{T}, \hat{\mathbf{u}}, \hat{\mathbf{v}}))}, \tag{6}$$

and obtain

$$\hat{\mathbf{v}}_{\mathcal{T}} = M(\mathcal{T})\hat{\mathbf{v}}_0. \tag{7}$$

The question of interest is whether the linear stability holds for the stationary and periodic solutions of the Navier–Stokes discrete system. Thus, we arrive at the eigenvalue problem for A and M. These matrices are real nonsymmetric and dense for high order methods. We do not discuss here the question of discretization size, but it must be noted that the high harmonics of the spatial Fourier series must be in a deep dissipation zone in order to achieve a good convergence of the eigenvalue problem. In this case, the size of either A or M is at least equal to the typical size for DNS. This exceeds any possible limits of explicit matrix storage. Thus, matrix-vector operations regarding (4) and (7) must be done without explicit calculation and storage of these matrices (the so-called matrix-free method).

3   Implicitly Restarted Arnoldi Method

The matrix-free approach is incorporated into the Arnoldi process with the operator $A \in \mathbb{R}^{N \times N}$. We use the IRA algorithm closely following [1,18]. We form the Krylov subspace

$$\mathcal{K}_m = \mathrm{span}\{\mathbf{v}, \mathbf{v}^{(1)}, \mathbf{v}^{(2)}, ..., \mathbf{v}^{(k)}, ..., \mathbf{v}^{(m-1)}\}, \tag{8}$$

where $\mathbf{v}^{(j)} = A^j \mathbf{v}$ and $(\mathbf{v}^{(l)}, \mathbf{v}^{(j)}) = \begin{cases} 0; l \neq j, \\ 1; l = j \end{cases}$.

IRA algorithm:

1. Initialization. Set $\mathbf{v}$ with random values, normalize the vector, select the number k of desired eigenvalues and the number of additional vectors p for the implicit procedure, $m = k + p$. Set the variables $s = 0$ and ε to some small predefined value.
2. Arnoldi step. For each element of (8) from s to $m - 1$, we build the Arnoldi algorithm with Gram–Schmidt process correction by calling

$$\mathbf{v}^{(j+1)} = A\mathbf{v}^{(j)}, j = s, ..., m - 1, \tag{9}$$

where the call is performed in a matrix-free way, applying (4) or (7) to the vector $\mathbf{v}^{(j)}$. At the end, this process generates the decomposition

$$AV_j = V_j H_j + \mathbf{w}_{j+1} H_{j+1,j} \mathbf{e}_j^T, \tag{10}$$

where H_j is an upper Hessenberg matrix of size $j \times j$, V_j are Arnoldi vectors of size $N \times j$, and $\mathbf{w}_{j+1}$ is the last Arnoldi vector. If the last value $H_{j+1,j} = 0$, then $\mathcal{K}_m$ is an invariant subspace of A. The process stops with $AV = VH$, and the eigenvectors (eigenvalues) of H are the selected eigenvectors (eigenvalues) of A. If the process does not stop (which is usually true), then continue to step 3.

3. Find estimates of the eigenvalues μ_m and eigenvectors of the Hessenberg matrix. Sort the estimates in a desired order (Largest Real (LR) or Largest Magnitude (LM) for our purposes). We designate the first estimates $\mu_0, ..\mu_{k-1}$ as the desired ones and the rest as undesired, and select shifts from the undesired ones using a function

$$\theta_j = \text{Shifts}_j(\mu_k, \mu_{k+1}, ..., \mu_{m-1}), j = k, ..., m-1, \tag{11}$$

which shall be defined later.

4. Perform the QR algorithm with shifts for H using polynomials with number of shifts $q = p$ (bulge-chasing):
$j = 1, .., q$

(a) $QR = (H - \theta_j E)$,

(b) $H = RQ + \theta_j E$.

Set $s = m - 1 - q$, with E being the identity matrix.

5. Apply shifts:
$\mathbf{h} = Q_{*,s}, \alpha_Q = Q_{s,s-1}, \alpha_H = H_{k+m-1,s-1}$,
$\mathbf{w} = \alpha_H \mathbf{h} + \alpha_Q \mathbf{w}$,
$Q_{a=s:(m+k-1),b=s:(m+k-1)} = \delta_{a,b}, V = VQ$.

6. At this point, we have the matrices V, H with additional last Arnoldi vector $\mathbf{w}_{j+1}$. Let the vector $\mathbf{x}$ contain the last elements of the eigenvectors $\mathbf{h}_l$ corresponding to the desired eigenvalue estimates $\mu_l, l = 0, .., k - 1$ of H: $x(l) = h_l(m)$. We denote $\mathbf{y}_l = V\mathbf{h}_l$ as a Ritz vector of A and Ritz value λ_l, both associated with $\mathcal{K}_m$. For these Ritz variables, we have [1]: $||A\mathbf{y}_l - \lambda_l\mathbf{y}_l||_2 = ||\mathbf{w}_{j+1}||_2|x_l|$. Thus, if

$$\max_{l=0,...,k-1} (||\mathbf{w}_{j+1}||_2|x_l|) \leq \varepsilon, \tag{12}$$

stop. Else go to step 2.

In the last step, we also check the convergence of the Ritz vectors by estimating the actual residual $||A\mathbf{y}_l - \lambda_l\mathbf{y}_l||_2$. This check is performed if and only if we have a successful convergence according to (12).

3.1   Filtering Polynomials

An exact filtering polynomial is used for the standard IRA method, i.e. one using the exact shifts $\theta_l = \mu_l$ in (11). In this case, the polynomial that is applied to the Arnoldi vectors as $\mathbf{w}_j^+ = P(\mathbf{A})\mathbf{w}_j$, where $\mathbf{w}_j^+$ is the vector in column j of V after step 5, is given in the form

$$P(\lambda) = \prod_{l=k}^{m-1} (\lambda - \mu_l). \tag{13}$$

It is known that IRA with exact shifts may fail to converge [19] for a special type of non-hermitian matrices.

Here we suggest an additional strategy, analogously to [20] where it was used in the Lánczos method. We define the undesired set $\Omega \subset \mathbb{C}^1$ with boundary $\partial \Omega$ formed by the elements μ_j, $j = k, ..., m - 1$, of H either by constructing the convex hull or by least-square ellipse fitting according to the algorithm proposed in [21]. After that, we place the zeros of the polynomial in Ω in such a way as to maximize the determinant of the Vandermonde matrix. This is achieved by means of either Fekete or Leja points in Ω. We consider Ω as a subset of $\mathbb{R}^2$, and apply a greedy algorithm, taken from [22], to generate Fekete or Leja points in $\Omega \subset \mathbb{R}^2$. It can be successfully applied through QR and LU factorizations. In addition, we can apply a certain "fix" to the polynomial. The "fix" is simply a user-specified point that fixes a boundary for the ellipse or the convex hull. If some spectral properties of an operator are known *a priori*, then one can facilitate by means of this "fix" the search of the desired eigenvalues in a region not included in the filter. This strategy turned out to be effective for Jacobi matrices by finding LR eigenvalues close to the imaginary axis.

3.2   Accelerating Techniques

The convergence of the eigenvalue estimates in the monodromy matrix calculations does not present any difficulty, since the IRA method is very good at finding eigenvalues close to the unit circle. However, it is known ([6,8,9,16]) that the unpreconditioned IRA method converges slowly or does not converge at all for the Jacobi matrix eigenvalues of Navier–Stokes stability problems. The use of either shift-and-invert or Cayley transformations requires the solution of

$$(A - \sigma E)v^{j+1} = v^j, \tag{14}$$

with σ being a shift that is inserted into IRA instead of (9) in a matrix-free way. As a result, one faces the same difficulties as in IRA, since Krylov-type methods are usually used for this problem (BiCGStab, GMRES etc.). The choice of preconditioners for matrix-free methods is a difficult task. We use instead transformation acceleration techniques that avoid the solution of (14). We consider an exponential matrix

$$J = e^A, \tag{15}$$

formed in a matrix-free way analogously to (6), i.e.

$$\int_{\hat{v}_0}^{\hat{v}_T(\epsilon)} d\hat{v} = - \underbrace{\int_t^{t+T(\epsilon)} (A(\tau, \hat{u})) \, d\tau}_{J(\hat{u}, \hat{v}))} . \tag{16}$$

The difference is that we use a frozen main solution $\hat{u}$ that is of interest for the stability, and the upper limit in the integral (16) is defined by the convergence in the norm $||\hat{v}^{n+1} - \hat{v}^{n-1}||_2 \leq \epsilon$, where n is the time step of numerical integration.

We need to stabilize the solution $\hat{u}$ of the main problem before replacing it in (16). For this purpose, we use a computationally cheap method outlined in [23].

Numerical integrations are performed using the Runge–Kutta 4 method with implicit diffusion, projection operators and explicit advection operator. Thus the stability of the integration is limited by the advection stability criterion (CFL number). The resulting integration is supplied to (9), which is called in each Arnoldi step. The implicit integration of diffusion allows for the use of larger time-steps, which leads to a better separation of eigenvalues around the unit circle. In order to obtain eigenvalues estimates for A, one can apply the transformation $R = V^H A V$, with the eigenvalues of R being the Ritz estimates of the Jacobi matrix. If one applies polynomial filtering, then this approach allows for a better estimation of the eigenvalues than the standard IRA does. If this method fails, then we apply the shift-and-invert method. In this case, the system (14) is solved by the GMRES method with a preconditioner that must be calculated additionally.

4   Implementation

The described IRA method is implemented in a multiGPU and CPU framework. As one can see, the method consists of some linear algebra operations. Let us denote the matrices $V \in \mathbb{R}^{N \times m}$, $H \in \mathbb{R}^{m \times m}$ and the orthogonal matrix $Q \in \mathbb{R}^{m \times m}$. We will use the vectors $\mathbf{v}, \mathbf{w} \in \mathbb{R}^N$, $\mathbf{x} \in \mathbb{R}^m$. Since matrices under consideration have either real or complex conjugate pairs of eigenvalues, we use real arithmetics in step 4 of the IRA algorithm. Either multiGPU or single GPU calculations are used for all bulky parts of the algorithm. The operations VQ, $V\mathbf{v}$, $\|\mathbf{v}\|_2 = \sqrt{\mathbf{v}^T \mathbf{v}}$, $(V, \mathbf{v}) = V^T \mathbf{v}$ and $(\mathbf{v}, \mathbf{w}) = \mathbf{v}^T \mathbf{w}$ operations of dimension N are distributed over multiple GPUs, so that $N_l = N/p$, where N_l is the local vector size in the g-th GPU, i.e. $V_l \in \mathbb{R}^{N_l \times m}, l = 1, \ldots, g, g \geq 1$. The operations $A\mathbf{v}$ and VQ are efficient and are executed locally on each GPU using CUBLAS or MAGMA. The operations $\sqrt{\mathbf{v}^T \mathbf{v}}$, $V^T \mathbf{v}$ and $\mathbf{v}^T \mathbf{w}$ are also executed with CUBLAS or MAGMA, but additionally require **MPI_Allreduce** with **MPI_SUM** operation. The effect of these operations will be investigated below.

We test two implementation strategies. The first strategy uses LAPACK on CPU with OpenMP for $QR \leftarrow H$, finding the eigenvalues of H and operations with $H, Q, \mathbf{x}$. Then CPU arrays Q, H and $\mathbf{x}$ are copied into the GPUs, and all the other computations are performed there. We also tested the application of MAGMA for QR and eigenvalue calculations, but the GPU MAGMA **magma_dgeqrf2_gpu** operation outperforms LAPACK on our hardware for matrix sizes starting at 600×600. In practice, $m \leq 150$, hence the reason why we do not compare the MAGMA approach here (see Sect. 5). However, it can be used in systems with more CPUs than GPUs and many CPU-GPU operations via PCI-buses. In this case, PCI transfers can be arranged in direct GPU mode.

The second strategy is based on the use of GPU-only calculations. In this case, the QR algorithm was implemented using Givens rotations on a GPU. Givens Q matrix is also used in the same manner during application to V matrix.

Operations with $\mathbf{x}$, H and Q are performed locally either on CPUs with further distribution to GPUs in the first strategy or on each GPU separately. In this case, the eigenvalues of H are compared between GPUs for consistency at each 10-th iteration of the method.

The Arnoldi process uses functor matrix-vector calls to (16), (6) or (4), which are executed alongside Navier–Stokes integration on a single or multiple GPUs. In the benchmark problems with predefined matrices, we replace the simple matrix-vector multiplication routine with CUBLAS.

5   Benchmark Problems

We use an 8-core (16 in hyper-threading) Intel Xeon E5-2640V2 Ivy Bridge and five k40 NVIDIA GPUs assembled in one chassis.

For the benchmark of convergence, we consider the following non-symmetric square real matrices from the SuiteSparse Matrix Collection [24]: **west0479**(479), **raefsky2**(3242), **S40PI_n**(2128), **GT01R**(7980), where matrix dimension is given in parentheses. The matrix **S40PI_n** has poorly separated eigenvalues with maximum real parts; the eigenvalues of the matrix **GT01R** are located on a strip stretched along the imaginary axis, with $\max(|\operatorname{Im}(\lambda(\mathbf{GT01R}))|) = 19\,553$. In addition, we also consider a Jacobi matrix constructed using exact calculations from stability of 2D (1790) and matrix-free 3D (10368) Kolmogorov flow problems [25]. In the 2D case, the solution is stable, with critical eigenvalues just before the Hopf bifurcation. In the 3D case, the solution is unstable, with main solution stabilized using the filtering approach [23]. The matrices are tested by the IRA and IRA+Leja algorithms with $\dim(\mathcal{K}_m) = 15$, residual tolerance ε set to $1 \cdot 10^{-8}$ in the L_∞-norm, and four LR targets against known spectra. The latter were obtained using the exact QR algorithm. The resulting convergence history of the residual in the L_∞-norm is presented in Fig. 1. The Leja-points polynomials can be suitable but not in all cases, as one can see for the **GT01R** matrix. The convergence of the 2D Jacobi matrix of the Kolmogorov problem is troublesome in both cases. However, this can be corrected by matrix exponential computations using (16) in this problem.

The 3D Jacobi matrix test case is more difficult, since the eigenvalues are located on a thin strip stretched along the imaginary axis, with clustering of complex conjugate pairs of small magnitude on the right half-plane (see Fig. 2). IRA as well as IRA+Leja algorithms fail to converge (Fig. 2). The latter can be adjusted by fixing the convex hull point near the imaginary axis. In this case, the convergence improves and the correct LR eigenvalues are found after 43 564 calls to the functor. The application of exponential transformations corrects this situation, what is shown in Fig. 3. The convergence was achieved by 255 calls to the functor (16) with a total of 15300 calls to the RK integration method. Thus, the application of the exponential matrix is more efficient.

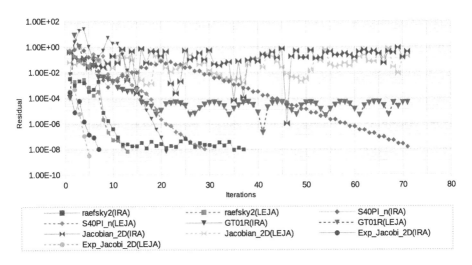

Fig. 1. Convergence history in the L_∞-norm for test matrices using different methods

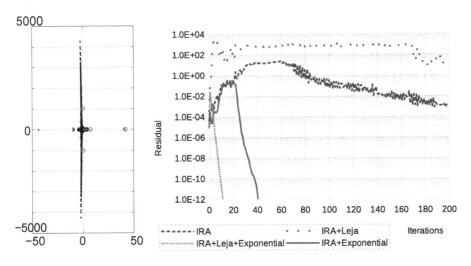

Fig. 2. Eigenvalues of the Jacobi matrix (10368) from the unstable solution of a 3D Kolmogorov flow problem with four LR leading eigenvalues found by the IRA algorithm, and residual convergence in the L_∞-norm for various algorithms

An additional benchmark is presented in [26], where a 3D Rayleigh–Benard convection stability problem is tested against an analytical neutral curve for a full resolution of 524 288 degrees of freedom, using the Fourier–Galerkin divergence-free method. Results show a good agreement with the analytical curve for different wave modes.

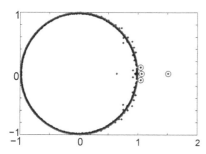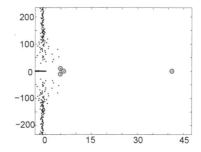

Fig. 3. Exponential transformation of the Jacobi matrix from the unstable solution of a 3D Kolmogorov flow problem with four LM and LR leading eigenvalues found by the IRA Exponential + Leja points algorithm. A zoom of found eigenvalues and exact spectra is given on the right

6 Parallel Efficiency

The two suggested strategies for operations with small dimensional arrays $Q, H, \mathbf{x}$ are tested on single-GPU calculation mode in our hardware. FLOPS count is performed for double precision using **nvprof** for the GPU part and **perf stat** for the CPU part as a function of the Krylov subspace dimension. All matrix-vector operations are executed using CUBLAS for GPUs and LAPACK for CPUs. The results of the LR Jacobi matrix analysis are shown in Fig. 4. Givens QR calculations are more efficient as the Krylov subspace dimension increases. Most of the performance drawbacks are due to low CPU FLOPS and **MPI_Isend,MPI_Irecv** during device-host transfers of the matrices H and Q. In practice, we take $m \leq 100$; therefore, we suggest using the Givens strategy in systems where the number of GPUs is much larger than that of CPUs. Performance improves for the calculation of exponential matrix or monodromy matrix analysis, since most of the computational time is spent on the highly efficient Arnoldi process. In these cases, the ratio of FLOPS drops to just below 2 for the maximum GPU loading.

For the Givens QR variant, we also investigate scaling on multiGPUs. We consider the Rayleigh–Benard convection problem described in [26] with Krylov subspace dimension 2: $m = 28$ and $m = 100$. The maximum size of the problem is scaled to fit in one GPU RAM. Results are shown in Fig. 5. The presented algorithm gives an acceleration close to linear. Scaling suffers when GPUs are not loaded; this, however, is a known fact. The increase in Krylov dimension reduces scalability only for small GPU loading; a normal GPU loading gives an irrelevant reduction of scalability. This is true, since for the Givens variant of implementation, it is almost perfectly localized. Thus, the influence of **MPI_Allreduce** is small for $m \leq 100$.

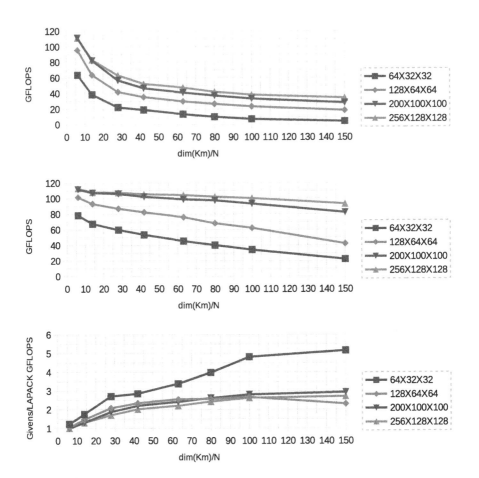

Fig. 4. Double precision FLOPS for LAPACK strategy (top) and Givens GPU strategy (middle), using the Jacobi matrix from a 3D Kolmogorov flow stability problem for different dimensions of the Krylov subspace

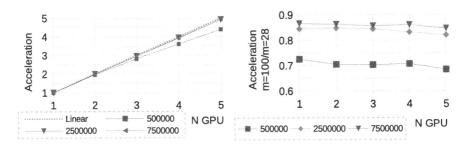

Fig. 5. Scaling on GPUs for $m = 28$, and ratios of accelerations for $m = 100$ to $m = 28$.

7   Application Examples

In this section, we demonstrate the application of the developed method to a bifurcation analysis in a 3D Kolmogorov flow problem described in [27]. In Fig. 6, we show the eigenvalues and eigenvectors for the supercritical pitchfork bifurcation of the first stationary bifurcation for Reynolds number $R = 5.3732$, and the monodromy matrix for the fold bifurcation with a period-3 cycle near $R = 11.1002$. The latter can be traced by the real eigenvalue passing through the point $(-1, 0i)$ on $\mathbb{C}^1$. By applying the IRA method, we were able to confirm the existence of a period-3 cycle and determine the values of the bifurcation parameters more precisely.

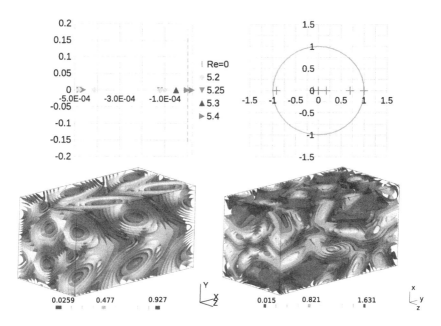

Fig. 6. Eigenvalues and leading eigenvector absolute values for the supercritical pitchfork bifurcation when $R = 5.2$–5.4 (on the left), and for the monodromy matrix when $R = 11.1$ with a period-3 cycle (on the right)

8   Conclusions

In this paper, we presented the modifications and implementation of the IRA method for CPU/multiGPU architecture adopted for fluid dynamics problems. These problems usually have nonsymmetric real matrices causing difficulties for the application of the IRA method. We tested different polynomial filters and matrix exponential calculation with the purpose of accelerating the convergence of the IRA method in case of a matrix-free approach. We successfully applied the Fekete or Leja-points strategy to the filtering polynomial.

We performed a benchmark analysis of the method for various nonsymmetric real matrices from the Matrix Collection and fluid dynamics stability problems. It was found that the use of the exponential matrix is essential for the convergence of the method in some test cases. It is desirable that the implementation of the proposed method takes GPU architectural features into consideration, namely small calculations (QR algorithm and eigenvalues of a Hessenberg matrix) must be performed locally on every GPU by the efficient Givens rotations method. This saves time for **MPI_Isend/recv** and decreases CPU use, which had a low efficiency in our hardware setup. But this strategy is only efficient if many GPUs are located in a single host chassis with one or two CPUs. In CPU-dominating architectures, this approach is not recommended, and increasing CPU use by calling LAPACK routines is suggested. However, the balance must be tested, since MPI operations may become a bottleneck in this case.

The use of MPI for the IRA method on multiple GPUs is efficient, and scalability does not decrease as the Krylov subspace dimension increases. This is achieved by a good algorithm decomposition in problem dimension (N in our case), which is natural for parallel algorithms in fluid dynamics problems, and the use of Givens rotations in local on-device calculations for small arrays.

The application of the method was demonstrated in the last section of the article. We emphasize that the method successfully detects unstable modes and allows for the determination of bifurcation points. We intend to apply the method to the problem of turbulent flow control. After subsequent testing and optimization, the method will be freely available from the author by request, under the terms of **GPLv3** license.

References

1. Saad, Y.: Numerical Methods for Large Eigenvalue Problems, Revised Edition (Classics in Applied Mathematics). SIAM, Philadelphia (2011). doi:10.1137/1.9781611970739
2. Lehoucq, R., Maschhoff, K.: Sorensen, D., Yang, C., ARPACK software, 1996–2016. http://www.caam.rice.edu/software/ARPACK
3. Sorensen, D.C.: Implicitly Restarted Arnoldi/Lanczos Methods For Large Scale Eigenvalue Calculations. NASA Contractor Report 198342 (1996). doi:10.1007/978-94-011-5412-3_5
4. Govaerts, W.J.F.: Numerical Methods for Bifurcations of Dynamical Equilibria. SIAM, Philadelphia (2000). doi:10.1137/1.9780898719543
5. Ericsson, T., Ruhe, A.: The spectral transformation Lanczos method for the numerical solution of large sparse generalized symmetric eigenvalue problems. Math. Comput. **35**, 1251–1268 (1980). doi:10.2307/2006390
6. Lehoucq, R.B., Scott, J.A.: Implicitly restarted Arnoldi methods and eigenvalues of the discretized Navier-Stokes equations. SIAM J. Matrix Anal. Appl. **23**, 551–562 (1997). doi:10.1137/s0895479899358595

7. Cliffe, K.A., Hall, E.J.C., Houston, P.: Adaptive discontinuous Galerkin methods for eigenvalue problems arising in incompressible fluid flows. SIAM J. Sci. Comput. **31**(6), 4607–4632 (2010). doi:10.1137/080731918

8. Burroughs, E.A., Romero, L.A., Lehoucq, R.B., Salinger, A.G.: Large scale eigenvalue calculations for computing the stability of buoyancy driven flows. Sandia Report SAND2001-0113 Unlimited Release (2001). doi:10.2172/782594

9. Henningson, D.S., Akervik, E.: The use of global modes to understand transition and perform flow control. Phys. Fluids **20**, 031302 (2008). doi:10.1063/1.2832773

10. Garnaud, X., Lesshafft, L., Schmid, P., Chomaz, J.-M.: A relaxation method for large eigenvalue problems, with an application to flow stability analysis. J. Comput. Phys. **231**, 3912–3927 (2012). doi:10.1016/j.jcp.2012.01.038

11. Tiesinga, G., Wubs, F.W., Veldman, A.E.P.: Bifurcation analysis of incompressible flow in a driven cavity by the Newton - Picard method. J. Comput. Appl. Math. **140**(1–2), 751–772 (2002). doi:10.1016/s0377-0427(01)00515-5

12. Barkley, D., Henderson, R.D.: Three-dimensional Floquet stability analysis of the wake of a circular cylinder. J. Fluid Mech. **322**, 215–241 (1996). doi:10.1017/s0022112096002777

13. Bres, G.A., Colonius, T.: Three-dimensional linear stability analysis of cavity flows. In: AIAA 2007–1126 45th AIAA Aerospace Sciences Meeting and Exhibit, 8–11 January 2007, Reno, Nevada (2007). doi:10.2514/6.2007-1126

14. Nebauer, J.R.A., Blackburn, H.M.: Floquet stability of time periodic pipe flow. In: Proceedings of International Conference on CFD in the Minerals and Process Industries, CSIRO, Melbourne, Australia, 10–12 December 2012

15. Baranyi, L., Darczy, L.: Floquet stability analysis of the wake of a circular cylinder in low Reynolds number flow. In: Proceedings of XXVI Conference in MicroCAD, University of Miskolc, Hungary (2012)

16. Dubois, J., Calvin, C., Petiton, S.G.: Accelerating The explicitly restarted arnoldi method with GPUs using an autotuned matrix vector product. SIAM J. Sci. Comput. **33**(5), 3010–3019 (2011). doi:10.1137/10079906x

17. Rantalaiho, T., Weir, W., Suorsa, J.: Parallel multi-CPU/GPU(CUDA)-implementation of the Implicitly Restarted Arnoldi Method. https://github.com/trantalaiho/Cuda-Arnoldi

18. Bujanovic', Z.: Krylov Type Methods for Large Scale Eigenvalue Computations. Ph.D. thesis, Zagreb (2011)

19. Embree, M.: The Arnoldi eigenvalue iteration with exact shifts can fail. SIAM J. Matrix Anal. Appl. **31**(1), 110 (2009). doi:10.1137/060669097

20. Breue, A.: New filtering strategies for implicitly restarted Lanczos iteration. Electron. Trans. Numer. Anal. **45**, 1632 (2016)

21. Fitzgibbon, A.W., Pilum, M., Fisher, R.B.: Direct least-squares fitting of ellipses. IEEE Trans. Pattern Anal. Mach. Intell. **21**, 476–480 (1996). doi:10.1109/34.765658

22. Bos, L., De Marchi, S., Sommariva, A., Vianello, M.: Computing multivariate Fekete and Leja points by numerical linear algebra. SIAM J. Numer. Anal. **48**(5), 1984–1999 (2010). doi:10.1137/090779024

23. Akervik, E., Brandt, L., Henningson, D.S., Hapffner, J., Marxen, O., Schlatter, P.: Steady solutions of the Navier-Stokes equations by selective frequency damping. Phys. Fluids **18**, 068102 (2006). doi:10.1063/1.2211705

24. SuiteSparse Matrix Collection. http://www.cise.ufl.edu/research/sparse/matrices/

25. Evstigneev, N.M., Magnitskii, N.A., Silaev, D.A.: Qualitative analysis of dynamics in Kolmogorovs problem on a flow of a viscous incompressible fluid. Differ. Equ. **51**(10), 1292–1305 (2015). doi:10.1134/s0012266115100055

26. Evstigneev, N.M.: Laminar-turbulent bifurcation scenario in 3D Rayleigh-Benard convection problem. Open J. Fluid Dyn. **6**, 496–539 (2016). doi:10.4236/ojfd.2016. 64035
27. Evstigneev, N.M., Magnitskii, N.A.: Nonlinear dynamics of laminar-turbulent transition in the generalized 3D Kolmogorov problem for the incompressible viscous fluid at symmetric solution subset. J. Appl. Nonlinear Dyn. (2017, accepted, to appear)

High-Performance BEM Simulation of 3D Emulsion Flow

Olga A. Abramova[1(✉)], Yulia A. Pityuk[1], Nail A. Gumerov[1,2],
and Iskander S. Akhatov[3]

[1] Center for Micro and Nanoscale Dynamics of Dispersed Systems, Bashkir State
University, Ufa, Russia
abramovacmndds@gmail.com
[2] Institute for Advanced Computer Studies, University of Maryland,
College Park, USA
[3] Skolkovo Institute of Science and Engineering (Skoltech), Moscow, Russia

Abstract. Direct simulations of the dynamics of a large number of
deformable droplets are necessary for a more accurate prediction of rhe-
ological properties and the microstructure of liquid-liquid systems that
arise in a wide range of industrial applications, such as enhanced oil
recovery, advanced material processing, and biotechnology. The present
study is dedicated to the development of efficient computational methods
and tools for understanding the behavior of three dimensional emulsion
flows in Stokes regime. The numerical approach is based on the accelerated boundary element method (BEM) both via an efficient scalable
algorithm, the fast multipole method (FMM), and via the utilization of
advanced hardware, particularly, heterogeneous computing architecture
(multicore CPUs and graphics processors). Example computations are
conducted for 3D dynamics of systems of tens of thousands of deformable
drops and several droplets with very high discretization of the interface in shear flow. The results of simulations and details of the method
and accuracy/performance of the algorithm are discussed. The developed
approach can be used for the solution of a wide range of problems related
to emulsion flows in micro- and nanoscales.

Keywords: Deformable droplets · Stokes flow · Boundary element
method · Fast multipole method · Graphics processors

1 Introduction

Detailed studies of the emulsion dynamics are very important for science and
many new technologies. Such dispersed systems appear in a wide range of industrial applications including crude-oil recovery, food processing, pharmaceutical

This study is supported in part by the Skoltech Partnership Program, grant RFBR
16-31-00029, grant of the President of Russia MK-3503.2017.8, and Fantalgo, LLC
(Maryland, USA).

© Springer International Publishing AG 2017
L. Sokolinsky and M. Zymbler (Eds.): PCT 2017, CCIS 753, pp. 317–330, 2017.
DOI: 10.1007/978-3-319-67035-5_23

manufacturing, biotechnology, etc. For many years progress in studying emulsion behavior was mainly empirical with just several examples of large-scale direct simulations (e.g. [25]). Nowadays, modern computational methods and powerful computer resources enable fast large-scale microfluid dynamics simulations, which makes them a valuable research tool. The main goal of this work is the development of efficient computational methods and tools for the investigation of the complex behavior of a three dimensional emulsion flow at low Reynolds numbers. In this study, we consider the 3D dynamics of deformable drops of an immiscible liquid in another liquid under imposed shear flow. The simulations are performed using the boundary element method (BEM) accelerated both via advanced scalable algorithms, particularly, the fast multipole method (FMM), and via the utilization of advanced hardware, particularly, graphics processors (GPUs) and multicore CPUs.

The BEM has been successfully applied for simulation of Stokes flows ([26]; see general overview and details of the BEM in the monograph of Pozrikidis [14]), but its application to simulation of large non-periodic systems is very limited. The major computational challenge is related to the solution of a large dense system of $3N$ algebraic equations for each time step, where N is the number of the surface discretization points. For systems of thousands of drops with hundreds of boundary elements per drop, N reaches a value of the order of millions. The direct methods of solution of algebraic systems, having computational complexity $O\left(N^3\right)$, become impractical. The use of efficient iterative methods reduces this complexity to $O\left(N_{\text{iter}}N^2\right)$, where $N_{\text{iter}} \ll N$ is the number of iterations, and $O\left(N^2\right)$ is the cost of a single matrix-vector product (MVP). For large N, this is not fast enough even when using high-performance computing, since the computational cost increases proportionally to the square of N. Thus, the key here is the application of fast algorithms for MVP. The use of the FMM for MVP reduces the computational complexity of the overall problem to $O\left(N_{\text{iter}}N\right)$ per time step, and potentially can handle direct simulations of large droplet systems with millions of boundary elements.

The FMM was first introduced by Rokhlin and Greengard [5] as a method for fast summation of electrostatic or gravitational potentials in two and three dimensions. The first application of the FMM for the solution of Stokes equations was reported by Sangani and Mo [19], who developed the method in the case of spherical rigid particles. The FMM accelerated BEM for domains of arbitrary shape was developed by Wang et al. [22]. In the work [24], the authors achieved substantial accelerations of droplet dynamics simulations via the use of multipole expansions and translation operators, which is very much in the spirit of the FMM, and can be considered as one- and two-level FMM. However, the $O\left(N\right)$ scalability of the FMM can be achieved only on hierarchical (multilevel) data structures, which were not implemented there.

Note that the FMM can be efficiently parallelized. The first implementation of the FMM on graphics processors was reported by Gumerov and Duraiswami [6], who showed that the use of a GPU for the FMM for the Laplace's equation in 3D can produce 30 to 60-fold accelerations, and achieved a time of 1 s for the MVP in a case of size 1 million on a single GPU. This approach was developed further and recent papers by Hu et al. [8,9] present scalable FMM algorithms and implementations of heterogeneous computing architectures that may contain many distributed nodes, each of them consisting of one multicore CPU and several GPUs. They performed the MVP for a system of size 1 billion achieving a time of about 10 s on a 32-node cluster, and also for systems of size 1 million in a time of the order of several tenths of a second on a single hetero-geneous CPU/GPU node. In the present study, we use a single heterogeneous node with one GPU and apply CPU/GPU parallelism. This realization enables computations in the case of 15 thousand dynamic droplets (more than 7 millions of unknowns) in a reasonable time.

2   Statement of the Problem

2.1   Governing Equations

The dynamics of deformable drops (fluid 2) of an immiscible liquid in another liquid (fluid 1) at low Reynolds numbers can be described by the Stokes equations for the motion of each fluid [7],

$$\nabla \cdot \boldsymbol{\sigma}_i = -\nabla p_i + \mu_i \nabla^2 \mathbf{u}_i = \mathbf{0}, \quad \nabla \cdot \mathbf{u}_i = 0, \tag{1}$$

where $\mathbf{u}$ and $\boldsymbol{\sigma}$ are the velocity and the stress tensors, μ is the dynamic viscosity, and p is the pressure, which includes the hydrostatic component, $i = 1, 2$. At the fluid-fluid interface (S), the boundary conditions for the velocity $\mathbf{u}$ and the traction $\mathbf{f}$ are

$$\mathbf{u}_1 = \mathbf{u}_2 = \mathbf{u}, \qquad \mathbf{f} = \boldsymbol{\sigma}_1 \cdot \mathbf{n}_1 - \boldsymbol{\sigma}_2 \cdot \mathbf{n}_2 = \mathbf{f}_1 - \mathbf{f}_2 = f\mathbf{n},$$

$$f = \gamma(\nabla \cdot \mathbf{n}) + (\rho_1 - \rho_2)(\mathbf{g} \cdot \mathbf{x}), \quad \mathbf{x} \in S, \tag{2}$$

where $\mathbf{n}$ is the normal to S pointing into fluid 1, ρ, γ and $\mathbf{g}$ are the density, the surface tension and the gravity acceleration, respectively. In the case of infinite domains, the condition $\mathbf{u}_1(\mathbf{x}) \rightarrow \mathbf{u}_\infty(\mathbf{x})$ should be imposed on the carrier fluid, where $\mathbf{u}_\infty(\mathbf{x})$ is a solution of the Stokes equations.

The dynamics of the fluid-fluid interface can be determined from the kine-matic condition

$$\frac{d\mathbf{x}}{dt} = \mathbf{u}(\mathbf{x}), \quad \mathbf{x} \in S, \tag{3}$$

where $\mathbf{u}(\mathbf{x})$ is the interface velocity determined from the solution of the elliptic boundary value problem stated above. Although the governing equations and boundary conditions are linear with respect to $\mathbf{u}$ and $\mathbf{f}$, the dynamics of the interface is a non-linear problem since $\mathbf{u}(\mathbf{x})$ depends on all the points of the surface.

2.2 Boundary-Integral Formulation

The problem is solved using the boundary element method, which is based on the integral equations for the determination of the velocity distribution over the boundary. Further, knowing this distribution, one can find the velocity field at any spatial point of the computational domain.

Then the boundary integral equations for the volume of fluid occupying the domain V bounded by the surface S can be written in the form

$$
\begin{aligned}
\mathbf{u}(\mathbf{y}) - \int_{S} \mathbf{K}(\mathbf{y}, \mathbf{x}) \cdot \mathbf{u}(\mathbf{x})\, dS(\mathbf{x}) &= \frac{1}{\mu} \int_{S} \mathbf{G}(\mathbf{y}, \mathbf{x}) \cdot \mathbf{f}(\mathbf{x})\, dS(\mathbf{x}), \quad \mathbf{y} \in V, \\
\frac{1}{2}\mathbf{u}(\mathbf{y}) - \int_{S} \mathbf{K}(\mathbf{y}, \mathbf{x}) \cdot \mathbf{u}(\mathbf{x})\, dS(\mathbf{x}) &= \frac{1}{\mu} \int_{S} \mathbf{G}(\mathbf{y}, \mathbf{x}) \cdot \mathbf{f}(\mathbf{x})\, dS(\mathbf{x}), \quad \mathbf{y} \in S, \quad (4) \\
- \int_{S} \mathbf{K}(\mathbf{y}, \mathbf{x}) \cdot \mathbf{u}(\mathbf{x})\, dS(\mathbf{x}) &= \frac{1}{\mu} \int_{S} \mathbf{G}(\mathbf{y}, \mathbf{x}) \cdot \mathbf{f}(\mathbf{x})\, dS(\mathbf{x}), \quad \mathbf{y} \notin S, V,
\end{aligned}
$$

where the fundamental solutions (Stokeslet and stresslet) are given by the second and third rank tensors

$$
\begin{aligned}
\mathbf{G}(\mathbf{y}, \mathbf{x}) &= \frac{1}{8\pi}\left(\frac{\mathbf{I}}{r} + \frac{\mathbf{rr}}{r^3}\right), \quad \mathbf{T}(\mathbf{y}, \mathbf{x}) = -\frac{3}{4\pi}\frac{\mathbf{rrr}}{r^5}, \\
\mathbf{K}(\mathbf{y}, \mathbf{x}) &= \mathbf{T}(\mathbf{y}, \mathbf{x}) \cdot \mathbf{n}(\mathbf{x}), \quad \mathbf{r} = \mathbf{y} - \mathbf{x}, \quad r = |\mathbf{r}|,
\end{aligned}
\quad (5)
$$

and $\mathbf{I}$ is the identity tensor. Thus, if $\mathbf{u}$ and $\mathbf{f}$ are known on the boundaries, the velocity field $\mathbf{u}(\mathbf{y})$ can be determined at the any spatial point.

The problem describes a flow in an unbounded domain. In this case, $\mathbf{u}_\infty$ is prescribed (e.g., a mean shear flow $\mathbf{u}_\infty = (Gz, 0, 0)$, G is a shear rate), and the boundary integral equation combining formulations inside and outside the drops can be written in the form (see [15, 23])

$$
\left.\begin{aligned}
\mathbf{y} \in V_1, & \quad \mathbf{u}(\mathbf{y}) - \mathbf{u}_\infty(\mathbf{y}) \\
\mathbf{y} \in V_2, & \quad \lambda \mathbf{u}(\mathbf{y}) - \mathbf{u}_\infty(\mathbf{y}) \\
\mathbf{y} \in S, & \quad \frac{1+\lambda}{2}\mathbf{u}(\mathbf{y}) - \mathbf{u}_\infty(\mathbf{y})
\end{aligned}\right\}
\quad (6)
$$
$$
= \int_{S} \left\{ -\frac{1}{\mu_1}\mathbf{G}(\mathbf{y}, \mathbf{x}) \cdot \mathbf{f}(\mathbf{x}) + (\lambda - 1)\mathbf{K}(\mathbf{y}, \mathbf{x}) \cdot \mathbf{u}(\mathbf{x}) \right\} dS(\mathbf{x}),
$$

where $\lambda = \mu_2/\mu_1$ is the viscosity ratio of the internal and external liquids, V_1 and V_2 are the domains occupied by the inner and outer fluids, respectively, and $\mathbf{f}(\mathbf{x})$ is determined according to Eq. (2). Hence, in this case, $\mathbf{f}(\mathbf{x})$ is completely determined by the surface, while $\mathbf{u}(\mathbf{x})$ can be found from the solution of the boundary integral equation (for points $\mathbf{y} \in S$), which is a Fredholm equation of the second kind with a singular kernel.

3   Numerical Technique

The numerical method is based on the discretization of droplet surfaces by triangular meshes. For accurate computation of boundary integrals and surface properties, such as the curvature, the mesh should be of a good quality, and for the method used for computation of the curvature, the valency of the mesh vertices should be not less than 5 (we used a slightly modified method of fitted paraboloid [26], which as we found provides good accuracy (no more than a few percent for the "worst" nodes)). The regular integrals over the patches were computed using second order accuracy formulae (trapezoidal rules). The collocation points for a droplet surface were located at the mesh vertices. The computation of singular integrals was performed based on the integral identities for the Stokeslet and stresslet integrals (see [1,14,15,26]), which allow to express these integrals via sums of regular integrals over the rest of the surface. Note that this can be done efficiently when using the FMM, since these sums are nothing but matrix-vector products (e.g., for problems of flow in infinite domains, this required four additional FMM runs per time step).

The boundary integral equations combined with the boundary conditions in discrete form result in a system of linear algebraic equations (SLAE) for each time step,

$$\mathbf{A}\mathbf{X} = \mathbf{b}, \tag{7}$$

where $\mathbf{A}$ is the system matrix, $\mathbf{X}$ is the solution vector, and $\mathbf{b}$ is the right-hand-side vector.

In the present study, the kinematic condition (3) is modified for mesh stabilization in the following way

$$\frac{d\mathbf{x}}{dt} = \mathbf{u}(\mathbf{x}) + \mathbf{w}(\mathbf{x}), \quad \mathbf{w} \cdot \mathbf{n} = 0, \quad \mathbf{x} \in S, \tag{8}$$

where $\mathbf{u}(\mathbf{x})$ is the velocity of the interface determined from the solution of the boundary integral equation via Eq. (6), and $\mathbf{w}(\mathbf{x})$ is a locally defined tangential velocity field that maintains the desired distribution of the point $\mathbf{x}^i$ on the drop surfaces [10]. A formula for the correction $\mathbf{w}(\mathbf{x})$ was proposed in [13]. In a simplified form, it was successfully applied in [10],

$$\mathbf{w}^i = \frac{N_{\Delta}^{\frac{3}{2}}}{300(1+\lambda)} (\mathbf{I} - \mathbf{n}_i\mathbf{n}_i) \sum_j \left(1 + |2k_j|^{\frac{3}{2}}\right) \Delta S_j \left(\mathbf{x}^j - \mathbf{x}^i\right), \tag{9}$$

where N_{Δ} is the number of points on the drop surface; k_j and ΔS_j are, respectively, the mean curvature and surface area associated with the point $\mathbf{x}^j$; $\mathbf{I}$ is the second-rank identity tensor. The summation extends over all the vertices $\mathbf{x}^j$ connected directly to $\mathbf{x}^i$, $i = 1, \ldots N_{\Delta}$.

For the integration of Eq. (8), we used the Adams–Bashforth–Moulton predictor-corrector scheme of the sixth order. This scheme requires two calls of the right-hand-side function per time step (therefore, Eq. (8) was solved two times per time step). It also requires an initialization, which was provided by a fourth order Runge–Kutta scheme. This requires more calls of the right-hand-side function, but it was needed just in a few initial steps.

The time step used in the explicit schemes should be sufficiently small. It was selected according to the stability condition for the drops, [23],

$$\Delta t \leq K \frac{\mu_1 \Delta x_{\min}}{\gamma}, \tag{10}$$

where $\Delta x_{\min}$ is the minimum length scale appearing in the computational problem, and $K = O(1)$ is some dimensionless factor.

4   FMM/GPU Accelerated BEM

Direct methods can be used for the solution of the SLAE (Eq. (7)) when the number of drops is small. However, as the problem size increases, these methods become impractical. This is due to the fact that the memory required for computations increases proportionally to the square of the number of mesh vertices. Furthermore, the computation time increases substantially, making computations impossible for systems with a number of drops $M \geq 10^2$. This issue can be fixed by applying iterative methods, which reduce time and memory costs significantly. The Krylov subspace projection methods are among the most efficient for the solution of SLAE. In this paper, we used the unpreconditioned general minimal residual method (GMRES) to solve the system [18]. In the computational tests, 5 to 7 iterations were usually sufficient to converge to a relative residual error $\sim 10^{-6}$. The number of iterations varies from step to step, since it depends on the distribution of the droplets in the system, but never exceeded 10 per solution.

In this method, the matrix-vector product (MVP) must be computed at each iteration, which presents the main computational complexity. The matrix-vector product specific for the present problem can be computed using the FMM [5]. The main advantage of this method is that the complexity of the MVP involving a certain type of dense matrices can be reduced from $O(N^2)$ to $O(N \log(N))$, or even $O(N)$. In this paper, we used the implementation technique of the FMM for the summation of the Stokeslets and stresslets (Eq. (5)) proposed in [20,22]. This is valid for a particular case of summation of the velocity fields of Stokeslets and stresslets, but is sufficient to compute the matrix-vector products for the BEM formulation used here. This approach is based on the summation of the fundamental solutions of the three-dimensional Laplace's equation. More details on the applied factorization of the respective velocity field can be found in [2].

In the present approach, GPU acceleration is used in the so-called heterogeneous FMM algorithm. The FMM is based on the formal decomposition of the matrix $\mathbf{A}$ in the MVP $\mathbf{AX}$ as

$$\mathbf{A} = \mathbf{A}_{\text{sparse}} + \mathbf{A}_{\text{dense}}, \tag{11}$$

where $\mathbf{A}_{\text{sparse}}$ is a sparse matrix that takes into account the interaction of particles (mesh nodes) residing in the neighbour boxes of the finest level of the octree data hierarchy, and $\mathbf{A}_{\text{dense}}$ is the dense matrix accounting for all other

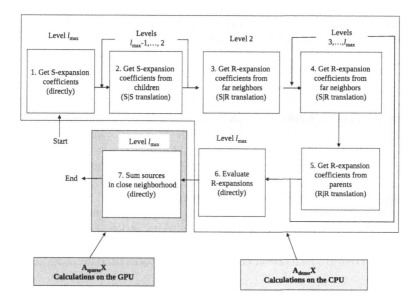

Fig. 1. Flow-chart of the heterogeneous FMM

interactions. The MVP $\mathbf{A}_{\text{sparse}}\mathbf{X}$ is performed directly, whereas $\mathbf{A}_{\text{dense}}\mathbf{X}$ is computed using generation of the multipole expansions for each source box at the finest level, multipole-to-multipole, multipole-to-local and local-to-local translation operators enabling both expansions for boxes of the coarser levels and evaluation of the local expansions. The flow-chart of the heterogeneous FMM is represented in Fig. 1. The features of the algorithm and implementations are presented in [2,6].

It was found in [6] that, owing to peculiarities of the GPU architecture, the MVP $\mathbf{A}_{\text{sparse}}\mathbf{X}$ can be performed very efficiently using the GPU (up to a 100-fold effective acceleration), while the GPU used for $\mathbf{A}_{\text{dense}}\mathbf{X}$ can be accelerated just a few times compared to a single core CPU. The latter effect is due to a relatively complicated data structure, which requires, first, an extensive random access to the GPU global memory (very expensive operations) and, second, limited local GPU memory, which is not sufficient for efficient storage of the translation operators. On the other hand, the availability of CPUs with P cores, such as those used in the present studies, enables relatively easy parallelization of the MVP $\mathbf{A}_{\text{dense}}\mathbf{X}$ via the Open MP. The efficiency of such parallelization is close to 100%, which means about a P-fold acceleration. It can be noticed that the actual acceleration of the total algorithm is much more than P-fold, owing to the fact that larger clusters of sources can be processed by the GPU more efficiently than on the CPU, resulting in a reduction of the depth of the octree and additional acceleration [6]. A careful tuning of the algorithm and the octree depth is based on the work load balance between the CPU and GPU. Using such an heterogeneous algorithm on a system with 8 to 12 CPU cores and one

GPU, the overall acceleration can rise by approximately 100 times compared to its value for a single-core CPU implementation.

Such a scheme was first proposed and tried in [8], where, additionally, the computation of the FMM data structure was performed on GPU with a new highly efficient algorithm, which provided up to a 100-fold acceleration of this step compared to the CPU realization. Such an acceleration is important in dynamic problems, especially in the case when the run part of the FMM is called one or two times for the same data structure. But in the present work, such a modification of the data structure algorithm is not necessary. Indeed, owing to the iterative process, the computation of the singular elements, and 3 to 4 calls of the FMM for the Laplace's equation to provide one call of the FMM for the Stokes equation, the number of "elementary" FMM calls on the same data structure can be 50 or so for a given source configuration (a fixed matrix $\mathbf{A}$). Hence, computation of the FMM data structure for a given mesh is amortized 50 times or so over the MVP computation, and the relative cost of the data structure generation compared to the total time becomes just a few percent.

The implemented FMM routine was tested in terms of accuracy and performance. The relative errors in the FMM MVP were of the order of 10^{-5} for the multipole series truncation number $p = 8$ (which means that $p^2 = 64$ terms in multipole expansions were used to represent the far fields of the singularities of the Laplace's equation). Numerical tests show that this is sufficient to solve the present problem. The speed of the FMM heavily depends on p, and perhaps a smaller p, which provides larger errors, can be used for the present problem. Indeed, all the errors of the numerical method, which include geometrical discretization errors that can be percent or fractions of percent, errors of quadratures, GMRES tolerance and errors of the time integrators, should be considered together, and more careful analysis is required to tune the entire method.

5   Numerical Results and Discussion

Calculations are performed on a workstation equipped with two Intel Xeon 5660, 2.8 GHz CPUs (12 physical + 12 virtual cores), 12 GB RAM, and one NVIDIA Tesla K20 GPU (5 GB of global memory). Several algorithm implementations were done, including CPU and CPU/GPU versions of the iterative algorithm with the FMM accelerated MVP and a conventional BEM in which the BEM matrices were computed and stored, and Eq. (7) was solved using standard general purpose linear algebra libraries providing $O\left(N^3\right)$ computational complexity. The latter implementation was developed for verification and validation purposes to ensure that the algorithms produce consistent results.

5.1   Validation Tests

The implemented methods were tested for the case of one fixed spherical droplet in an unbounded flow. The obtained results were compared against the analytical solution for a droplet in a Stokes flow [7]. The relative error in the L_∞-norm was defined as $\delta(\mathbf{u}^*) = \|\mathbf{u}^* - \mathbf{u}\|_\infty / \|\mathbf{u}\|_\infty$, where $\|\mathbf{u}\|_\infty = \max|u_i|$, $\mathbf{u}^*$ is the numerical result, $\mathbf{u}$ is the analytical solution, $i = 1, ..., N_\Delta$. Here the relative error was about 1 to 1.5% for $N_\Delta = 642$, $\lambda = 2.5$, and the error decreases as the number of mesh points increases.

The calculations for an isolated drop in a shear flow at several viscosity ratios and capillary numbers $\mathrm{Ca} = \mu_1 aG/\gamma$, where a is the undeformed droplet radius, were also compared against the experimental results of Torza et al. [21], Rumscheidt and Mason [17], small-deformation computations by Cox [3], and numerical results of Rallison [16], and Kennedy et al. [11]. Usually, the inclination angle α and the deformation $D = (c-b)/(c+b)$ are used to describe the changes of the droplet form, where c and b are the maximum and minimum distances from the center of the droplet to its surface, and α is the angle between $\mathbf{u}_\infty$ and the direction of the maximum droplet elongation. Figure 2 shows the steady drop deformation D and the inclination angle α as functions of Ca for $\lambda = 3.6$. The results obtained at small values of Ca for the deformation are in excellent agreement with the small-deformation theory [3]. Also, for each Ca, our results for both the drop inclination angle and the deformation are in good agreement with numerical computations [11,16] and experimental data [17,21].

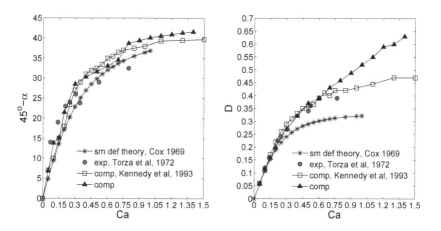

Fig. 2. Steady drop deformation D and $45° - \alpha$ as functions of Ca, $\lambda = 3.6$

5.2 Large Scale Numerical Tests

We conducted simulations of the motion of two close droplets in a shear flow $(Gz, 0, 0)$ with a high mesh resolution of the surfaces. Similar cases for other physical parameters have been considered in various works [4,12,13,24,26]. Furthermore, in these works the maximum number of mesh points on the droplet surface did not exceed $N_\Delta = 138\,240$. In the present work, we considered the motion of two initially spherical droplets with equal radii and number of mesh

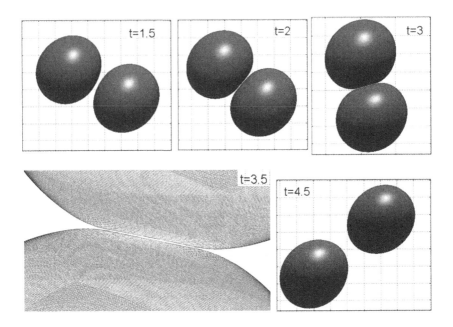

Fig. 3. Dynamics of two droplets in a shear flow, Ca = 0.05, $\lambda = 1$

points $N_\Delta = 163\,842$ (which corresponds to $327\,680$ triangular elements). In this case, Ca = 0.05, $\lambda = 1$, the distance between droplets centers was $\Delta x = 2.4a$, $\Delta y = 0$, and $\Delta z = 0.7a$. The integration time step Δt was chosen according to the stability condition (10) and was varied depending on droplet distribution.

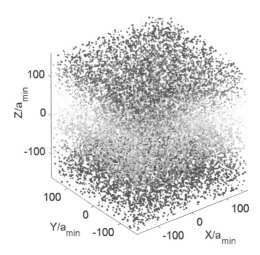

Fig. 4. A snapshot of the initial spatial distribution of 15342 drops

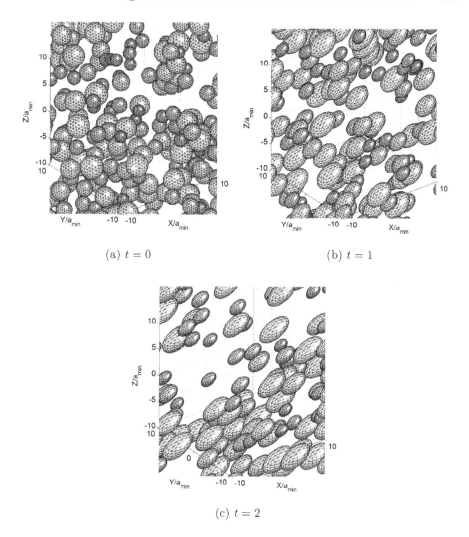

(a) $t = 0$

(b) $t = 1$

(c) $t = 2$

Fig. 5. 15 342 drops in a shear flow, $N_\Delta = 162$, $0.25 \leq \text{Ca} \leq 0.5$, $\lambda = 1.5$, $\alpha = 8.8 \cdot 10^{-3}$ (a fragment of the computational domain)

At the initial moment, $\Delta t = 5 \cdot 10^{-4}$, then, as droplets entered the near-contact zone, the time step decreased to $\Delta t = 2 \cdot 10^{-5}$.

Figure 3 shows the droplet dynamics at different non-dimensional times. Note that droplets affect each other in the time interval $2 \leq t \leq 4$. The minimum distance between droplet surfaces was $h_{\min} \sim 10^{-2} \cdot a - 10^{-3} \cdot a$.

Test computations were also conducted for a polydisperse emulsion with initially spherical drops and randomly uniform spatial distributions. The droplet radii were varied from $a_{\min}$ to $a_{\max}$, $a_{\max}/a_{\min} = 2$, $0.25 \leq \text{Ca} \leq 0.5$, and $\lambda = 1.5$. The results are presented for the non-dimensional time $t = t_{\text{nondim}} = \gamma t_{\text{dim}}/(\mu_1 a)$, where $a = a_{\min}$ is the minimal initial droplet size.

Figure 4 shows the initial distribution of $M = 15\,342$ spherical drops in a cube with edge size $300a_{min}$ and volume fraction $\alpha = 8.8 \times 10^{-3}$. Each drop was discretized by means of a triangular mesh with $N_\Delta = 162$ nodes. Thus, the total number of computational points in this case was $N = M \times N_\Delta = 2\,485\,404$. Note that this leads to the solution of a linear system with $\sim 7.5 \cdot 10^6$ independent variables for each time step. Thus, in the conventional BEM, the memory required to store the system matrix was ~ 445 TB.

Fragments of the computational domain $(-10a_{min}; 10a_{min})^3$ are shown in Fig. 5 for the case of 15 342 droplets at dimensionless time $t = 0$, $t = 1$, $t = 2$. The integration time step $\Delta t = 0.01$, which corresponds to $K \ll 1$ in Eq. (10). During the simulation time, the drops elongate downstream and the effect of gravity is not significant in this case. Here the average angle and deformation are the same as for single-drop computations, and are consistent with the respective experimental data and computational results. The run time for one FMM call with 15 342 droplets was about 7 s; for one time step, it was 4 min, and for 100 time steps, about 7 h.

In Fig. 6, the computational wall-clock times for one FMM call, for one time step and for one hundred time steps are shown as functions of the problem size. It can be seen that these functions are close to linear. This allows one to estimate the computational times for larger scale problems.

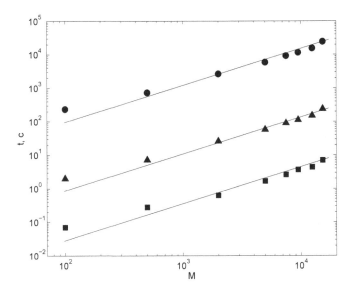

Fig. 6. Wall-clock time: one matrix-vector product (■), one time step (▲), 100 time steps (●)

6   Conclusions

In this paper, we developed and tested an approach based on the boundary element method for 3D problems, accelerated both via an advanced scalable algorithm (FMM) and via the utilization of a heterogeneous computing architecture (multicore CPUs and graphics processors). The verification and validation of the developed codes showed that the algorithms produce consistent results. Example computations were conducted for 3D dynamics of systems of tens of thousands of deformable drops and systems of several droplets, with very high discretization of the interface in a shear flow. We also studied the accuracy/performance of the method. The results of this study show that the software developed by us enables direct simulation of droplet systems with dynamic deformable interfaces discretized by millions of boundary elements on personal supercomputers. The presented algorithms can be mapped onto heterogeneous computing clusters [9], which should both accelerate computations and allow for the treatment of larger systems. The developed method can become a valuable research tool for investigations into the dynamics of emulsions.

References

1. Abramova, O.A., Itkulova, Y.A., Gumerov, N.A.: Modelirovanie trehmernogo dvijeniya deformiruemih kapel v stoksovom regime metodom granichnih elementov. Comput. Contin. Mech. **6**(2), 214–223 (2013)
2. Abramova, O.A., Itkulova, Y.A., Gumerov, N.A., Akhatov, I.S.: Three-dimensional simulation of the dynamics of deformable droplets of an emulsion using the boundary element method and the fast multipole method on heterogeneous computing systems. Vychislitelnye Metody i Programmirovanie **14**, 438–450 (2013)
3. Cox, R.G.: The deformation of drop in general time-dependent fluid flow. J. Fluid Mech. **37**(3), 601–623 (1969). doi:10.1017/s0022112069000759
4. Cunha, F.R., Almeida, M.H.P., Loewenberg, M.: Direct numerical simulations of emulsion flows. J. Braz. Soc. Mech. Sci. Eng. **25**(1), 30–40 (2003). doi:10.1590/s1678-58782003000100005
5. Greengard, L., Rokhlin, V.: A fast algorithm for particle simulations. J. Comput. Phys. **73**, 325–348 (1987). doi:10.1006/jcph.1997.5706
6. Gumerov, N.A., Duraiswami, R.: Fast multipole methods on graphics processors. J. Comput. Phys. **227**(18), 8290–8313 (2008). doi:10.1016/j.jcp.2008.05.023
7. Happel, J., Brenner, H.: Low Reynolds Number Hydrodynamics. Springer, Dordrecht (1981). doi:10.1007/978-94-009-8352-6. 553 pages
8. Hu, Q., Gumerov, N.A., Duraiswami, R. Scalable fast multipole methods on distributed heterogeneous architectures. In: Proceedings of Supercomputing 2011, Seattle, Washington (2011). doi:10.1145/2063384.2063432
9. Hu, Q., Gumerov, N.A., Duraiswami, R.: Scalable distributed fast multipole methods. In: Proceedings of the 2012 IEEE 14th International Conference on High Performance Computing and Communications, UK, Liverpool, pp. 270–279 (2012). doi:10.1109/ipdpsw.2014.110

10. Itkulova, Y.A., Solnyshkina, O.A., Gumerov, N.A.: Toward large scale simulations of emulsion flows in microchannels using fast multipole and graphics processor accelerated boundary element method. In: Proceedings of ASME 2012 International Mechanical Engineering Congress Exposition, IMECE 2012, USA, Houston (2012). doi:10.1115/imece2012-86238

11. Kennedy, M.R., Pozrikidis, C., Skalak, R.: Motion and deformation of liquid drops, and the rheology of dilute emulsions in simple shear flow. Comput. Fluids **23**(2), 251–278 (1994). doi:10.1016/0045-7930(94)90040-x

12. Loewenberg, M., Hinch, E.J.: Collision of two deformable drops in shear flow. J. Fluid Mech. **338**, 299–315 (1997). doi:10.1017/s0022112097005016

13. Loewenberg, M., Hinch, E.J.: Numerical simulation of a concentrated emulsion in shear flow. J. Fluid Mech. **321**, 395–419 (1996). doi:10.1017/s002211209600777x

14. Pozrikidis, C.: Boundary Integral and Singularity Methods for Linearized Viscous Flow. Cambridge University Press, Cambridge (1992). doi:10.1017/cbo9780511624124. 259 pages

15. Rallison, J.M., Acrivos, A.: A numerical study of the deformation and burst of a viscous drop in an extensional flow. J. Fluid Mech. **89**(1), 191–200 (1978). doi:10.1017/s0022112078002530

16. Rallison, J.M.: A numerical study of the deformation and burst of a viscous drop in general shear flows. J. Fluid Mech. **109**, 465 (1981). doi:10.1017/s002211208100116x

17. Rumscheidt, F.D., Mason, S.G.: Particle motions in sheared suspensions XII. Deformation and bust of fluid drops in shear and hyperbolic flow. J. Colloid Interface Sci. **16**, 238 (1961). doi:10.1016/0095-8522(61)90003-4

18. Saad, Y.: Iterative Methods for Sparse Linear System, 2nd edn. SIAM, Philadelphia (2000). doi:10.1137/1.9780898718003. 534 pages

19. Sangani, A.S., Mo, G.: An O(N) algorithm for Stokes and Laplace interactions of particles. Phys. Fluids **8**(8), 1990–2010 (1996). doi:10.1063/1.869003

20. Tornberg, A.K., Greengard, L.: A fast multipole method for the three-dimensional Stokes equations. J. Comput. Phys. **227**(3), 1613–1619 (2008). doi:10.1016/j.jcp.2007.06.029

21. Torza, C., Cox, R.G., Mason, S.G.: Particle motions in sheared suspensions XXVI. Streamlines in and around liquid drops. J. Colloid Interface Sci. **35**, 529 (1972). doi:10.1016/0021-9797(71)90211-6

22. Wang, H., Lei, T., Li, J., Huang, J., Yao, Z.: A parallel fast multipole accelerated integral equation scheme for 3D Stokes equations. Int. J. Numer. Methods Eng. **70**, 812–839 (2007). doi:10.1002/nme.1910

23. Zinchenko, A.Z., Davis, R.H.: An efficient algorithm for hydrodynamical interaction of many deformable drops. J. Comput. Phys. **157**, 539–587 (2000). doi:10.1006/jcph.1999.6384

24. Zinchenko, A.Z., Davis, R.H.: A multipole-accelerated algorithm for close interaction of slightly deformable drops. J. Comput. Phys. **207**, 695–735 (2005). doi:10.1016/j.jcp.2005.01.026

25. Zinchenko, A.Z., Davis, R.H.: Large-scale simulations of concentrated emulsion flows. Philos. Trans. R. Soc. Lond. A **361**, 813–845 (2003). doi:10.1098/rsta.2003.1178

26. Zinchenko, A.Z., Rother, M.A., Davis, R.H.: A novel boundary-integral algorithm for viscous interaction of deformable drops. Phys. Fluids **9**(6), 1493–1511 (1997). doi:10.1063/1.869275

Optimization of Drop Characteristics in a Carrier Cooled Gas Stream Using ANSYS and Globalizer Software Systems on the PNRPU High-Performance Cluster

Stanislav L. Kalyulin[1]([✉]) [iD], Evgenya V. Shavrina[1], Vladimir Y. Modorskii[1] [iD],
Konstantin A. Barkalov[2] [iD], and Victor P. Gergel[2] [iD]

[1] Perm National Research Polytechnical University, Perm, Russia
{ksl,modorsky}@pstu.ru
[2] Lobachevsky State University of Nizhny Novgorod, Nizhny Novgorod, Russia
{konstantin.barkalov,victor.gergel}@itmm.unn.ru

Abstract. We describe in this article the optimization calculations of spray droplets in a gas injected through a nozzle into a work area, as a part of a research icing on model objects in a small-size climatic wind tunnel. Calculations were performed in a three-dimensional formulation. It is assumed that the drop has some speed, temperature and diameter as it enters the gas flow, which has a specified speed and temperature, so that certain temperature limits are attained when it interacts with a remote obstruction. We determined the maximum gas flow temperature, which corresponds to the minimum of cooling energy consumption. The optimization was carried out using the Globalizer software (Lobachevsky State University of Nizhny Novgorod). Also, we could solve the integration issue between Globalizer and ANSYS Workbench 13.0. ANSYS was employed as a tool to calculate optimization criteria values, whereas Globalizer was used as an optimal parameter search tool. Calculations were performed on the PNRPU high-performance cluster (with a peak performance of 24 TFLOPS).

Keywords: Small-size climatic wind tunnel · Global optimization · Multiextremal functions · Parallel algorithms · Droplets flying · Numerical simulation · Gas flow

1 Introduction

An energy-efficient [2] closed-loop small-sized climatic wind tunnel (CWT) is being developed at Perm National Research Polytechnical University (PNRPU) [1] with a view to continue scientific investigations into the field of aircraft

The study was supported by the grants of the Russian Science Foundation: project 14-19-00877 (V. Modorskii, S. Kalyulin and E. Shavrina, Sects. 1, 2 and 5) and project 16-11-10150 (K. Barkalov, V. Gergel, Sects. 3 and 4).

L. Sokolinsky and M. Zymbler (Eds.): PCT 2017, CCIS 753, pp. 331–345, 2017.
DOI: 10.1007/978-3-319-67035-5_24

aerodynamics. This installation which will provide an opportunity to conduct aerodynamic experiments with simulated in-flight icing conditions at flow Mach numbers as high as $M = 0.8$ and stagnation temperatures up to $30\,^\circ\text{C}$ [3].

As it is known, tests in wind tunnels can help to solve the following problems [4]:

1. Investigation of the shape influence of the gas-streamlined object on its aerodynamic characteristics depending on flow velocity and the body position in space.
2. Investigation of air machines: gas turbines, compressors, propellers, windmills, fans, etc.
3. Investigation of engine characteristics (turbojet, ramjet, etc.).
4. Investigation of aircraft flight dynamics.
5. Investigation of the influence of aerodynamic forces on the elasticity of aircraft structures (e.g., research into aircraft wing flutter).
6. Physical investigations related to air flow under different conditions (research into the boundary layer, supersonic flows, spatial streams, etc.).
7. Methodical and scientific investigations related to the creation of wind tunnels as physical facilities, development of test methods in tunnels and processing of the results obtained.

The article discusses the design of an energy-efficient closed-loop small-sized CWT for full-scale icing simulation. The modeling of this process requires structural elements such as air-cooling system, injectors for water supply to the tunnel, a device to heat them, and a dehumidifying system (described in the patent [5]).

A distinctive feature of the design is its high energy efficiency, which will allow for testing under icing conditions at a much lower cost than already existing large testing facilities. This economy is achieved through its ability to consume only the energy required in operating mode to maintain a predetermined level of air flow and by reducing the characteristic dimensions of the CWT [3].

According to studies by Russian scientists [6], the contour of the CWT must be open, so that moisture does not accumulate. The working part is enclosed as its length is enough to equalize the drop temperature to air temperature. Owing to the fact that large CWTs are energy-intensive installations, small-size models are becoming more popular. However, scale model sizes less than 1 : 8 lead to distortion of the icing shape, which is unacceptable. Thus, there is a limitation on the minimum diameter of the CWT.

Given the large time and material costs that are necessary for the preparation and execution of physical experiments in an energy-efficient closed-loop small-sized CWT, numerical simulation of experiments has become the most popular method for studying icing processes. We propose the joint use of numerical and physical experiments.

Ice formation processes during field experiments (spraying water with subsequent freezing) determine three tasks that must be accomplished for the numerical simulation of processes occurring in a CWT:

1. Simulation of cooling gas flowing in the CWT.
2. Simulation of drop disintegration to determine correctly the size of ice particles.
3. Simulation of the "water-ice" phase transition.

ANSYS CFX numerical experiments concerning spraying of water particles
are possible with the use of different mathematical models of various degrees
of accuracy. In most cases, however, significant computational resources are
required.

A modeling of icing was performed in [7], but the method suggested by the
authors does not allow to model the characteristics of the surface roughness, nor
complex-shaped ice build-ups (such as the unevenness of the "lobster tail") in
icing areas at small slip angles of the drops. According to [8], small build-ups
of this kind have a more significant impact than the "horn-shaped" ones. Also,
that paper discusses a technique that allows to estimate the probability of such
ice build-ups. It is not possible, however, to obtain sufficiently accurate data on
the nature of the build-ups in such processes as flow separation from a wing.

Icing simulation in ANSYS FENSAP-ICE has not gained much popularity
due to requirements to conclude additional license agreements and an insufficient
number of validation studies. Nevertheless, a solution example of the problem of
icing in ANSYS FENSAP-ICE in conjunction with ANSYS CFX was presented
by researchers from Louisiana in [9]. ANSYS CFX and ANSYS FLUENT gas
dynamics packages are usually used in this regard in research practice.

As a rule, information on temperature, humidity, speed and other gas dynamics parameters are used to predict icing, since a direct modeling of the "water-
ice" phase transition is not possible in these packages. Besides, investigations on
gas dynamics problems performed abroad with the use of the ANSYS CFX and
ANSYS FLUENT packages relied on subsequent accurate gas dynamics parameters to solve the problem of icing by semi-empirical or analytical methods [10].

The statement of the problem of numerical simulation depends on the type
of icing, which can be divided into four categories [6, 11, 12]:

1. The roughness of the ice surface surpasses the height of the local boundary
   layer, affecting not only the transition of the boundary layer to a turbulent
   state, but also its separation downstream.
2. The icing is characterized by grooves along the stream. In this case, the area
   of flow separation primarily depends on the attack angle. In combination with
   its roughness, this type of icing mainly affects the increase in resistance.
3. In some cases, the droplets spreading over the surface crystallize immediately.
   Then the ice begins to grow in a vicinity of the critical point adopting the
   shape of a "horn". In other cases, crystallization is delayed; a film of water
   creeps over the surface, and ice forms along two areas on the side of the
   critical point, forming two "horns". The icy "horns" create a large separated
   region which affects aerodynamics. This is accompanied by the separation of
   flow depending on horns shape, position, and angle of attack. It significantly
   reduces the aircrafts load-bearing properties.

4. When a thermal anti-icing system is working, a film of water moves across the surface to the outer boundary of the heated region and then freezes in the form of a so-called barrier ice. An obstacle is formed which leads to the appearance of a large separation region in front and behind the obstacle. This significantly affects the aerodynamic characteristics depending on the position and geometry of the obstacle [12].

The nature of these phenomena and the conditions producing a particular type of icing have not been thoroughly studied until now.

Furthermore, the estimation of the surface tension of small droplets with a typical diameter of 20 μm have shown that the excess pressure, which could affect the freezing point of water and the crystallization process, is so small that it practically does not affect the freezing temperature of water [6]. The surface tension only has an effect on the spreading of drops when they hit the surface and the disintegration of the liquid into drops at being expelled from the injectors [13]. These processes are poorly understood and will be the subject of forthcoming experimental studies in CWT.

In this regard, mathematical modeling of icing at the moment is not perfect. Therefore, physical experiments cannot be completely excluded from consideration. Thus, to maximize at present the effectiveness of research into icing processes, both numerical and physical experiments are equally needed.

Investigations are frequently limited to finding one or several solutions that satisfy the requirements of technical specifications. Searching for an optimal solution is usually not possible, since it requires large computational resources.

Therefore, the numerical search for optimal solutions in an acceptable time frame is one of the most important tasks [14,15]. It increases the quality of decisions and enhances the benefits of numerical calculations, enabling the identification of new dependencies.

2   Gas Dynamics Problem

2.1   A Physical Model of the Gas Dynamics Problem

In the physical statement of the problem, it is assumed that a drop with a given velocity, temperature and diameter falls into a gas dynamic flow having some speed and temperature, so that, upon contact with a distant obstacle, certain temperature limits are attained [16]. The distance to the obstacle is equal to 2 m. The maximum temperature of the carrying gas dynamic flow is determined, and it corresponds to the minimum energy consumption of the energy-efficient closed-loop small-sized CWT when it cools.

The calculation is carried out in a three-dimensional formulation. The internal CWT volume is modeled, and the two-phase environment, gravity and interaction with the cooling gas dynamic flow are taken into account.

2.2   Mathematical Model of the Gas Dynamics Problem

We applied the finite volume method for the numerical simulation of the drop flight from the site of injection through the nozzle to the working area of the CWT. This implies that the field of the gas dynamic flow is approximated by a finite number of calculation points.

A mathematical model is developed in accordance with the adopted physical model. This model is based on the mass, momentum and energy conservation laws, and is includes state equations of an ideal compressible gas and turbulence, as well as with initial and boundary conditions. The mathematical model of the gas dynamics problem includes the following relations [17]:

- The equation of mass conservation for the gas:

$$\frac{\partial}{\partial t}(\rho_{air}) + \bar{\nabla} \cdot (\rho_{air} V_{air}) = Q_{mass}^{drop}, \tag{1}$$

where ρ_{air} is the air density, V_{air} is the air velocity vector, and Q_{mass}^{drop} is a source term expressing the increase in weight caused by evaporation of water from the drop surface.

- The equation of momentum conservation for the gas:

$$\frac{\partial}{\partial t}(\rho_{air} V_{air}) + (\rho_{air} V_{air} \cdot \bar{\nabla}) \cdot V_{air} = -\bar{\nabla} P + \bar{\nabla} \tau_{air} + \rho_{air} g + Q_{force}^{drop}, \tag{2}$$

where P is pressure, g is the gravity vector, τ_{air} is the shear stress at the wall, and Q_{force}^{drop} is a source term that expresses the force with which the drop acts on the air.

- The equation of energy conservation for the gas:

$$\frac{\partial}{\partial t}(\rho_{air} H_{air}) + \bar{\nabla} \cdot (\rho_{air} V_{air} H_{air}) = \frac{\partial P}{\partial t} \cdot \bar{\nabla} \cdot \left(\left(\frac{\lambda_{air}}{c_{p_{air}}} + \mu_t \right) \bar{\nabla} H_{air} \right) + Q_{energy}^{drop}, \tag{3}$$

where H_{air} is the total enthalpy of the air, λ_{air} is the coefficient of thermal conductivity of the air, $c_{p_{air}}$ is the air heat capacity, μ_t is turbulent dynamic viscosity, Q_{energy}^{drop} is a source term expressing energy transfer between the phases.

- The equation of mass conservation for the steam:

$$\frac{\partial}{\partial t}(\rho_{air} Y_{steam}) + \bar{\nabla} \cdot (\rho_{air} V_{air} Y_{steam}) = \bar{\nabla} \cdot \left(\left(\rho_{air} D + \frac{\mu_t}{S_{c_t}} \right) \bar{\nabla} Y_{steam} \right) + Q_{mass}^{drop}, \tag{4}$$

where $Y_{steam} = 1 - Y_{air}$ is steam mass concentration, Y_{air} is air mass concentration, D is the diffusion coefficient, and S_{c_t} is the Schmidt turbulent number ($S_{c_t} = 1$).

- The equation of state:

$$P = \frac{\rho_{air} R_0 T_{air}}{M}, \tag{5}$$

where $R_0 = 8314.41$ J/kmol $\cdot$ K is the universal gas constant, M is the molecular weight, and T_{air} is the air temperature.

• The equation of the drop motion dynamics:

$$\frac{\partial V_{\text{drop}}}{\partial t} = \frac{\pi d_{\text{drop}}^2}{8 m_{\text{drop}}} C_D \rho_{\text{drop}} |V_{\text{air}} - V_{\text{drop}}| (V_{\text{air}} - V_{\text{drop}}) + g \left(1 - \frac{\rho_{\text{air}}}{\rho_{\text{drop}}} \right), \quad (6)$$

where d_{drop} is the drop diameter, m_{drop} is the drop weight, ρ_{drop} is the drop density, V_{drop} is drop velocity, and C_D is the coefficient of resistance.
• The equation of mass conservation for the drop:

$$\frac{\partial m_{\text{drop}}}{\partial t} = -m^* \pi d_{\text{drop}}^2, \quad (7)$$

where m^* is the steam injection parameter per drop surface unit.
• The equation of energy conservation for the drop:

$$\frac{\partial T_{\text{drop}}}{\partial t} = \left(\left[\text{Nu} \frac{\lambda_{\text{air}}}{d_{\text{drop}}} (T_{\text{air}} - T_{\text{drop}}) \right] - m^* q(T_{\text{drop}}) \right) \frac{6}{d_{\text{drop}} \rho_{\text{drop}} c_{p_{\text{drop}}}}, \quad (8)$$

where $c_{p_{\text{drop}}}$ is the drop heat capacity, T_{drop} is the drop temperature, and Nu is the Nusselt number.
• The equation of turbulent energy:

$$\frac{\partial}{\partial t} (\rho_{\text{air}} K) + \bar{\nabla} \cdot (\rho_{\text{air}} V_{\text{air}} K) = \bar{\nabla} \cdot \left(\left(\mu + \frac{\mu_t}{\sigma} \right) \bar{\nabla} K \right) + \mu_t G - \rho_{\text{air}} \varepsilon, \quad (9)$$

where σ is a constant, G determines the speed of turbulent energy generation, ε is the rate of turbulent energy dissipation, and K is turbulent energy.
Given that the computational domain has a cylindrical geometry, the gas dynamic flow has a high speed, and the length is small and equals the distance from the injection of droplets through the nozzles to the barrier (experimental model), it is advisable to use the $K - \varepsilon$ turbulence model, which is applicable when the influence of inertial forces is large compared to viscosity forces.
• The equation of turbulent energy dissipation rate:

$$\frac{\partial}{\partial t} (\rho_{\text{air}} \varepsilon) + \bar{\nabla} \cdot (\rho_{\text{air}} V_{\text{air}} \varepsilon) = \bar{\nabla} \cdot \left(\left(\mu + \frac{\mu_t}{\sigma_\varepsilon} \right) \bar{\nabla} \varepsilon \right) + C_1 \frac{\varepsilon}{K} \mu_t G - C_2 f_1 \rho_{\text{air}} \frac{\varepsilon^2}{K},$$
$$(10)$$

where σ_ε, C_1 and C_2 are constants.

2.3 Parameters and Criteria for Optimization of Gas Dynamics Calculation

The problem is solved in ANSYS CFX, as the initial and boundary conditions are given. We studied the solution convergence. Next, input and output parameters were parameterized in CFX and transferred to the ANSYS Workbench, a step that was necessary for the iterative start of ANSYS in the Globalizer optimization program.

The following ranges of input parameters were set:

1. $T_{air} = -30 \ .. \ 0\,°C$ for gas dynamic flow temperature;
2. $V_{air} = 10 \ .. \ 270$ m/s for gas dynamic flow rate;
3. $T_{drop}|_{l=0} = +5 \ .. \ +10\,°C$ for the drop temperature at the initial moment of contact with the air flow;
4. $V_{drop} = 10 \ .. \ 270$ m/s for the drop velocity at the initial time of contact with the air flow.

The average drop diameter for injecting was taken equal to 20 μm, and did not change during computations.

We adopted the following output criteria for the optimization algorithm:

1. $T_{drop}|_{l=0} = -0.5 \ .. \ +0.5\,°C$ for the drop temperature at the moment when it reaches the obstacle $(L = 2$ m$)$;
2. $T_{air} \rightarrow$ max for the initial temperature of the gas dynamic flow.

The input parameter T_{air} is simultaneously an output criterion that approaches the maximum.

As we know, when integrating the multi-criteria parallel optimization IOSO PM program complex with ANSYS, it is possible to set the input parameter as an output optimization criterion only with additional program code associated with syntactic IOSO PM parameters, i.e. it is not foreseen within the core function of the optimization complex.

As regards the drop freezing, the air temperature cannot physically be above $+0.5\,°C$, otherwise the drop will not be able to cool down to the temperature defined by the output criterion.

The optimization problem was solved with the generalized global optimization algorithm implemented in Globalizer. To allow for sorting the results of the Globalizers gas dynamics solution in ANSYS CFX, we wrote the respective functions, and parameterized the total count time (TotalTime) and the time step (TimeSteps).

The description of both the optimization algorithm and the Globalizer program system is given in the next section.

3   Global Optimization with Non-convex Constraints

3.1   Problem Statement

Let us consider the N-dimensional global optimization problem

$$\varphi(y^*) = \min \{\varphi(y) : y \in D, \ g_i(y) \leq 0, \ 1 \leq i \leq m\}, \tag{11}$$

with search domain

$$D = \{y \in R^N : -2^{-1} \leq y_j \leq 2^{-1}, 1 \leq j \leq N\}. \tag{12}$$

This problem statement covers a large class of problems since the hyperinterval

$$S = \left\{ y \in R^N : a_j \le y_j \le b_j, 1 \le j \le N \right\}$$

can be reduced to the hypercube (12) by linear transformation. The objective function $\varphi(y)$ (henceforth denoted by $g_{m+1}(y)$) and the left-hand sides $g_i(y)$, $1 \le i \le m$, of the constraints satisfy the Lipschitz conditions with constants L_i, $1 \le i \le m+1$, respectively, and may be multiextremal. It is assumed that the functions $g_i(y)$ are defined and computed only at the points $y \in D$ satisfying the conditions

$$g_k(y) \le 0, \ 1 \le k < i. \tag{13}$$

Employing the continuous single-valued Peano curve $y(x)$ (*evolvent*) that maps the unit interval $[0, 1]$ of the x-axis onto the N-dimensional domain (12), it is possible to find the minimum in the problem (11) by solving the one-dimensional problem

$$\varphi(y(x^*)) = \min \left\{ \varphi(y(x)) : x \in [0, 1], \ g_i(y(x)) \le 0, \ 1 \le i \le m \right\},$$

where, as it follows from (13), the functions $g_i(y(x))$ are defined and computed in the domains

$$Q_1 = [0, 1], \ Q_{i+1} = \left\{ x \in Q_i : g_i(y(x)) \le 0 \right\}, \ 1 \le i \le m.$$

These conditions allow for the introduction of a classification of the points $x \in [0, 1]$ according to the number $\nu(x)$ of the constraints computed at this point. The *index* $\nu(x)$ can also be defined by the conditions

$$g_i(y(x)) \le 0, \ 1 \le i < \nu, \ g_\nu(y(x)) > 0,$$

where the last inequality is inessential if $\nu = m + 1$.

The considered dimensionality reduction scheme juxtaposes a multidimensional problem with lipschitzian functions to a one-dimensional problem where the corresponding functions satisfy the uniform Hölder condition (see [18]), i.e.

$$|g_i(y(x')) - g_i(y(x''))| \le H_i |x' - x''|^{1/N}, \ x', x'' \in [0, 1], \ 1 \le i \le m+1.$$

Here N is the dimensionality of the initial multidimensional problem and the coefficients H_i are related with the Lipschitz constants L_i of the initial problem by the inequalities $H_i \le 2L_i\sqrt{N+3}$.

Thus, a *trial* at a point $x^k \in [0, 1]$ executed at the k-th iteration of the algorithm will consist in the following sequence of operations:

- Determine the *image* $y^k = y(x^k)$ in accordance with the mapping $y(x)$.
- Compute the values $g_1(y^k), ..., g_\nu(y^k)$, where the index $\nu \le m$ is determined by the conditions

$$g_i(y^k) \le 0, \ 1 \le i < \nu, \ g_\nu(y^k) > 0, \ \nu \le m.$$

The occurrence of the first violation of the constraint terminates the trial. In the case when the point y^k is a feasible one, i.e. when $y(x^k) \in Q_{m+1}$, the trial includes the computation of the values of all the functions of the problems and the index is assumed to be $\nu = m + 1$. The pair of values

$$\nu = \nu(x^k), \quad z^k = g_\nu(y(x^k))$$

is a *trial result*.

3.2   Generalized Global Search Algorithm

The rules that determine the work of the index method are the following.

The first trial is executed at an arbitrary internal point $x_1 \in (0,1)$. The selection of the point x^{k+1}, $k \geq 1$, of any next trial is determined by the following rules.

Rule 1. Renumber the points $x^1, ..., x^k$ of the preceding trials by the lower indices in ascending order of coordinate values, i.e.

$$0 = x_0 < x_1 < \cdots < x_k < x_{k+1} = 1,$$

and assign to them the values $z_i = g_\nu(y(x_i))$, $\nu = \nu(x_i)$, $1 \leq i \leq k$, computed at these points. The points $x_0 = 0$ and $x_{k+1} = 1$ are introduced additionally, while the values z_0 and z_{k+1} are not defined.

Rule 2. Classify the indices i, $1 \leq i \leq k$, of the trial points according to the number of the problem constraints fulfilled at these points, by constructing the sets

$$I_\nu = \{i : 1 \leq i \leq k, \ \nu = \nu(x_i)\}, \ 1 \leq \nu \leq m + 1,$$

containing the numbers of all the points x_i, $1 \leq i \leq k$, with the same values of ν. The end points $x_0 = 0$ and $x_{k+1} = 1$ are interpreted as the ones having indices equal to zero. An additional set, $I_0 = \{0, k + 1\}$, corresponds to them.

Determine the maximum value of the index:

$$M = \max \{\nu(x_i), \ 1 \leq i \leq k\}.$$

Rule 3. Compute the current lower estimates,

$$\mu_\nu = \max \left\{ \frac{|z_i - z_j|}{(x_i - x_j)^{1/N}}, \ i, j \in I_\nu, \ i > j \right\},$$

for the unknown Hölder constants H_ν of the functions $g_\nu(y), 1 \leq \nu \leq m + 1$. If a set I_ν contains less than two elements, or if μ_ν is equal to zero, then assume $\mu_\nu = 1$.

Rule 4. For all nonempty sets I_ν, $1 \leq \nu \leq m + 1$, compute the estimates

$$z_\nu^* = \begin{cases} -\epsilon_\nu, & \nu < M, \\ \min\{g_\nu(y(x_i)) : i \in I_\nu\}, & \nu = M, \end{cases}$$

where the nonnegative numbers $(\epsilon_1, ..., \epsilon_m)$ are parameters of the algorithm.

Rule 5. For each interval (x_{i-1}, x_i), $1 \le i \le k+1$, compute the *characteristics* $R(i)$:

$$R(i) = 2\Delta_i - 4\frac{z_i - z_\nu^*}{r_\nu \mu_\nu}, \quad \nu = \nu(x_i) > \nu(x_{i-1}),$$

$$R(i) = \Delta_i + \frac{(z_i - z_{i-1})^2}{r_\nu^2 \mu_\nu^2 \Delta_i} - 2\frac{z_i + z_{i-1} - 2z_\nu^*}{r_\nu \mu_\nu}, \quad \nu = \nu(x_i) = \nu(x_{i-1}), \quad (14)$$

$$R(i) = 2\Delta_i - 4\frac{z_{i-1} - z_\nu^*}{r_\nu \mu_\nu}, \quad \nu = \nu(x_{i-1}) > \nu(x_i),$$

where $\Delta_i = (x_i - x_{i-1})^{1/N}$. The values $r_\nu > 1$, $1 \le \nu \le m+1$, are parameters of the algorithm. An appropriate selection of r_ν allows to consider the product $r_\nu \mu_\nu$ as an estimate of the Hölder constants H_ν, $1 \le \nu \le m+1$.

Rule 6. Find the interval (x_{t-1}, x_t) with the maximum characteristic

$$R(t) = \max\{R(i) : 1 \le i \le k+1\}. \quad (15)$$

Rule 7. Make the next trial at the midpoint of the interval (x_{t-1}, x_t) if the indices of the points x_{t-1} and x_t are not the same, i.e.

$$x^{k+1} = \frac{x_t + x_{t-1}}{2}, \quad \nu(x_{t-1}) \neq \nu(x_t).$$

Otherwise, make the trial at the point

$$x^{k+1} = \frac{x_t + x_{t-1}}{2} - \frac{\text{sign}(z_t - z_{t-1})}{2r_\nu} \left[\frac{|z_t - z_{t-1}|}{\mu_\nu}\right]^N, \quad \nu = \nu(x_{t-1}) = \nu(x_t). \quad (16)$$

We may take as termination condition the inequality $\Delta_t \le \epsilon$, where t is from (15) and $\epsilon > 0$ is the predefined accuracy.

Various modifications of this algorithm and the corresponding theory of convergence are given in [18–21].

The algorithm considered above is very flexible and allows for an efficient parallelization for shared memory, for distributed memory, and for accelerators [22–25].

4   Integration of Globalizer and ANSYS Workbench

This section contains a brief explanation of the Globalizer software system. The system expands the family of global optimization software that has been consistently developed at Nizhny Novgorod State University during the past several years.

The development of the system was based on the generalized global search algorithm that was described in the previous section. A major advantage of Globalizer is that the system is designed to solve time-consuming multiextremal optimization problems. The system efficiently uses modern high-performance computer systems to obtain the optimal solution within a reasonable time and at a reasonable cost.

The structural components of the Globalizer are the following blocks.

- Block 0 consists of the procedures for computing the function values (objective and constraints) for the optimization problem that is being solved; this is an external block with respect to the Globalizer system.
- Blocks 1–3 form the optimization subsystem and solve unconstrained (Block 1), constrained (Block 2) and multicriteria (Block 3) global optimization problems.
- Block 4 is a subsystem for accumulating and processing search information.
- Block 5 contains the dimension reduction procedures based on evolvents; the optimization blocks solve the reduced (one-dimensional) optimization problems. Block 5 provides for interaction between the optimization blocks and the initial multidimensional optimization problem.
- Block 6 is responsible for managing the parallel processes while performing the global search (determining the optimal configuration of parallel processes, distributing the processes between computer elements, balancing computational loads, etc.).
- Block 7 is a common managing subsystem, which fully controls the entire computational process when solving global optimization problems.

In order to integrate Globalizer with ANSYS Workbench 13.0, we need to provide the call of an external calculator of function values (i.e. ANSYS) as Block 0. It is necessary to transfer the input parameter values of the ANSYS Workbench project, update the project and receive the output parameter values. The transfer of the parameters between the ANSYS Workbench and Globalizer was implemented by means of an IronPython script.

In the conducted experiments, computations were performed on a single cluster node. The parallelization of computations of the problem function values relied on ANSYS tools. The computation costs of executing the optimization algorithm in Globalizer (the choice of the next trial points, processing the trial results, etc.) were significantly less than those of computing the problem function values in ANSYS. Therefore, it was not necessary to parallelize the operations of the optimization module itself.

It is important to note that the integration of ANSYS and Globalizer allows for larger scale experiments to be performed with the use of several cluster nodes as well. In this case, search trials at different points of the search domain will be performed at all the nodes that are employed at each iteration of the method. Therefore, the results of the computations can be processed in Globalizer both synchronously and asynchronously. In the first case, the system waits until the computations on all nodes employed are completed. Then the processing of the results is performed, and the next iteration is started. This regime is preferred when computing time is the same at any point of the search domain. In the second case, the results obtained at one of the nodes are processed immediately, and the next trial begins at this node. The asynchronous regime provides for the full load of the nodes to be utilized in the cases when trial times are different at different points of the search domain. The experiments performed earlier on test problems have demonstrated this approach to be promising [26]. Performing experiments of this kind on the solution of applied problems will be the subject of further investigations.

5   Optimization Results

The Globalizer system allowed to construct a field of input parameter values that satisfy the above-mentioned (Sect. 2.3) restrictions of the gas dynamics problem (Fig. 1).

The total time spent on finding the solution of the optimization problem at the PNRPU high-performance computing complex was 48 h.

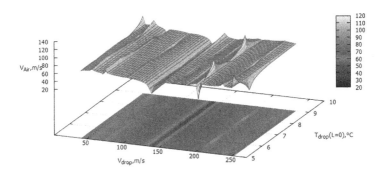

Fig. 1. Field of input parameter values that satisfy the output criteria of the optimization problem: V_{drop} — drop velocity, $T_{drop}|_{l=0}$ — drop temperature at the time of injection, V_{air} — air velocity

It was assumed that the energy-efficient modes of the CWT are those in which the gas dynamic flow temperature in the working section approaches the maximum at the drop temperature $+0.5\,^{\circ}\mathrm{C}$ when it reaches the barrier at a 2 m distance from the injection point:

$T_{air} \to$ max;

$T_{drop}|_{l=2\,m} = 0.5\,^{\circ}\mathrm{C}$.

However, according to expert estimates, the ranges of permissible drop speed V_{drop} and temperature $T_{drop}|_{l=0}$ at the time of injection are the following:

$V_{drop} = 5\,..\,270\,\mathrm{m/s}$;

$T_{drop}|_{l=0} = +5\,..\,+10\,^{\circ}\mathrm{C}$.

As the optimization computations show, the values of drop speed and temperature in the energy-efficient modes of the CWT can vary within the following ranges (Fig. 1):

$V_{drop} = 42.37\,..\,237.63\,\mathrm{m/s}$;

$T_{drop}|_{l=0} = +5\,..\,+9\,^{\circ}\mathrm{C}$.

An analysis of the results of the computational optimization experiment (Fig. 1) showed that the range of drop velocities V_{drop} and drop temperatures

$T_{\text{drop}}|_{l=0}$ corresponding to these modes of operation is narrower than the recommended pre-peer evaluations.

These findings identify the speed (by 26.32%) and temperature (by 20%) injection modes that produce icing in the work area based at the specified distance of the experimental model to the injection zone.

Figure 2 shows that the variation of only two parameters, such as the speed of the gas dynamic flow (V_{air}) and the drop temperature at the time of injection through the nozzles ($T_{\text{drop}}|_{l=0}$), allows to achieve the desired value of the drop temperature ($+0.5\,^{\circ}$C) upon reaching the barrier (experimental model) when meeting the output optimization criterion $T_{\text{air}} \rightarrow +0.5\,^{\circ}$C. This may help to achieve energy-efficient modes in the CWT.

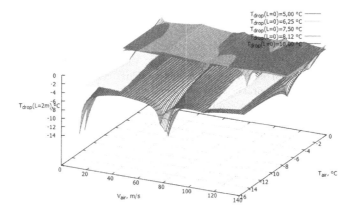

Fig. 2. Dependence of V_{air} (gas dynamic flow rate) and T_{air} (gas dynamic flow temperature) on $T_{\text{drop}}|_{l=2\text{ m}}$ (drop temperature when reaching the obstacle ($L = 2$ m)) for different values of $T_{\text{drop}}|_{l=0}$ (drop temperature at time of injection)

For example, in the given operation mode of the studied experimental model, by varying only technological parameters such as temperature and velocity of drops at the initial moment of contact with the gas flow, one can achieve an energy-efficient simulation of ice formation, except for a few bottlenecks in the drop velocity modes. Moreover, modes can be specified by applying a greater number of iterations (ANSYS starts) in the Globalizer optimization algorithm, which, of course, requires longer computation times.

6 Conclusions

1. As a result of the solution to the optimization problem, a combined region of speed and temperature parameters has been detected, as well as the air flow velocity at which the gas dynamic flow temperature in the working section approaches the maximum. This region is energy efficient, since it allows for icing without achieving significant negative airflow temperatures.

2. The possibility of implementing an energy efficient operation mode of the small-sized closed-loop CWT has been demonstrated by changing only technological parameters such as the initial drop temperature and velocity.
3. The integration of the Globalizer optimization system and the engineering software package ANSYS has been realized for the first time.
4. The Globalizer software package helps to find the best solutions at reasonable time and material costs.
5. The application of the Globalizer optimization software system made it possible to identify a fairly wide range of parameter combinations that allows to maintain the energy-efficient operation mode of the small-sized closed-loop CWT with sufficient accuracy for engineering practice ($\pm 1\,^{\circ}\text{C}$).

References

1. Modorskii, V.Y., Shevelev, N.A.: Research of aerohydrodynamic and aeroelastic processes on PNRPU HPC system. In: Fomin, V. (ed.) ICMAR 2016, AIP Conference Proceedings, vol. 1770, art. no. 020001 (2016). doi:10.1063/1.4963924
2. Shmakov, A.F., Modorskii, V.Y.: Energy conservation in cooling systems at metallurgical plants. Metallurgist **59**(9–10), 882–886 (2016). doi:10.1007/s11015-016-0188-8
3. Kalyulin, S.L., Modorskii, V.Y., Paduchev, A.P.: Numerical design of the rectifying lattices in a small-sized wind tunnel. In: Fomin, V. (ed.) ICMAR 2016, AIP Conference Proceedings, vol. 1770, art. no. 030110 (2016). doi:10.1063/1.4964052
4. Afanasiev, V.A., Barsukov, V.S., Gofin, M.Y., Zakharov, Y.V., Strelchenko, A.N., Shalunov, N.P.: Experimental testing of spacecraft. MAI, Moscow (1994). (in Russian)
5. Goryachev, A.V., Mezhzil, E.K., Petrov, S.B., Syrov, V.A., Harlamov, A.V., Chivanov, S.V.: A way of ground testing objects of aircraft, subject to icing, and a device for its implementation. Patent RF, no. 2345345 (2007)
6. Klemenkov, G.P., Prihodko, Y.M., Puzyrev, L.N., Haritonov, A.M.: Modelling of aircraft icing processes in aeroclimatic tubes. Thermophys. Aeromeh. **15**(4), 563–572 (2008). (in Russian)
7. Alekseenko, S.V., Prihodko, A.A.: Numerical simulation of icing cylinder and profile. Models review and calculation results. TsAGI Sci. J. **44**(6), 25–57 (2013). (in Russian)
8. Prihodko, A.A., Alekseenko, S.V.: Numerical simulation of icing aerodynamic surfaces in the presence of large overcooled water drops. JETP Lett. **40**(19), 75–82 (2014). (in Russian)
9. Hannat, R., Morency, F.: Numerical validation of conjugate heat transfer method for anti-/de-icing piccolo system. J. Aircr. **51**(1), 104–116 (2014). doi:10.2514/1.c032078
10. Villalpando, F., Reggio, M., Ilinca, A.: Prediction of ice accretion and anti-icing heating power on wind turbine blades using standard commercial software. Energy **114**, 1041–1052 (2016). doi:10.1016/j.energy.2016.08.047
11. Lynch, F.T., Khodadoust, A.: Effects of ice accretions on aircraft aerodynamics. Prog. Aeosp. Sci. **37**(8), 669–767 (2001). doi:10.1016/s0376-0421(01)00018-5
12. Bragg, M.B., Broeren, A.P., Blumenthal, L.A.: Iced-airfoil aerodynamics. Prog. Aeosp. Sci. **41**(5), 323–362 (2005). doi:10.4271/2003-01-2098

13. Modorskii, V.Y., Sipatov, A.M., Babushkina, A.V., Kolodyazhny, D.Y., Nagorny, V.S.: Modeling technique for the process of liquid film disintegration. In: Fomin, V. (ed.) ICMAR 2016, AIP Conference Proceedings, vol. 1770, art. no. 030109 (2016). doi:10.1063/1.4964051

14. Gaynutdinova, D.F., Modorsky, V.Y., Masich, G.F.: Infrastructure of data distributed processing in high-speed process research based on hydroelasticity tasks. In: Sloot, P. (ed.) YSC, Procedia Computer Science, vol. 66, pp. 556–563 (2015). doi:10.1016/j.procs.2015.11.063

15. Modorskii, V.Y., Gaynutdinova, D.F., Gergel, V.P., Barkalov, K.A.: Optimization in design of scientific products for purposes of cavitation problems. In: Simos, T.E. (ed.) ICNAAM 2015, AIP Conference Proceedings, vol. 1738, art. no. 400013 (2016). doi:10.1063/1.4952201

16. Modorskii, V.Y., Sokolkin, Y.V.: Dynamic behavior of a thick-walled cylinder under pressurization. Izv. Vyss. Uchebnykh Zaved. Aviats. Tek. **4**, 14–16 (2002)

17. Kozlova, A.V., Modorskii, V.Y., Ponik, A.N.: Modeling of cooling processes in the variable section channel of a gas conduit. Rus. Aeronaut. **53**(4), 401–407 (2010). doi:10.3103/s1068799810040057

18. Strongin, R.G., Sergeyev, Y.D.: Global Optimization with Non-Convex Constraints. Sequential and Parallel Algorithms. Kluwer Academic Publishers, Dordrecht (2000). doi:10.1007/978-1-4615-4677-1

19. Barkalov, K.A., Strongin, R.G.: A global optimization technique with an adaptive order of checking for constraints. Comput. Math. Math. Phys. **42**(9), 1289–1300 (2002)

20. Gergel, V.P., Strongin, R.G.: Parallel computing for globally optimal decision making on cluster systems. Future Gener. Comput. Syst. **21**(5), 673–678 (2005). doi:10.1016/j.future.2004.05.007

21. Sergeyev, Y.D., Strongin, R.G., Lera, D.: Introduction to Global Optimization Exploiting Space-Filling Curves. Springer, New York (2013). doi:10.1007/978-1-4614-8042-6

22. Barkalov, K., Ryabov, V., Sidorov, S.: Parallel scalable algorithms with mixed local-global strategy for global optimization problems. In: Hsu, C.-H., Malyshkin, V. (eds.) MTPP 2010. LNCS, vol. 6083, pp. 232–240. Springer, Heidelberg (2010). doi:10.1007/978-3-642-14822-4_26

23. Barkalov, K., Gergel, V., Lebedev, I.: Use of xeon phi coprocessor for solving global optimization problems. In: Malyshkin, V. (ed.) PaCT 2015. LNCS, vol. 9251, pp. 307–318. Springer, Cham (2015). doi:10.1007/978-3-319-21909-7_31

24. Barkalov, K., Gergel, V.: Parallel Global Optimization on GPU. J. Glob. Optim. **66**(1), 3–20 (2016). doi:10.1007/s10898-016-0411-y

25. Barkalov, K., Gergel, V., Lebedev, I.: Solving global optimization problems on GPU cluster. In: Simos, T.E. (ed.) ICNAAM 2015, AIP Conference Proceedings, vol. 1738, art. no. 400006 (2016). doi:10.1063/1.4952194

26. Gergel, V., Sidorov, S.: A two-level parallel global search algorithm for solution of computationally intensive multiextremal optimization problems. In: Malyshkin, V. (ed.) PaCT 2015. LNCS, vol. 9251, pp. 505–515. Springer, Cham (2015). doi:10.1007/978-3-319-21909-7_49

Author Index

Printed in the United States
By Bookmasters